Cities in Global Transition

Raymond Charles Rauscher • Salim Momtaz

Cities in Global Transition

Creating Sustainable Communities in Australia

Springer

Raymond Charles Rauscher
School of Environmental & Life Sciences
University of Newcastle
East Gosford, NSW, Australia

Salim Momtaz
School of Environmental & Life Science
University of Newcastle
Ourimbah, NSW, Australia

ISBN 978-3-319-39864-8 ISBN 978-3-319-39865-5 (eBook)
DOI 10.1007/978-3-319-39865-5

Library of Congress Control Number: 2016952547

Printed on acid-free paper

This Springer imprint is published by Springer Nature
The registered company is Springer International Publishing AG Switzerland

There are many people who contribute to making cities work, and this book is dedicated to a few of those people, as noted below. Cities change constantly, and into the 2000s, most cities are feeling the effects of global factors. One major factor is population movements between countries, usually from one city to another. Recent civil unrest in a clutch of nations has seen mass international migration of cultural groups across the globe (many group members becoming refugees). As this book focuses on Australia and its planning of cities, most chapters have been dedicated to an Australian(s) who made a contribution to a subject or locality the chapter addresses. A biographical note for each identity appears within that chapter, with a list of those individuals hereunder:

Chapter 1: Cadigal and Wangal Aboriginal Peoples
Chapter 2: Urban Planning – Denis Winston
Chapter 3: Marrickville – Mary Reibey
Chapter 4: Ashfield – Pamela Travers
Chapter 5: Leichhardt – Nick Origlass

Chapter 6: Burwood – Sir Charles Lloyd Jones
Chapter 7: Strathfield – Miriam Hyde
Chapter 8: Canada Bay – Edith Walker
Chapter 9: Sydney Inner West – Maybanke Anderson, Augustus Alt, Henry Parkes, William Arnott, Eleanor Dark, and Thomas Walker
Chapter 10: Parramatta – Dowell Philip O'Reilly
Chapter 11: Auburn – Jack Lang
Chapter 12: Holroyd – Betty Cuthbert
Chapter 13: Greater Parramatta – John William Lewin, David Lennox, and Allan Cunningham
Chapter 14: Hurstville – Hugh Patrick and Jack Lindwall
Chapter 15: Kogarah – Myles Dunphy and Eddie Charlton
Chapter 16: Rockdale – Norm Provan and Fleur Mellor
Chapter 17: St. George – Miles Franklyn, Sumner Locke, Christina Stead, and H.V. Evatt

Book Abstract

This book examines the planning of cities in global transition (e.g., population movements), looking at how sustainable urban planning (SUP) practices can be applied to plans and development within these cities. Australia's Greater Sydney is taken as a case example, with focus on municipalities and districts (groups of municipalities). The subjects of global transition and SUP are introduced in Chap. 1. How Greater Sydney approaches planning is outlined in Chap. 2. Here, three case study districts are selected for critiquing the extent SUP is being applied. The districts, beyond the city of Sydney, are Sydney Inner West, Greater Parramatta, and St. George. A methodology to assess SUP practices within each of the municipalities (twelve in all) within the three districts is then outlined. Three key SUP principles are selected from international protocols (i.e., United Nations) to apply to the municipalities and districts. These principles are central to all chapters and include (1) developing sustainable urban centers, (2) developing sustainable transport, and (3) protecting the natural environment.

In determining how each municipality is applying SUP practices, the first question is as follows: what is the pattern of development and population change (e.g., global population movements) the municipality is experiencing? In addressing this, four phases of municipal development are examined (early, traditional, new, and future). From here, key plans of the municipal council are critiqued to determine the extent of SUP practices being applied. The next question is as follows: what is the impact of Greater Sydney planning and development on the municipality (including recent proposals for council amalgamations)? Examples of state plans are critiqued and an overall conclusion is reached on whether each municipality is meeting (and can meet in the future) the challenges of planning and development with SUP practices.

Moving from municipal- to district-level planning and development, the first question is as follows: what is the extent the state is applying SUP practices? Examples of state plans and developments for each district are critiqued under the adopted three SUP principles. From these critiques, overall conclusions are reached on the extent of SUP practices being addressed by the state within each district. The final chapter (Chap. 18) reviews all conclusions (Chaps. 3, 4, 5, 6, 7, 8, 9, 10, 11, 12, 13, 14, 15, 16, and 17) with a focus on directions for planning and developing a sustainable Greater Sydney. The final question in this chapter is as follows: how can cities anywhere in the world apply SUP practices in planning and development, in particular to address factors of global transition impacting on those cities?

Acknowledgments

We acknowledge first of all the contributions of residents, administrators, and decision-makers (past and present) of the Greater Sydney (i.e., within municipalities and districts that the book focuses on). These districts (and municipalities within) are Sydney Inner West (Marrickville, Ashfield, Leichhardt, Burwood, Strathfield, and Canada Bay), Greater Parramatta (Parramatta, Auburn, and Holroyd), and St. George (Hurstville, Kogarah, and Rockdale). It is the citizens of those areas of Greater Sydney who often initiate planning ideas and follow through the governance levels to gain results. We thank the New South Wales (NSW) state government and various institutions associated with the planning and development of these municipalities and districts.

Secondly, specific individuals and organizations (and their members) we wish to thank for assistance rendered (either directly or indirectly) include (a–z) Abnett (Bob), town planner, Queensland; Australian Institute of Architects (NSW); Australian Institute of Planning (NSW); Cook and Archies Cafe, Surry Hills; Fenner (Rolfe), town planner (Australian Local Govt. Assoc., Canberra); Ferrandiz (Susan), historian (past Slippery Rock University, USA); Gallery Cafe, Surry Hills (Mark and Pearl Williams); Henry George Society, Redfern (special acknowledgment to the late Henry George, NYC, USA); (late) Iceton (Dr. Ned) (Social Developers Network (SDN) (Australia); McKay (Darren), lecturer at Newcastle University; and Sydney University, School of Architecture and Planning.

Thirdly, on the book advancement side (a–z), thanks go to Habitat Association for Arts and Environment (HAAE) Inc. for offering technical advice, in particular to Margaret O'Toole, art director of HAAE, who assisted with proofreading the book and offering artistic advice. Other directors of HAAE to thank include John Christoforou, Ruth Dickson, David Holland, Megan Hitchens, Diane Rauscher, and public officer Kevin Armstrong. Thanks also go to IT support from Kevin and Anna Rochford, with photos review by Brooklyn Rochford and Hugo Wheelahan. In addition, we thank the University of Newcastle School of Environment and Life Sciences (SELS), in particular for secretarial assistance at the Ourimbah Campus from Pam, Nicole, and Fiona.

Fourthly, thanks go to book supporters (a–z): Martin (Joyce, Richard, Richie, and Ray), Rauscher (Barbara, Kim, Philip, and Artie Jr.), Rauscher (Grace (late) and Henry (late)), Riccio (Gracie (late), Frankie (late), and Joseph (late)), and Wheelahan (Maree and Michael).

Fifthly, we wish the mayors and staff of all the municipalities (as studied) the continued success in leading their councils into still further achievements in sustainable urban planning (SUP) practices (especially in the face of global transition factors). We wish success to the following at NSW government level: State Premier Mike Baird, parliamentary members, and the Greater Sydney Commission. May they meet success in applying the principles of sustainability for Greater Sydney (and regions adjacent and beyond Greater Sydney covering NSW). We hope this book makes a contribution toward SUP practices beyond NSW to the wider Australia and within other countries.

Contents

Part IV Planning in St George District

Author Biographies

Raymond Charles Rauscher is a conjoint lecturer at the University of Newcastle, Australia, and a director of Habitat Association for Arts and Environment Inc. Delving into the subject of sustainable urban planning (SUP), Ray completed a PhD (2009) at the University of Newcastle. Ray has at various times lived and worked within the three Greater Sydney districts that the book focuses on – Sydney Inner West, Greater Parramatta, and St. George. Being born in Brooklyn, New York City (1943), he graduated from the City College of New York (Bachelor of Engineering, Civil, 1966). Later, he completed a master's degree (commencing) at the University of Michigan (research submission on Detroit entitled *A Solution to the Urban Crisis: Proposal for the Creation of Region Serving New Cities* (unpublished 1969)). A masters of town and country planning was completed at Sydney University (1971) with the thesis *Community Response to a Redevelopment Proposal* (University of Sydney Library Microfilm Dept.). The thesis covered planning conflicts in Erskineville and Newtown (Sydney) and measures to resolve these. Wanting to research further, Ray published (co-author Salim Momtaz) *Sustainable Communities: A Framework for Planning* (Springer 2014). Continuing to study urban change, he published (co-author Salim Momtaz) *Brooklyn's Bushwick: Urban Renewal in New York, USA* (Springer 2014). Ray then brought together long-term research (1970–2013) of the city of Sydney inner city areas, resulting in the publication (co-author Salim Momtaz) *Sustainable Neighbourhoods in Australia: City of Sydney Urban Planning* (Springer 2015). Preparing for this book (*Cities in Global Transition*) meant numerous field trips into the Greater Sydney districts, as he witnessed municipal and state efforts to adopt SUP practices in the face of global transition (e.g., population movements and settlement changes). On a community note, since 1973, Ray has been a member of the Australian-based Social Developers Network (SDN) (http://www.ned.org.au/sdn/sdnblurb.htm). The network commenced at the time of (late) PM Gough Whitlam (1972–1975) and (late) Tom Uren (minister for planning and regional development under the PM). Both of these renowned Australians introduced environmental and social bases to urban planning (Ray worked in a federal program at the time). A number of those Whitlam government planning initiatives have proven pivotal to development of all three districts that the book addresses. Contact for Ray Rauscher: ray.r@idl.net.au

Salim Momtaz is an associate professor at the University of Newcastle, Australia. He teaches in the area of sustainable resource management. He received a PhD from the University of London under a Commonwealth Scholarship. A geographer, environmental scientist, and social planner by training, Salim's research interests include environmental planning and governance, social adaptation to climate change, social impact assessment, community engagement, and urban governance. His recent published books include *Experiencing Climate Change in Bangladesh*, Elsevier (2016); *Sustainable Neighbourhoods in Australia: City of Sydney Urban Planning*, Springer (2015); *Brooklyn's Bushwick: Urban Renewal in New York, USA,* Springer (2014); *Sustainable Communities: A Framework for Planning*, Springer (2014); and *Evaluating Environmental and Social Impact Assessment in Developing Countries*, Elsevier (2013). Contact for Salim Momtaz: salim.momtaz@newcastle.edu.au http://www.newcastle.edu.au/profile/salim-momtaz

List of Abbreviations

197-a	City of New York community planning charter
ABS	Australian Bureau of Statistics
CB	Community board
CCP	Cities for Climate Protection
DPE	Department of Planning and Environment
DCP	Development Control Plan
ESD	Ecologically sustainable development
GSC	Greater Sydney Commission
GSR	Greater Sydney Region
GUT	Global urban transition
ICLEI	International Council for Local Environmental Initiatives
LA21	Local Agenda 21
LEP	Local Environment Plan
LG	Local government
LGA	Local government area
NGO	Nongovernment organization
SIW	Sydney Inner West
ST	Sustainable transport
SUC	Sustainable urban centers
SUP	Sustainable urban planning
UN	United Nations
USWIM	Urban Stormwater Integrated Management
WWF	World Wildlife Fund

List of Addendums

List of Information Boxes and Tables

List of Plates

List of Tables

Part I
Global Urban Transition and Sustainable Urban Planning

Part I of the book addresses planning of global cities and the applications of sustainable urban planning (SUP) practices. Firstly (Chap. 1), a background on globalization and a review of global cities is presented. Further, the theory and practice of applying SUP to urban areas (including global cities) around the world is examined. The planning of Greater Sydney is presented in Chap. 2. In this chapter a methodology for evaluating the extent of SUP practices being applied at municipal and district levels (groups of municipalities) is outlined. Three case study districts are nominated, being: Sydney Inner West; Greater Parramatta; and, St George. Finally, key planning principles are adopted and an outline is provided on how these principles will be applied to the case study municipalities and districts.

Chapter 1
Planning for Sustainable Cities

Dedication *This chapter is dedicated to the Cadigal and Wangal Aboriginal peoples (Plate 1.1), first inhabitants of the Sydney Districts of: Sydney Inner West, Greater Parramatta, and St George. A brief summary of early Aboriginal groups of Greater Sydney is provided in Chap. 2.*

Plate 1.1 Cadigal aboriginal peoples (Source Annandale Street Wall Art, Artist Unknown 2015)

Abstract The chapter addresses planning of global cities and the applications of sustainable urban planning (SUP) practices. A background on theory and practice of SUP is presented, particularly applicable to municipalities and districts (group of municipalities) undergoing rapid change (focusing on the effects of global transition). The chapter covers definitions and examples of sustainability and ecologically sustainable development (ESD). Reviews and critiques of writers and key documents on the subject of SUP are presented. Recent (1980s–2016) planning schools of thought (i.e., compact cities, new urbanism, smart growth, eco-city planning and SUP) are also commented on. Finally, examples of current worldwide approaches to planning municipalities and districts using SUP practices is canvassed.

R.C. Rauscher, S. Momtaz, *Cities in Global Transition*,
DOI 10.1007/978-3-319-39865-5_1

1.1 Cities in Global Transition

Looking at global cities worldwide, many of these cities are part of metropolises and conglomerations. Greater Sydney is a global city (Plate 1.2) in the Asia Pacific Area. Other global cities in this region include (right to left, bottom to top): Auckland, Jakarta, Singapore, Kuala Lumpur, Bangkok, Mumbai, Hong Kong, Shanghai, Seoul, Tokyo and Beijing.

The phenomena of global cities becoming world metropolises (note populations) is illustrated in Table 1.1.

Planning for the share size of these world metropolises challenges concepts of sustainability. Reading down the list is the city that is the subject of this book, Greater Sydney (# 59) (note Melbourne, the second Australian city on the list at # 68). There are a number of questions, raised now at the outset that this book on the subject of cities in global transition, including: (1) What is a global city? (2) What is the meaning of cities in global transition? (3) What is the extent of impacts of global cities (social, environmental and economic? (4) Can urban development in global cities be assessed against sustainability principles (e.g., UN environmental protocols)? (5) Is there some universal practices of sustainable urban planning (SUP) that can be applied to cities and city districts? (6) Can specific plans and developments likely to impact on a city district be measured and appropriate government policies be adopted to prevent (or minimize) any impact? (7) Can other cities learn lessons from the experience of one city and translate these lessons into improved governance planning policies? These questions are addressed throughout the book and revisited in the final chapter of the book Chap. 18).

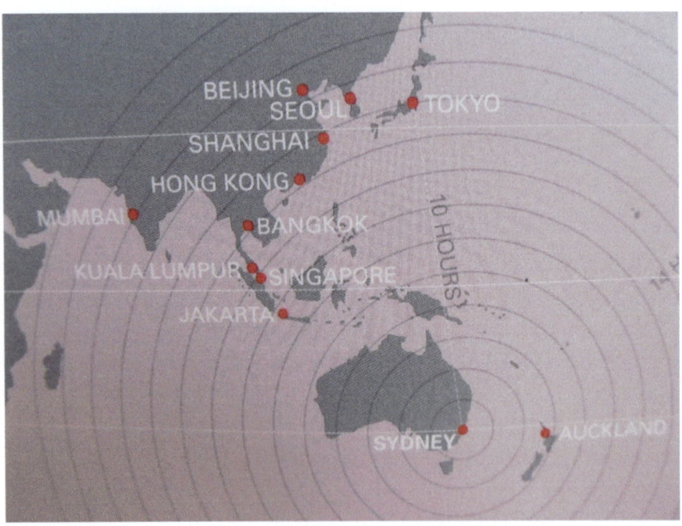

Plate 1.2 Sydney as a global city in Asia Pacific Area (Source NSW State Archives 2014)

Table 1.1 Major world metropolises and populations

1 Tokyo Japan 36,923,000	24 Shenzhen China 10,630,000	47 Belo Horizonte. Bra. 5,767,414
2 Seoul South Korea 25,620,000	25 Jakarta Indonesia 10,075,310	48 Atlanta US 5,614,323
3 Shanghai China 24,750,000	26 Lima Peru 9,886,647	49 Singapore, Sing. 5,535,000
4 Guangzhou China 23,900,000	27 Chicago US 9,554,598	50 Barcelona Spain 5,445,616
5 Delhi India 21,753,486	28 Karachi Pakistan 9,339,023	51 Munich Germany 5,203,738
6 Mexico City Mexico 21,339,781	29 Nagoya Japan 9,107,000	52 Stuttgart Germany 5,200,000
7 Beijing China 21,148,000	30 Chennai India 8,917,749	53 Ankara Turkey 5,150,072
8 Lagos Nigeria 21,000,000	31 Bangalore India 8,728,906	54 Hamburg Germany 5,100,000
9 São Paulo Brazil 20,935,204	32 Bangkok Thailand 8,305,218	55 Pune India 5,049,968
10 Mumbai India 20,748,395	33 Bogotá Colombia 7,878,783	56 Berlin Germany 5,005,216
11 New York US 20,092,883	34 Hyderabad India 7,749,334	57 Guadalajara Mexico 4,796,050
12 Osaka Japan 19,342,000	35 Hong Kong China 7,298,600	58 Boston US 4,732,161
13 Dhaka Bangladesh 17,151,925	36 Taipei Taiwan 7,045,488	59 Sydney Australia 4,667,283
14 Kolkata India 14,617,882	37 Dallas–Fort Worth US	60 San Francisco US 4,594,060
15 Tehran Iran 14,595,904	38 Santiago Chile 6,683,852	61 Surat India 4,585,367
16 Istanbul Turkey 14,377,018	39 Luanda Angola 6,542,944	62 Phoenix US 4,489,109
17 London UK 14,031,830	40 Madrid Spain 6,378,297	63 Monterrey Mexico 4,477,614
18 Los Angeles US 13,262,220	41 Ahmedabad India 6,352,254	64 Inland Empire US 4,441,890
19 Buenos Aires Arg. 13,074,000	42 Houston US 6,490,180	65 Rome Italy 4,321,244
20 Paris France 12,005,077	43 Philadelphia US 6,051,170	66 Detroit US 4,296,611
21 Rio de Janeiro Br. 11,973,505	44 Washington.DC US 6,033,737	67 Milan Italy 4,267,946
22 Manila Philippines 11,855,975	45 Toronto Canada 6,055,724	68 Melbourne Australia 4,246,345
23 Rhine-Ruhr Ger. 11,470,000	46 Miami US 5,929,819	

Source: UN Global Population Report 2015

The concept of global cities stretches back to Patrick Geddes when he used the term in 1915 (century ago) to describe the city's influence in finances and trade around the world. Later, Canadian philosopher Marshall McLuhan popularized the term Global Village, beginning in 1962. His view suggested that globalization would lead to a world where people from all countries would become more integrated and aware of common interests and shared humanity. By 2015, the Institute of Urban Strategies (IUS 2015) had completed a study on global cities and adopted 70 indicators for these cities, including for example: economy, research and development, cultural interaction, livability, environment and accessibility. Some common global city activities within these indicators include: international financial services (including real estate, stock exchange and banking); head offices of multinational corporations; major manufacturing centers with ports; centers for new ideas and innovation; media and communications on a global network basis; and, high quality educational facilities. There has been, however, debate (pros and cons) on the advantages and disadvantages of globalization of cities. To look closer at this debate, a sample of four (4) aspects of globalization (focusing on global cities) are addressed next. These include aspects of: (1) population growth; (2) Social, cultural and economic factors; (3) environmental and human resources; and, (4) urban planning.

1.1.1 Population Growth

Global cities normally experience rapid and continued population increases (natural and from immigration). Current projections, by the UN based Food and Agriculture Organization (FAO), show a continued increase in projected world population (but a steady decline in the population growth rate) (UN FAO 2015). The FAO projects that the world population will hover about nine billion by 2050 (vs. current population of seven billion). The FAO warns, however, that due to a decline in fossil fuels (thus impacting on transportation and food production) the world could see a period of contraction and economic delocalization in coming decades. An important factor, in looking at planning of global cities, the UN notes megacities (cities having a population in excess of ten million) grew in number from 3 megacities in 1973 to 24 by 2013, with estimates of up to 27 megacities by 2025. The FAO notes further that India's and China's high economic growth is not sustainable. The FAO writes that the world's ecological capacity is simply insufficient to satisfy, in a sustainable way, the ambitions of China, India, Japan, Europe and the United States as well as the aspirations of the rest of the world. The FAO thus indicates that cities within urban agglomerations need to examine their sustainability. In Asia, for example, the urban agglomerations include: Osaka, Karachi, Jakarta, Mumbai, Shanghai, Manila, Seoul, and Beijing (each already home to over 20 million people). Looking further, Delhi and Tokyo are forecast to approach or exceed 40 million people, each within the coming decade. Outside Asia, the FAO reports that Mexico City, São Paulo, New York, Lagos, Los Angeles, and Cairo are, or soon will be, home to over 20 million people. Looking back, there has been a rapid urbanization of the world's population over the twentieth century. The UN reported in 2005 within the *Revision of the UN World Urbanization Prospects Report* (UN 2005):

> The global proportion of urban population rose dramatically from: 13 % (220 million) in 1900 to 29 % (732 million) in 1950 and, 49 % (3.2 billion) in 2005. The same report projected that the figure is likely to rise to 60 % (4.9 billion) by 2030. It is expected that by 2050, the global urban population (that is living in cities or urban regions) will be at 6.4 billion). (*Source* UN 2005)

1.2 Social, Cultural and Economic Factors

There are a number of social, cultural and economic characteristics related to globalization of cities, for example: cosmopolitanism; gentrification; and, pros and cons of globalization. Firstly, cosmopolitanism is the proposal that all ethnic groups belong to a single community based on a shared morality. K.A. Appiah (1997) writes:

> A person who adheres to the idea of cosmopolitanism in any of its forms is called a cosmopolitan. A cosmopolitan community might be based on an inclusive morality, a shared economic relationship, or a political structure that encompasses different nations. The cosmopolitan community is one in which individuals from different places (e.g., nation-states) form relationships based on mutual respect. There is the possibility of a cosmopolitan community in which individuals from varying locations (physical, economic, etc.) enter relationships of mutual respect, despite their differing beliefs (religious, political, etc.).

Moving to urban gentrification, this term denotes socio-cultural changes in an area, such as wealthier people buying housing property in a less prosperous community. One description of this in Wikipedia notes:

> Consequent to gentrification, the average income increases and average family size decreases in the community. This often results in the informal economic eviction of the lower-income residents, because of increased rents, house prices, and property taxes. This type of population change reduces industrial land use when it is redeveloped for businesses and housing (though industry is often already moving further away from these centers). In addition, new businesses, catering to a more affluent base of consumers moving into these areas. This further increases the appeal to more affluent migrants and decreases the accessibility to less wealthy natives.

Finally, on the question pros and cons of globalization, there are two sides to this debate. Amartya K. (1970) notes there is a broad-range of ideologies (leading to social movements) on the issue of globalization. Proponents of economic growth (expansion and development), in general, view globalizing processes as desirable (or necessary) to the well-being of human society There are, however, also international political factors of globalization that can lead to conflicts. K Sorrells (2013) points out that the forces of globalization have led to the spread of western-style democracy, but with some backlashes. Sometimes, on the downside, he argues, this spread has been accompanied by an increase in inter-ethnic tension and violence.

This can occur where different political views are expressed (i.e., free market economic policies, or democratic processes of universal suffrage). This has in some regions of the world led to an escalation in militarization (i.e., civil conflicts over imposition of political systems, including socialistic vs. democratic ones).

1.3 Environmental and Human Resources

There are a number of environmental and human resource questions that arise as a result of urban globalization. The UN (2005) notes:

> The travel time between distances is shrinking between continents and countries due to globalization, causing developing and developed countries to find new ways to solve problems on a global rather than regional scale. Agencies like the United Nations now must be the global regulators of pollution, whereas before, regional governance tackled this problem. Action has been taken by the United Nations to monitor and reduce atmospheric pollutants through: the Kyoto Protocol, the UN Clean Air Initiative, and studies of air pollution and public policy. Global traffic, production, and consumption are causing increased global levels of air pollutants. The northern hemisphere, for example, has been the leading producer of carbon monoxide and sulfur oxides. In addition, on the human resources side, the global workforce (international labor), reflects a new division of labor that has been emerging since the late 1970s. The global economic factors driving the rise of multinational corporations – namely, cross-border movement of goods, services, technology and capital – are changing ways of thinking about labor and the structure of today's workforce. With roots in the social processes surrounding the shift to standardization and industrialization, post-industrial society in the Western world has been accompanied by industrialization in other parts of the world, particularly in Asia. (*Source* UN 2005)

Helena Norberg-Hodge (1992), the director and founder of Local Futures International Society for Ecology and Culture, criticizes globalization in many ways. In her book *Ancient Futures*, she writes:

> Centuries of ecological balance and social harmony are under threat from the pressures of development and globalization. The standardization and rationalization of globalization does not always yield the expected growth outcomes. Although globalization takes similar steps in most countries, it might not be effective in certain countries and globalization has actually moved some countries backward instead of developing them. (*Source* Hodge 1992)

Finally, new factors of environmental concern can arise, for example the recent alarm over the existence of urban heat islands within cities. The process is defined as follows:

> An urban heat island is formed when industrial and urban areas produce and retain heat. Much of the solar energy that reaches rural areas is consumed by evaporation of water from vegetation and soil. In cities, where there is less vegetation, the exposure is greater. Most of the sun's energy is instead absorbed by buildings and asphalt. This leads to higher surface temperatures. Vehicles, factories and industrial and domestic heating and cooling units release even more heat. As a result, cities are often 1 to 3 °C (1.8 to 5.4 °F) warmer than surrounding landscapes. Impacts also include reducing soil moisture and a reduction in reabsorption of carbon dioxide emissions.

1.4 Urban Planning

When the residential areas shift outward, the process is called suburbanization. A number of researchers and writers (i.e., Garreau 1991) suggest that suburbanization has gone so far to form new points of concentration. These are often outside the core of cities, both in developed and developing countries. This networked, poly-centric form of concentration is considered by some as an emerging pattern of urbanization. It is called variously by terms such as exurbia, edge city network city (Batten 1995), or postmodern city (Dear 2000). Dear points out that Los Angeles is the best-known example of this type of urbanization. Interestingly, Dear points out, in the United States this process has reversed as of 2011, with re-urbanization occurring (due to chronically high transport costs).

Moving to planned urbanization, this process is based on an advanced plan being adopted. This plan can be prepared for military, aesthetic, economic or urban design reasons. UN agencies, for example, prefer to see urban infrastructure installed before urbanization occurs (UN 2005). Landscape planners are responsible for landscape infrastructure (public parks, sustainable urban drainage systems, greenways etc.). This can be planned before urbanization takes place, or afterward to revitalize an area and create greater livability within a region. As the population continues to grow and urbanize at unprecedented rates, new urbanism and smart growth techniques are needed to create successful transitions into developing sustainable cities. Smart growth and new urbanism's principles (discussed later in this chapter), for example include walkability, mixed-use development, comfortable high-density design, land conservation, social equity, and economic diversity. Finally, mixed-use communities work to fight gentrification with affordable housing. This promotes social equity, decreases automobile dependency (thus lowering use of fossil fuels), and promoting a localized economy. An examination of sustainable urban planning (SUP) is addressed next.

1.5 Sustainable Urban Planning (SUP)

The subject of sustainable urban planning (SUP) has become a major focus of town planning in most countries, developed and developing. An approach to the planning for sustainable communities is contained in *Sustainable Communities: A Framework for Planning* (Rauscher and Momtaz 2014a). The authors note there has been a call for governments throughout the world to respond to environmental issues, especially within cities and including: "need to renew older cities; impact of climate change; carbon emissions; pressures to accommodate new immigrant residents; and, depleting resources (e.g., peak oil)". All of these issues are relevant to municipalities and metropolitan districts (as well as cities and regions overall). The authors point out that governments, in response, are trying to develop policies to address these concerns, often adopting strategies aimed to achieve sustainable communities (social/cultural, environmental, and economic). The question of how to achieve these communities at municipal and district levels is addressed in coming chapters with an examination of Greater Sydney.

Addressing SUP further, Rauscher and Momtaz (2014a) note that current planning practices are moving towards a more coordinated and holistic framework in incorporating sustainability principles in cities. These principles will be examined more closely latter in this chapter. Today (especially since the early 2000s) a growing portion of the development industry accepts evaluation of development proposals under SUP policies, including ecologically sustainable development (ESD) principles. The more education, bonuses and tax breaks that the industry gets, the more interest there is in adopting ESD principles in development. One common development concern within government, for example, is the fact that 50 % of world's oil resources ('peak oil') has been expended. Thus, car dependent cities are now needing to examine alternatives to motor transport, including more cycling and walking. Also, thinking sustainably, cities are looking towards alternative transport and new fuels. The municipalities within Greater Sydney's Districts of Sydney Inner West, Greater Parramatta, and St George, for example, have been expanding their cycle networks and participating in extensions to the light rail system within their jurisdictions (to be addressed in coming chapters).

As worldwide expressions of concern about deteriorating environments continue in 2016, the writings on the subject provide valued backgrounds leading to solutions. Most writers agree that environmental problems are a causal factor of uncontrolled urban growth. Greater Sydney (incorporating the City of Sydney and the whole of the metropolitan region), for example, experiences this phenomenon of ongoing major growth. The result is ever outward expansion as well as urban consolidation (normally medium and high density development). Key advances in sustainability applications in most world cities have been translated into ESD related documents (in the form of protocols, acts and reports). These documents usually refer to sustainability components, including: sustainability principles; sustainability goals; and, indicators of sustainability. The documents exist today at all levels of government (municipal, district, city, regional, state, national, and international), as well as being drawn up by non-government organizations. International ESD related documents (particularly those applying to the natural and built environments) and being interpreted in planning cities around the world. These documents have increased in numbers, in application and subject area since the early 1970s. The central international protocol, Agenda 21 (UN 1992), relates to ESD and is a global action plan for sustainability. The document lays out key procedures for governments to adopt ESD strategies from city level to neighborhood level. In moving beyond Agenda 21, the UN developed *Local Agenda 21* (UN 1993) from one chapter Agenda 21. Local Agenda 21 is a mechanism to encourage greater involvement by local authorities in delivering Agenda 21 programs. In 1997 *Local Agenda 21 Model Communities Program* (UN 1997) was adopted. This program aids local government in implementing the Local Agenda 21 program. The Model Communities Program documented those local government areas that had adopted Local Agenda

21 planning processes for sustainable development. The take up of the Local Agenda 21 by a number of Greater Sydney councils will be addressed in coming chapters.

Authorities, having adopted ESD principles (partly using the above agreements), normally move on to ESD goals. Low (2000) argues that adopting ESD principles and goals is crucial to confronting cities that are depleting resources without limit. He spells out how cities can adopt these principles and goals, and as a result ensure renewal of cities and preservation of resources for future generations. Equipped with ESD goals, authorities are in a position to adopt the key ESD components for measurement (e.g., 'indicators of sustainability'). These indicators enable ESD goals to be measured and monitored at municipal and district. The application of indicators of sustainability by municipalities in Greater Sydney will be examined in upcoming chapters. By 2002, the United Nations adopted the *Political Declaration and Plan of Implementation (Johannesburg)* (UN 2002). This was an important document for all governments around the world to address as it incorporated steps for authorities to adopt sustainability goals in their planning.

Generally, the NGO sector has been increasingly pro-active in alerting governments to act on sustainability planning. Key NGO documents, for example, have contributed to the debate around national actions on climate change (Stern 2006). The take up by the NGO sector of Al Gore's (2007) *An Inconvenient Truth* is a measure of this sector's work in educating the public on environmental issues. The worldwide education of 'climate change educators' (trained under Al Gore) is further testimony to take up actions. It is the climate change concerns, incidentally, that continue today (2016) to drive public environmental interest (especially from 2003 onwards). In addition, by the early 2010s the mayors of a number of cities, agreed to cooperate on CO2 emissions. The mayors adopted a maximum future 2° world temperature change to avert the predicted greater disasters if temperature changes are greater than 2°. In late 2015 the Paris Agreement on the subject reinforced this stance by also adopting the maximum of 2° change. The coming chapters, in looking at case study districts of Greater Sydney, examine various municipal greenhouse gas reduction programs.

Key urban planning schools of thought, Rauscher and Momtaz (2014a) note, have influenced the take up by authorities of ESD criteria. Some of these schools of thought include: *compact cities, new urbanism*; *smart growth*; and, *eco-city planning and eco-villages*. Compact cities aim for higher density development around transport nodes. This includes the slowing of urban sprawl by increasing allowable urban densities in inner city areas as well as growth centres. This was a reversal of past strategies (after the Second World War) that encouraged urban sprawl. It is under compact cities that ESD principles gained attention, including within case study and districts to be examined in coming chapters. *New urbanism* emphasizes building neighborhoods with a diversity of residential, commercial and light industrial land uses in close proximity. The notion of *smart growth* followed new urbanism as a complementary school of planning thought in the early 1990s in the United States (Urban Land Institute 1995) and cities in Australia too. This school aims to limit urban sprawl through improved land use and transport policies. Older inner city areas often reflect the aims of smart growth.

A further land use urban planning school of thought is *eco-city and eco-village planning*. Walker (1997) argues that a range of tools are being applied within the concept of eco-cities, providing authorities adopt the appropriate strategies. Upcoming chapters examine the extent of municipalities within case study districts taking up an eco-city principle of biodiversity protection. There has been emerging since the 1990s a *sustainable urban planning* (SUP) school of thought. This school of thought partly relies upon frameworks being developed to guide ESD based urban planning. Looking at the application of SUP, the City of New York adopted a charter for community engagement to create local plans as outlined in *Brooklyn's Bushwick – Urban Renewal in New York, USA* (Rauscher and Momtaz 2014b). Aware of this community planning interest as early as 1963, thc City of New York wrote into the City Charter the establishment of Community Boards. The mechanisms for more cooperative planning between Greater Sydney's municipalities and districts and the NSW State government will be examined in upcoming chapters, particularly Chap. 18. Planning sustainable cities and sustainable city districts is examined next.

1.6 Sustainable Cities and Sustainable City Districts

There are a number of authors who have challenged authorities to develop new frameworks for planning cities and city districts, often beyond the main city CBD. In *The City Reader* (fifth edition) (Richard T. LeGates 2011) a number of writers on the subject of sustainable urban planning (SUP) reinforce the importance of this approach to city planning. One city that has taken up this approach is Melbourne, under *The Urban Design Charter* (Victoria 2014). Here is a commitment by the Victorian government to make cities and towns in Victoria more liveable through good urban design. The Charter identifies a range of design principles, such as: sense of place, continuity and change, safety, inclusiveness and interaction. Moving to another city (Tokyo) looking at achieving sustainability principles, Marieluise Jonas (Jovis 2015) emphasizes the design of vacant spaces can be done in a creative way. He argues that a city has unlimited access to spaces that could be more efficiently used. Taking this approach, the City of Sydney announced in 2015 that large swaths of its South Sydney heavy industry lands would be earmarked for affordable housing and commercial and light industry for a population increase of up to 100,000. Finally, Rachel Keeton (2011) in *Rising in the East* argues for completely planned and built forms for cities. She notes that throughout the Asian continent, much of the recent urbanization has come in the guise of New Towns. The challenge for existing urban areas (including older metropolitan districts) undergoing renewal is to find the sites to create these 'new towns in town'.

Master planning of cities and city districts is an approach increasingly favoured by authorities to ensure sustainability principles are included. Leonard Schenk (2013) emphasizes that master planning needs to be functional and still result in urban designs that create a vibrant city or city district. Moving to the development of new cities, Daan Roggeveen (2012) looks to China for master planning that addresses big populations and ethnic diversity. This approach applies to the growing multi-cultural composition of older metropolitan districts in Sydney (and other global cities). In contrast Robert Stern (2013) emphasizes the re-emergence of design principles of the planning of garden cities. Stern addresses bringing changes in the transition of older suburbs into metropolitan communities with added open spaces and natural lands. Addressing the ecologies of cities and master planning in Australia, Timothy Beatley (2008) emphasizes ecological assets and green urbanism. Examples of Australian capital cities (and districts within those cities) moving to more sustainable practices are citied. Still within Australia, Ken Yeang (2010), in *Dictionary of Ecodesign: An Illustrated Reference,* outlines the means to achieve a sustainable future is through 'bio-integration', design of human interventions into the natural environment. Yeang calls for wider scale regional and urban planning to enable engineering and architecture disciplines to carry through the designs (thus reflecting these plans) in older districts and new ones. He emphasizes that master planning is the opportunity to redress current environmental challenges. Finally, Theodore Spyropoulos (2013) states that it is vital master planning needs to cope with cultural complexities in city building. He coins the phrase 'adaptive ecologies' to develop a model of urban planning.

Having addressed the importance of master planning in planning cities districts sustainably, other writers address how urban areas can be distinctly different from each other (and still incorporating sustainability principles). In *The Spirit of Cities: Why the Identity of a City Matters in a Global Age* Daniel Bell (2011) calls for the revival of the classical idea that a city expresses its own distinctive ethos or values. Districts outside the CBD of a main city (such as districts beyond the City of Sydney to be introduced in Chap. 2) have the potential to achieve this concept. Gleeson (2014), in his book *The Public City* on the importance of civic and institutional spaces in a city, argues for creation of the 'public city'. A useful guide to create this pubic city is the *Urban Code* (Mikoleit 2011). Mikoleit here offers lessons to read the city, and thus know its requirements at the pedestrian level. To achieve changes in any city, given the challenges noted earlier in this chapter, takes wide community input. In *We Own The City: Enabling Community Practice in Architecture and Urban Planning* (Miaza 2014), Miaza promotes community ownership of urban spaces. He presents instances of improved urban areas through greater involvement of local communities. Agreeing with this approach is Matthew Carmona (2012) in *Capital Spaces: The Multiple Complex Public Spaces of a Global City.* Carmona notes that the development market has too often been left to create urban public space to its own interests. In these instances, the author argues, the public sector has been relegated to the edges of the space-making process. The result, it is argued, fails to meet the needs of a wider public and of the diversity of users that make up that community. Reinforcing this is Farrell (2014) in *Tangled Bank,* where he encourages planners and architects to follow the biologists. He thus argues for urban design to learn from nature's patterns and to compliment these patterns (not opposed to them).

Moving from planning metropolitan districts with a difference in design, the accommodation of human needs within these districts is paramount in ensuring sustainability. Marusic (2014) in *Human Cities: Celebrating Public Space* argues that public spaces need to be designed to encourage interactions among people. He then presents tools that can be used within urban planning to achieve better public space design. Following this theme, Vittorio Aureli in *City as Project* (2013) draws attention to cities (and districts) being shaped not only by physical forces, but also by cultural ones. He argues that planners need to appreciate that urban places should reflect these cultural aspirations, which is turn lead to specific architectural designs. Finally, in an attempt to also ensure adequate housing is provided, Alejandro Aravena (2012) in *Elemental: Incremental Housing and Participatory Design Manual* sets the goal (in Santiago) of planners designing urban places to first meet the social needs of the community. He argues for community participatory building and for projects devoted to meeting basic infrastructure and transportation needs.

Moving from these concepts of planning cities to the tools in building cities, a number of authors offer advice here. Ken Yeang (2010) (noted earlier) provides a whole dictionary on eco-designing. He outlines over 1500 descriptions of eco-design terms to assist planners and city administrators. Moving again to Asia, many developments there over the last several decades have gone from green design to sustainable communities. Nirmal Kishnani (editor) (2013), notes some of these initiatives in *Greening Asia – Emerging Principles for Sustainable Architecture*. He examines a number of cities attempting to lessen the impact of development. Richard Hyde (2012), in *Bioclimatic Housing: Innovative Designs for Warm Climates*, takes the challenge of climate change protection in designing cities. He focuses on the growing interest in the relationship between nature and design within cities. Hyde (under a long term International Energy Agency project) offers low-energy architectural solutions in housing design within many countries, including Australia. Finally, Chris Van Uffelen (2015), in *Pedestrian Zones: Car Free Urban Spaces*, emphasizes that it is the urban areas where pedestrians congregate that need the most attention to bring about sustainable cities. Uffelen argues that pedestrians need to have right of way over vehicles. He provides many examples where cities around the world have turned over sections of downtowns to open plazas (including many Australian cities). The next Chap. 2 will introduce the planning of Sydney's metropolitan districts.

1.7 Summary

The chapter addressed planning of global cities and the applications of sustainable urban planning (SUP) practices. First, a background on theory and practice of SUP was presented, focusing on municipalities and districts that are undergoing rapid change (especially within global cities). The chapter covered definitions of sustainability and ecologically sustainable development (ESD), with examples given.

Reviews and critiques of writers and key documents on the subject of SUP were then presented. Recent (1980s–2016) planning schools of thought (i.e., compact cities, new urbanism, smart growth, eco-city planning and SUP) were also commented on and further examples given. Finally, current worldwide approaches to planning municipalities and districts using SUP practices were canvassed.

References

Amartya K (1970) Collective choice and social welfare. Holden-Day, San Francisco

Appiah KA (1997) Cosmopolitan patriots. Crit Inq 23(3):617–639. Spring, NY

Aravena A (2012) Elemental: incremental housing and participatory design manual. Hatje Cantz, New York

Aureli PV (ed) (2013) The city as a project. Ruby Press, New York

Batten D (1995) Network cities – creative urban agglomerations in the 21st century. Sage, New York

Beatley T (2008) Green urbanism down under. Island Press, Washington, DC

Bell DA (2011) The spirit of cities: why the identity of a city matters in a global age. Princeton University Press, New York

Carmona M (2012) Capital spaces: the multiple complex public spaces of a global city. Routledge, New York

Dear M (2000) Postmodern urban conditions. Wiley, New York

Farrell T (2014) The city a Tangled bank. Wiley, New York

Garreau J (1991) Edge city: life in the new frontier. Doubleday, New York

Gleeson B (2014) The public city. Melbourne University Press, Melbourne

Gore A (2007) An inconvenient truth. Wiley, New York

Hyde R (ed) (2012) Bioclimatic housing: innovative designs for warm climates. Routledge, New York

Institute of Urban Strategies (2015) Indicators for global cities. Author, Tokyo

Jonas M (2015) Tokyo void: possibilities in absence. Jovis, New York

Keeton R (2011) Rising in the East. Sun, New York

Kishnani N (2013) Greening Asia – emerging principles for sustainable architecture. Bci publications, Singapore

LeGates RT (2011) The city reader, 5th edn. Routledge, New York

Low N (ed) (2000) Consuming cities: the urban environment in the global economy after the Rio declaration. MIT Press, Cambridge, MA

Marusic B (ed) (2014) Human cities: celebrating public space. Stichting Kunstboek, Oostkamp

Miaza F (2014) We own the city: enabling community practice in architecture and urban planning. Valiz/Trancity, New York

Mikoleit A (2011) Urban code. MIT Press, Boston

Norberg-Hodge H (1992) Ancient futures: learning from Ladakh. Sierra Club, San Francisco

Rauscher RC, Momtaz M (2014a) Sustainable communities: a framework for planning. Springer, Amsterdam

Rauscher RC, Momtaz M (2014b) Brooklyn's bushwick – urban renewal in New York, USA. Springer, Amsterdam

Roggeveen D (2012) Go west project: China's new megacities. David McManus, New York

Schenk L (2013) Designing cities – basics, principles, projects. Birkhauser, New York

Sorrells K (2013) Intercultural communication: globalization and social justice. Sage, Thousand Oaks

Spyropoulos T (2013) Adaptive ecologies. Architectural Association Publications, New York
Stern R (2006) International climate change final report. Her Majesty's Treasury, London
Stern RA (2013) Paradise planned: the garden suburb and the modern city. Stern Architects, New York
United Nations (1992) Agenda 21. United Nations, New York
United Nations (1993) Local agenda 21. United Nations, New York
United Nations (1997) Local agenda 21 model communities program. United Nations, New York
United Nations (2002) Political Declaration and Plan of Implementation (PDPI). United Nations, New York
United Nations (2005) Revision of the UN world urbanization prospects report. United Nations, New York
United Nations Food and Agriculture Organization (2015) Global population report. United Nations, New York
Urban Land Institute (1995) Smart growth. Cambridge Press, New York
Van Uffelen C (2015) Pedestrian zones: car free urban spaces. Braun Publishers, New York
Victoria (2014) The urban design charter. Victoria, Australia, Melbourne
Walker B (1997) Sustainable cities: concepts and strategies for eco-city development. Eco Home Publications, Los Angeles
Yeang K (2010) Dictionary of ecodesign: an illustrated reference. Routledge, New York

Chapter 2
Planning Greater Sydney and Sydney Districts

Dedication *This chapter is dedicated to Denis Winston (1908–1980) (Plate 2.1), architect and town planner, who was the foundation chair of the school of Town and Regional Planning (first in Australia, at the University of Sydney in the early 1960s). Winston provided Sydney with an academic basis for sound planning and the training of professional town planners to undertake the challenges facing a growing Greater Sydney, including the emergence of sustainable urban planning (SUP) principles (Chap. 1).*

Plate 2.1 Denis Winston (Source Australian Dictionary of Biography 2016)

Abstract In this chapter a methodology for evaluating the extent of sustainable urban planning (SUP) practices being applied at municipal and district (groups of municipalities) levels is outlined. The chapter commences with an introduction to the planning of Greater Sydney. Three case study districts are nominated, being: Sydney Inner West; Greater Parramatta; and, St George. A methodology is adopted for assessing the extent each municipality within the three districts is applying SUP practices (as outlined in Chap. 1). Key planning principles are then adopted and an outline given on how these principles will be applied to case study districts

R.C. Rauscher, S. Momtaz, *Cities in Global Transition*,
DOI 10.1007/978-3-319-39865-5_2

(and municipalities within). These principles are: (1) Developing Sustainable Urban Centers; (2) Developing Sustainable Transport; and, (3) Protecting the Natural Environment. The process of State initiated local government reform and council amalgamations (given connections to planning Greater Sydney) are also reviewed at this stage. Finally, the chapter notes that conclusions reached in examining the selected districts (and municipalities within) will be used in the final chapter (Chap. 18). That chapter considers Greater Sydney planning reforms that are needed for municipal and district planning to ensure a sustainable Greater Sydney in the future. Finally, understanding these reforms for Greater Sydney, this final chapter will also comment on the means of achieving a sustainable city anywhere in the world.

2.1 Greater Sydney Planning

The Greater Sydney Region (sometimes called Greater Metropolitan Region (GMR), or Sydney Metropolitan Area (SMA) is herein called Greater Sydney (located on the east coast of Australia). Other urban areas follow the coast line of Australia, north and south of Greater Sydney and around a major portion of Australia. Greater Sydney stretches from: the ocean (east) to the Blue Mountains (west); and, north to Broken Bay leading into the Central Coast (Gosford and Wyong) and south to the Illawarra Region (City Wollongong and other centres).

Those Greater Sydney districts can be viewed in the Plate 2.2, showing all districts.

The Greater Sydney districts relative to this book (as case studies) include: Inner West (center and green) to the left of Central Sydney (Purple); Parramatta (red) to the west and next to Inner West; and, St George (orange) to the south (lower). Greater Sydney's Outer Growth Regions (bottom and clockwise) include: Sutherland Shire; South West (includes cities of Liverpool, Fairfield, Campbelltown, and Wollondilly and Shire of Camden); Outer West (includes cities of Blacktown, Penrith and Blue Mountains); and, North West (includes the City of Hawkesbury and Shire of The Hills).

There are separate regions beyond Greater Sydney within commuter distance (1–2+ hours), including: to the north is the Central Coast Region as noted above (Gosford and Wyong); further north is the Lower Hunter Region (containing cities of Newcastle, Lake Macquarie, Cessnock, and Maitland and Shire of Port Stephens); and, finally to the south is the Illawarra Region as noted above (containing cities of Wollongong and Shellharbour, Shire of Wingecarribee and Town of Kiama). Hence, Greater Sydney has a number of regions (containing cities, shires and towns) in north and south directions. Greater Sydney and these regions as noted (and remainder of urban and rural regions of the State of New South Wales) come under the State planning authority centered on the Dept. of Planning and Environment (with Sydney and regional offices).

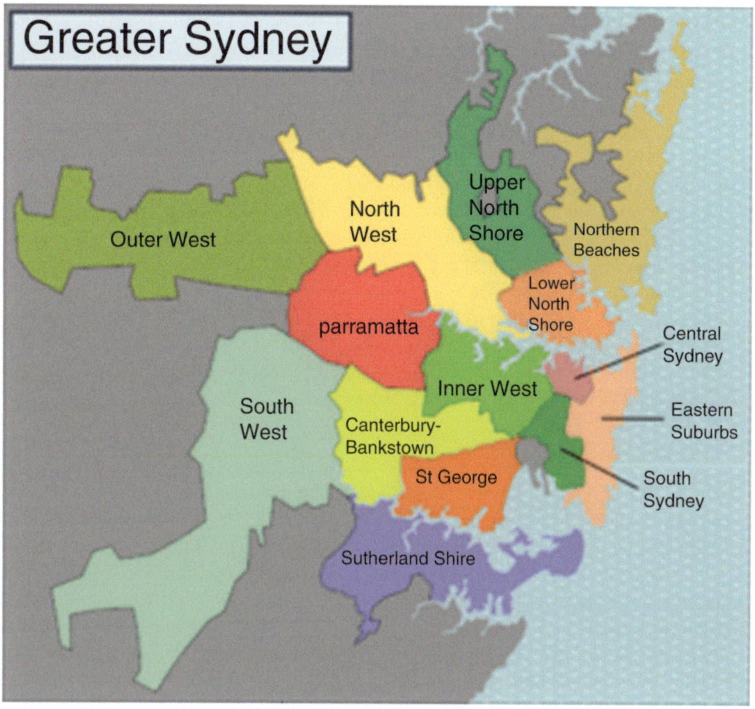

Plate 2.2 Greater Sydney and Districts (Source *Wikipedia Travel*. 2016)

The first people to inhabit Greater Sydney (45,000–50,000 years ago) area were indigenous Australians, having migrated from northern Australia. The Cadigal and Wangal Aboriginal clans were part of the first inhabitants of a number of Greater Sydney districts (i.e. Sydney Inner West, Greater Parramatta and St George) (these districts to be introduced later in this chapter). The Cadigal and Wangal Aboriginal clans (of the Dharug language group) are part of the Eora nation. A historical note on the Aboriginal people in Wikipedia reads:

> The earliest British settlers called them Eora people. "Eora" is the term the indigenous population used to explain their origins upon first contact with the British. Its literal meaning is "from this place". Prior to the arrival of the British there were 4000 to 8000 native people in Sydney from as many as 29 different clans. The first meeting between the native people and the British occurred on 29 April 1770 when Lieutenant James Cook landed at Botany Bay. Cook was on a mission of exploration and continued north along the east coast of Australia and claiming the new land he had discovered for Britain. Development has destroyed much of the city's history (of that era) including that of the first inhabitants.

The chapters ahead commence with European settlement and phases of development from the 1800s. Common to these areas is the Aboriginal communities' ancestry. There has been a resurgence of renewing the culture of the Aboriginal people of

Greater Sydney since the early 1960s. This movement is evident and strong in all of the local government areas surveyed in coming chapters. A brief snippet of the Cadigal Aboriginal people follows.

> The Cadigal, sometimes spelled Cadigal, are a group of Aboriginal Australians who originally inhabited the area that they called 'Cadi'. Cadigal territory lies south of Port Jackson covering today's Sydney CBD and stretches from South Head to Petersham with part of the southern boundary lying on the Cooks River. The Cadigal were coastal people who were dependent on the harbor for providing most of their food. The occupation of the Sydney area by the British and the subsequent introduction of European diseases including smallpox decimated the Eora people and their neighbors. The disastrous 1789 smallpox epidemic is estimated to have killed about 50 % of Sydney's indigenous population, and it has been claimed that only three Cadigal people were left alive by 1791. Part of the Cadigal territory is in the Marrickville local government area of Sydney. In 1994 the Marrickville Aboriginal Consultative Committee was established.

A look at the Wangal Aboriginal people follows.

> The Wangal Aboriginal people were believed to have a territory encompassing modern-day Concord and stretched east to the swampland of Long Cove Creek (now known as Hawthorne Canal). The land was heavily wooded at the time with tall eucalypts covering the higher ground and a variety of swampy trees along Iron Cove Creek… Wangal country was known as 'wanne' and it originally extended from the suburbs of Balmain and Birchgrove in the east to Silverwater and Auburn in the west. The northern boundary was the Parramatta River. Neighboring Dharug bands were the Cadigal to the east, the Wallumattagal on the northern shore of the Parramatta River and the Bediagal to the south.

Acknowledging the early Aboriginal people, the settlement of Sydney commenced with the arrival of the first fleet from England in 1788 under Captain Cook. Early Sydney was built with a significant input from convicts as transported from England. The municipal council of Sydney was incorporated in 1842 and became Australia's first city. By 1850 (final year of convict transportation) Sydney had a population of 35,000. With the discovery of gold in 1851 within parts of the colony (under England) the population increased dramatically, seeing Sydney at 200,000 by 1871. Following the depression of the 1890s, the six colonies agreed to form a federated nation of The Commonwealth of Australia (1901). Sydney, with a population of 481,000, then became the state capital of New South Wales. The population continued to boom to reach one million in 1925. The Great Depression of the 1930s had a severe effect on Sydney's economy, as it did with most cities throughout the industrial world. Post World War Two saw a mass migration to Australia, with increases in all ethnic groups including, for example: Greek, Lebanese, Italian, Czech, Lithuanian, Polish, German, Ashkenazi Jews, Serbian, Macedonian, Russian, Sri Lankan, Filipino, Korean, and Fijian communities… By 1950 there were 1.7 million people living in Sydney, increasing to almost three million by 1975. Sydney's population at the time of the 2011 census was 4,391,674 (ABS 2011). At this census it was recorded that there were 1,503,620 people born overseas. The overseas born accounted for 42.5 % of the population of the City of Sydney and

34.2 % of the population of Greater Sydney). On a world comparative basis this placed the City of Sydney as the seventh highest city with this proportion of overseas born. It has been forecasted by the Australian Bureau of Statistics (ABS) that the population of Greater Sydney will grow to between 8 and 8.5 million by 2061 (ABS 2011). Despite this increase, the ABS predicts that Melbourne will replace Sydney as Australia's most populous city by 2053. Finally, in terms of density of urban areas, the four most densely populated suburbs in Australia are located in Sydney with each having more than 13,000 residents per square kilometer (33,700 residents per square mile) (ABS 2011).

Greater Sydney is now a metropolitan area built on a spread of local government areas (LGAs) (43 currently in 2016). Most of the areas of Greater Sydney were developed in the 1800s (as ‘villages’, ‘neighborhoods’ and ‘suburbs’ grew) (Plate 2.3). These LGAs are beyond the ‘City of Sydney’ which is the core municipality of Greater Sydney. The City of Sydney planning is outlined in *Sustainable Neighbourhoods in Australia: City of Sydney Urban Planning* (Rauscher and Momtaz 2015).

Plate 2.3 Early Sydney Suburbs. (Source Spearritt 1988)

The *draft Metropolitan Strategy for Sydney* (NSW 2012a) sets out a plan for Greater Sydney's future, including affordable homes, jobs and transport. This Strategy as developed the NSW State Government (herein call 'the State') was superseded by *A Plan for Growing Sydney* (NSW 2014a) (herein referred to as 'the Plan'). This Plan is an overarching document supported by the principle of subregions of Greater Sydney accommodating future growth. The Plan is thus a guide for Greater Sydney absorbing a further 1.6 m people by 2031 (bringing the population to 6 m). Some to the key local government growth areas of Greater Sydney include (not inclusive, clockwise from top): Macquarie Park; Chatswood; St. Leonards; North Sydney; City of Sydney; Bondi Junction; Green Square; Randwick; Kogarah; Hurstville; Bankstown; Burwood; Sydney Olympic Park; Rhodes; Parramatta; and Blacktown. The Plan, in examining where growth will occur, notes:

> Long term planning in Western Sydney will be necessary to create a setting for jobs, infrastructure and services to meet the needs of current and future populations. The growth is thus targeted towards strategic centres across Western Sydney. This will include taking advantage of opportunities arising from new infrastructure investment such as Badgerys Creek Airport, the South West Rail Link, North West Rail Link and upgrades to Northern Road, Elizabeth Drive and Bringelly Road. (*Source* NSW Govt. 2014a)

Among the State aims within the Plan is 'to create a more resilient city, with connected green spaces, infrastructure and housing'. Important planning principles within the Plan include, for example:

> *Principle 1: Stronger economic development in strategic centers and transport gateways*
> For example, llocating jobs in around 30 to 40 large centers across Sydney provides the greatest benefits to the city's overall productivity
>
> *Principle 2: Connecting centers with a networked transport system*
> For example, to ensure public transport network connects people to jobs, education facilities, health centers and hospitals, and sporting, cultural and entertainment facilities

The State, under the Plan, proposes 'subregions' as the link between strategic planning directions and detailed planning controls at the municipal level. The State is careful to note (given the reforms of local government the State has foreshadowed in 2015) (Chap. 18):

> The State will also deliver planning outcomes across local council boundaries, where coordination between State agencies and/or local government is required. Better planning outcomes around major infrastructure projects will significantly benefit from improved coordination between local councils and State agencies. Subregional plans will build on the actions set out in *A Plan for Growing Sydney*. Councils, the community, the Greater Sydney Commission and the NSW Government will work together to finalize and implement these plans. Sydney's six sub-regions are proposed as: Central; West Central; West; North; South West; and, South. (*Source* NSW 2014a)

The Plan addresses further the concept of subregions:

1. The State had developed the sub-regional boundaries in consultation with stakeholders and local government.
2. The sub-regions reflect the economic and community relationships between Local Government Areas, allowing more cohesive and integrated planning.

3. Priorities for each sub-region in Sydney are set out, along with further investigations that are needed to shape sub-regional plans (in a sub-regional planning process). (Source NSW 2014a)

Within the State's 'Central Subregion' are the municipalities (a–z) of Ashfield, Botany Bay, Burwood, Canada Bay, Leichhardt, Marrickville, Randwick, Strathfield, Sydney (City of), Waverley, and, Woollahra. The drawing up of sub-regions by the State is also reviewed in respect to the State's foreshadowing amalgamations of councils. This question of nominated sub-regional boundaries and the role of municipal councils in planning within sub-regions is further addressed in Chap. 18 (Directions for Planning Sustainable Cities).

At the same time as establishing subregions the State proposes the establishment of the Greater Sydney Commission (under the State) to help plan for the future of Greater Sydney. The State urban planning legal framework (incorporating municipal and district level planning) comes under the *Environmental Planning and Assessment Act 1979* (NSW 1979). This Act, however, is subject to changes under draft reforms to the NSW Parliament (2013). After an exhibition period and debate in Parliament, the reforms were subject to considerable opposition. By late 2015, the State (holding office to 2019) gave indications of other means of planning reform. The State, for example, aims for the Commission to be 'independent' and notes:

> In partnership with councils, the community and stakeholders, the Commission will oversee the implementation of *A Plan for Growing Sydney*. It will look at the 'big picture' of what's needed to build vibrant communities so Sydneysiders get the most out of their neighborhoods, their region and their city. This will ensure everyone has access to the things they value – things like parks, hospitals, public transport, jobs and housing. Finally, the Commission will be located within the NSW Planning and Environment cluster and will have a Board with independent, state agency and local government representatives. The chair of the Board will report directly to the Minister for Planning. (*Source* NSW 2014a)

The State outlines the work of the Commission as follows:

> The Commission will work with councils, state agencies, the community and stakeholders in each subregion to build vibrant communities with access to the services they want and need. The Commission will take on a coordination role to help each subregion:
>
> 1. Identify places for housing and jobs which are close to transport and services
> 2. Identify new and improved services, such as public transport, that will be essential as communities grow
> 3. Improve local environments and open spaces
> 4. Help create well-designed neighborhoods and suburbs.
>
> The Commission will monitor the progress to make sure the right homes, jobs, infrastructure and services are being delivered when and where they are needed. This will be done in close consultation with communities and everyone will get a chance to have a say. (*Source* NSW 2014a)

Planning at the municipal and district levels of Greater Sydney is introduced next.

2.2 Municipal and District Planning

Councils have had the task of providing planning input on plans as noted above. This includes, for example, input on: making and reviewing public submissions; implementing State planning policies at the local level; joint partnership on service or infrastructure projects with State or private companies; and, monitoring State or private development projects affecting a council. Other Council planning responsibilities under State legislation that also faces the Council includes: engaging the municipality community (e.g. residents, business, services, development interests, and institutions) on State planning policies and development proposals; statutory plan making (e.g. adopting local environment plans (LEPs); and, reporting back to the State on a range of planning matters (i.e. State proposed roads, state of environment reporting and infrastructure projects).

Local councils are instruments of the NSW Government and are not formally recognized under the Australian constitution. Most planning statutes (as noted above) are adopted by the NSW Parliament, with implementation guidelines formulated for local government councils. Councils, for example, have the responsibility to implement aspects of State generated plans affecting local government. This includes, for example, councils making submissions to the State on planning matters generated by the State; implementing State planning policies at the local level; joint partnerships with the State on service or infrastructure projects; and, monitoring State or private development projects affecting the local government area.

The 43 councils (prior to 12th May 2016 *Local Government (Council Amalgamations) Proclamation 2016*) (NSW 2016) within Greater Sydney and 119 councils (prior to 12th May 2016) outside Greater Sydney (covering the regional areas of the State) make up the State. In many instances the councils of Greater Sydney and regional areas have achieved 'city' status. For the purposes of this book the term 'municipalities' and 'councils' is used to refer in general to all Greater Sydney councils. The State (overseeing councils as noted) reviews on an ongoing basis the operations and performance of councils. Commencing in 2009, the Minister for Local Government introduced the *Local Government Amendment (Planning and Reporting) Bill* (NSW 2009) to Parliament which requires all NSW councils to produce a long term Community Strategic Plan. This required councils to produce 4 year delivery programs and annual operational plans linked to longer term planning outcomes of those councils. The new framework encouraged councils to integrate their various plans and prioritize their works and services based on community priorities (identified through consultation and engagement). The State at the time noted:

> The clear goal is to turn community aspirations into reality, to ensure Council's priorities are set according: to residents' needs; to guarantee transparency in the decision-making process; and, to promote cooperation between all government levels, community groups and business. (*Source*: NSW 2009)

By 2011, the State commenced looking at changes in the way councils in the State operated. In that year the State called representatives from all 152 NSW councils to come together in Dubbo for the Destination 2036 workshop. Looking at some of the conclusions reached, representatives at the workshop agreed that (NSW 2012b):

a. existing communities of interest should be preserved
b. there is no particular structural reform model that fits all Local Government areas
c. long term financial sustainability must be assured (funding reforms are essential) (*Source* NSW 2012b)

The selection of case study districts is examined next.

2.3 Selecting Case Study Sydney Districts

Having examined State planning (including A *Plan for Growing Sydney* (2014a); Greater Sydney Commission; and, amalgamation of councils) the selection of Sydney case study districts is addressed next. The districts selected (Plate 2.4) are Sydney Inner West (center at Burwood, middle of the map); Greater Parramatta (center on Parramatta, left-top of map); and, St George (center at Hurstville, bottom-center of map).

Looking more closely at the first two districts of Sydney Inner West District and Greater Parramatta District (Plate 2.5) shows the connection to the City of Sydney

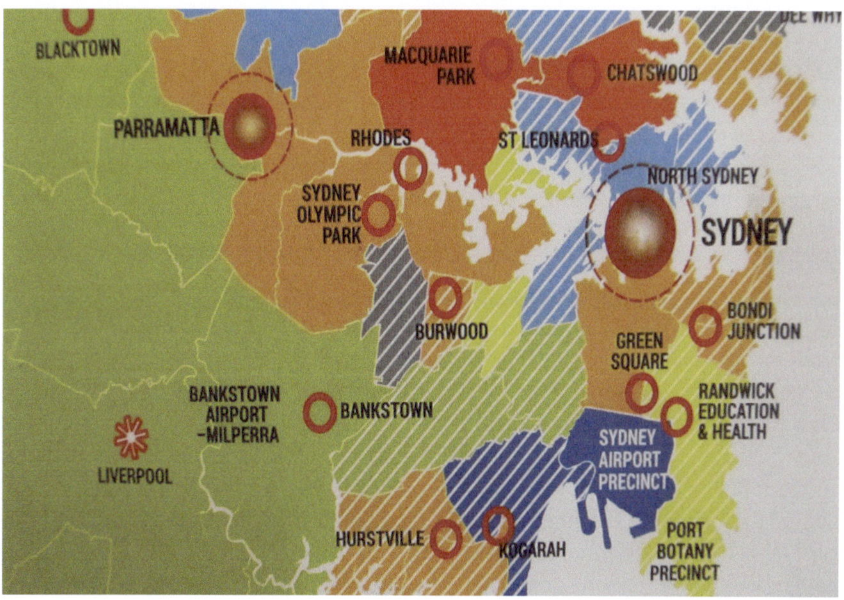

Plate 2.4 Key Greater Sydney Growth Areas. (Source NSW 2014a)

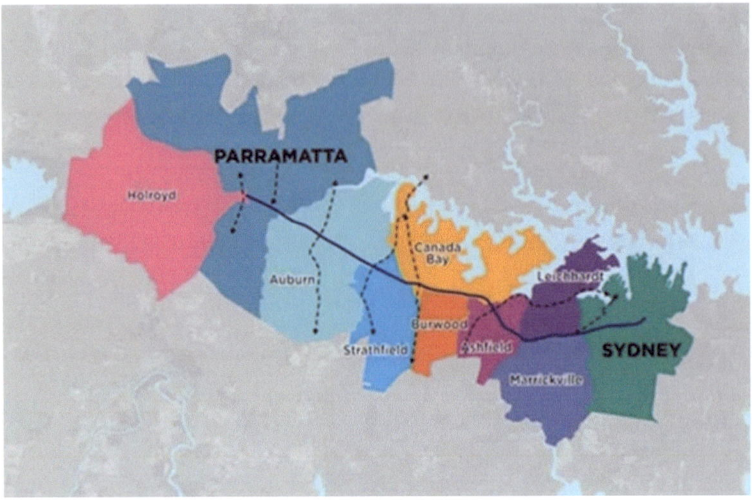

Plate 2.5 City of Sydney via Inner West to Parramatta

(right). These two districts share a common transport link (Parramatta Rd) along the centre spine) and waterway (Parramatta River) bordering on the north (top).

Sydney Inner West (from City of Sydney, right) consists of the municipalities of Marrickville (blue); Ashfield (burgundy); Leichhardt (purple); Burwood (dark orange); Strathfield (blue); and, City of Canada Bay (light orange). Greater Parramatta consists of the cities of Parramatta (dark blue); Auburn (light blue); and, Holroyd (pink). The St George District consists of the cities of Hurstville, Kogarah and Rockdale. Each of the municipalities and cities will be examined in coming chapters. These districts were selected as they form the main 'metropolitan' part of Greater Sydney. This is in contrast to the outer Greater Sydney growth districts such as Blacktown, Northwest Hills and Liverpool districts. The three selected case study districts are distinctively different from each other in history and development, thus making useful comparisons. Each district as a whole entity will be examined within the Chaps. 9, 13 and 17, applying key planning principles to the respective districts.

Each municipality (within the three districts) will be examined (acknowledging the 12th May 2016 *Local Government (Council Amalgamations) Proclamation 2016*) for its application of sustainable urban planning (SUP) practices (as outlined in Chap. 1). This is done under a 'qualitative research' method (relying on field work and interpretation). This includes studying the plans adopted by councils and noting if these plans and their implementation reflect SUP practices. The critiques of these plans was supplemented by author (Rauscher) qualitative research work in each of the case study municipalities. This research methods included: (a) photographing; (b) critiquing reports; and, (c) interpreting development and population changes (12 municipalities contained within 3 districts). The research included, for example, interpreting phases of development (early, traditional, new and future) and population changes. The key planning principles adopted are: (1) Developing

Sustainable Urban Centers; (2) Developing Sustainable Transport; and, (3) Protecting the Natural Environment. These planning principles will be applied initially at the municipal level and later at the district level. From these assessments of the degree of SUP practices (municipal and district levels) the conclusions reached will be used in the final chapter (Chap. 18). That chapter will consider what reforms are needed for Greater Sydney municipalities and the State to move further in applying SUP practices at municipal and district levels.

Looking back, the subject of local government reform (including urban planning functions) commenced under the current State government in 2009. To consider the importance of this reform (and how the reform is likely to affect urban planning at local and State levels) a timeline (2009–2015) and background to reform proposals has been prepared (Addendum 2.1 at the end of the Chapter). In 2012 the State's *Destination 2036* (NSW 2012b) was released, a strategic plan and delivery program for councils. This plan mirrors key elements of the integrated planning and reporting frameworks noted above. The plan provides a process and a forum for councils to explore the issues of changing population, technologies and economies. In addition, the plan (the State notes) considers and develops structures and approaches to councils in NSW that will allow councils to meet the needs and expectations of communities. The subject of local government reform and amalgamations (related to planning Greater Sydney) is examined next.

2.4 Local Government Reform and Amalgamations

The process of NSW local government reform and amalgamations affects most councils in the Greater Sydney region, and is important and topical in looking at the three districts as selected above. The reform process was stepped up in later in 2012 with the creation of an Independent Steering Committee (ISC). The ISC would also look at potential council mergers. A merger is a political and administrative process of two or more councils (elected administrative bodies) to amalgamate into a single entity. The State wanted to build on the work of *Destination 2036* (NSW 2012b) and to develop a specific Action Plan for reform and amalgamations. This process from 2012–2016 is briefly outlined next (referring again to Addendum 2.1). The State wanted to build on the work of *Destination 2036* (NSW 2012b) (#1) and to develop a specific Action Plan. Out of this Action Plan an Independent Review Commission was formed. In July 2012, under the Commission, an Independent Review Panel produced a paper (*Strengthening Your Community*) (NSW 2012c) (#2), asking for local government feedback. Leichhardt Council (selecting a sample council from Sydney Inner West district) responded in its paper *Leichhardt Council's Feedback to the Strengthening Your Community (*Leichhardt 2012a) (#3). In November 2012, the Independent Review Panel produced a second paper *The Case for Sustainable Change* (NSW 2012c) (#4) about reforms for local councils. Leichhardt Council again responded with a paper *Leichhardt Council's Feedback to the Case for Sustainable Change Paper* (Leichhardt 2012b) (#5).

Moving further, the Independent Steering Committee released the *Destination 2036 Outcomes Report* (NSW 2012e) (#6), including a *Draft Vision for NSW Government* (NSW 2012f) (#7) for comment. To continue to follow the trend of councils see, as an example, the Leichhardt Council submission to the Outcomes Report (*Leichhardt Council's Outcomes Report Submission* (Leichhardt 2012c) (#8). Finally, the Independent Steering Committee then produced the *Destination 2036 draft Action Plan* (NSW 2012g) (#9) for comment and again councils made submissions (accessible at the NSW Division of Local Government).

Given these earlier documents on council reform, in 2014 there was considerable discussion in NSW (as generated by the State) about these reforms. Firstly, the Independent Local Government Review Panel's report *Revitalizing Local Government* (NSW 2014b) (#10) was released. In addition the *Taskforce Report on a New Local Government Act* (NSW 2014c) (#11) was released. In September 2014 NSW Premier Mike Baird and Minister for Local Government Paul Toole announced funding of $153 million for Sydney councils to assist those who decide to merge. All NSW councils were requested to undertake a self-assessment, prepare a road map under the State's document *Fit for the Future* (NSW 2014d) (#12). Councils were asked to make submissions to the State by 30 June 2015. The Panel's recommendations and the final report, *Local Government Act Taskforce Report* (NSW 2014e) (#13), were provided to the State Government for consideration at the end of 2014.

The Independent Review Panel's paper (*Future Directions for NSW Local Government*) (NSW 2014f) (#14) recommended, for example, that Leichhardt Council be amalgamated with the City of Sydney, Woollahra, Waverly, Randwick, Botany and Marrickville Council. Leichhardt Council objected to this Panel's initial recommendations (*Leichhardt Council's Submission on the Future Directions Paper*) (Leichhardt 2014a) (#15). A subsequent proposal (Plate 2.6) recommended

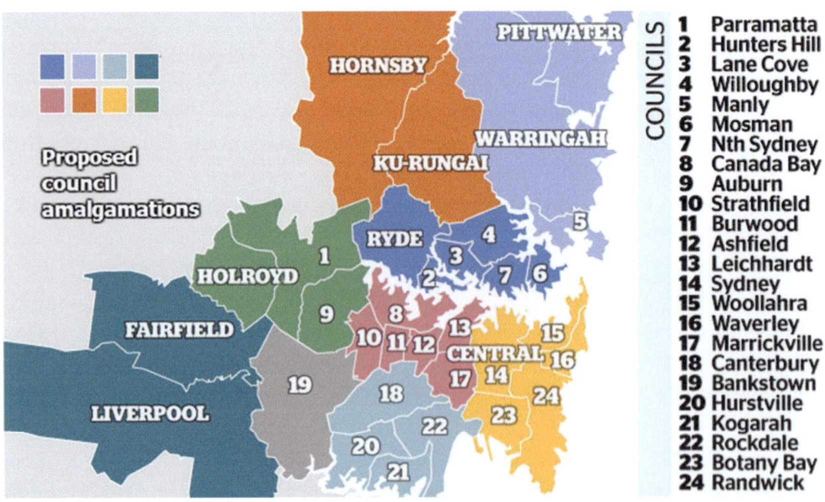

Plate 2.6 Initial Local Government Review Panel Suggested Citywide Council Mergers. (Source NSW Government 2014)

the City of Sydney be amalgamated with these councils, but without Leichhardt and Marrickville.

The City of Sydney commented in May 2015 (Mayor Clover Moore Newsletter):

> The Government's preferred outcome — a mega-council the size of Tasmania, made up of the City, Randwick, Waverley, Woollahra and Botany — would be a shockingly wasteful and disruptive exercise. The City has a 10-year, $1.95 billion plan to invest in new infrastructure. This includes $220 million for light rail and $440 for Green Square. This work would be put at risk by an amalgamation. (Source: City of Sydney 2015)

The *Revitalizing Local Government Final Report* (NSW 2014g) (#16) made 65 recommendations around local government reform. The Review Panel recommended, for example, that Leichhardt amalgamate with Ashfield, Burwood, Canada Bay, Marrickville and Strathfield (Plate 18.1). The Panel requested evidenced based responses to this proposal. Leichhardt Council response was very clear, in that it supported structured and systematic local government reform centered on resource sharing/strategic alliances. Council, however, firmly rejected any proposals involving a merger with neighboring councils (a trend reflected in most council responses). A detailed response to both reports was considered by Leichhardt Council meeting (*Leichhardt Council's Submission to the Revitalizing Local Government Report*) (Leichhardt 2014b) (#17). Having reviewed the documents and timeframe of the local government reform and amalgamations processes (taking Leichhardt Council as an example) the wider scope is now address, looking at: (1) amalgamations and other councils; (2) local government act changes 2015; (3) local government update 2016; and, (4) broader local government reforms.

1. Amalgamations and Other Councils

Looking beyond Leichhardt (taken above as a case example) Strathfield Council (within Sydney Inner West district) argued that any amalgamation of Council with other councils would directly impact Strathfield residents. Council noted, for example, the lower residential rates in comparison to all other Sydney Inner West councils, estimating resident rates would rise by 16 % with amalgamation. Further, looking beyond Strathfield, Canada Bay City Council (in Sydney Inner West district) noted on the subject of amalgamations in 2014:

> The Mayors and General Managers of the City of Canada Bay, Ashfield, Burwood, Leichhardt and Marrickville councils have signed a Memorandum of Understanding (MOU) to work together to prepare responses to the State Government. Our community deserves to be fully informed as this process unfolds. We simply do not know as the State Government has not provided this information.

In 2014, the NSW Government announced one of the most significant reform proposals in our State's history, named *Fit for the Future* (2014d), with the intention to reduce the number of Sydney Metropolitan councils from 41 to 18. The reform is based on the final recommendations of the Independent Local Government Review Panel and the Local Government Acts Taskforce which were commissioned to carry out reviews of the local government sector. The *Fit for the Future* package strongly indicates that Sydney Metropolitan councils, notably the inner and eastern councils, will need to meet the 'scale and capacity' (i.e. size) outlined in the Panel's recommendations to ensure financial sustainability and more effective governance. In particular, *Fit for the Future* supports the Panel's recommendation to merge Burwood, Ashfield, Canada Bay, Leichhardt, Marrickville and Strathfield Councils to form one Inner West Council (Plate 2.7).

By 30 Jun 2015 (deadline for submissions to the State) the Sydney Inner West councils of Burwood, Canada Bay and Auburn indicated a willingness to consider an amalgamation. Strathfield, in response, continued to argue for no amalgamation of that Council with any neighbouring council. Burwood Council has lodged a joint merger proposal with Auburn City and Canada Bay Councils to the Independent Pricing and Regulatory Tribunal (IPART) as part of the NSW Government's *Fit for the Future* reforms. The decision to explore a merger comes following the State Government's strong indication that 'no-change' is not an option for councils. The proposal focuses on a new entity to possibly be known as Sydney Olympic Park City Council. Burwood Council voted to adopt an alternative merger (Plate 2.8) and explains:

> Following extensive community consultation, independently commissioned research and discussions with neighbouring councils, Burwood Council voted to adopt an alternate merger proposal including Auburn City Council, Burwood Council and City of Canada Bay Council. At the Council Meeting of 9 June 2015. Auburn City and City of Canada Bay Councils also voted in favour of this proposal at Council Meetings in May and June 2015. The decision to merge with Auburn City and City of Canada Bay Councils was based on communities of interests and financial long term benefits. In the event that the NSW Government does not proceed with its reform agenda, each Council has opted to stand-alone under their existing structures.

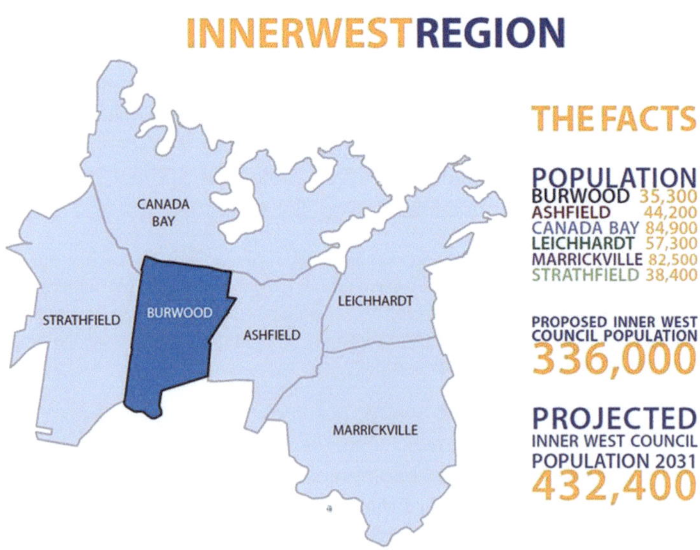

Plate 2.7 Local Government Review Panel Suggested Inner West Council Mergers. (Source NSW Government 2014)

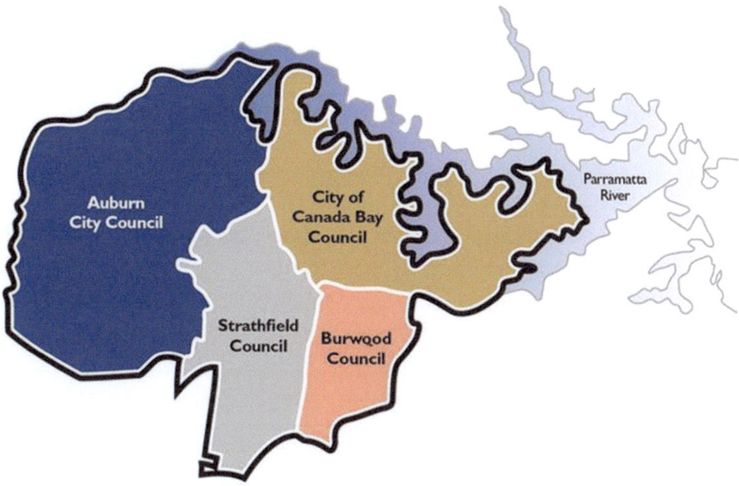

Plate 2.8 Burwood Council Proposed Council Mergers (Source: Burwood Council 2015)

Council notes the benefits of the merger plan as follows:

> The merger plan: Meets estimated minimum population size for Councils; Protects local town centres and villages; Same level of services provided; Combined financial strength which meets *Fit for the Future* benchmarks including low debt; Cultural synergies – cosmopolitan communities; Investment in community facilities and infrastructure; Strong connectivity between the Councils though Parramatta Road and extensive river frontage; Powerful local economies that build on existing industry and business, strengthened employment opportunities; Key strategic centres, including Burwood, Rhodes and a city based around Sydney Olympic Park; and, Equal footing of merged Councils.

Looking beyond Sydney Inner West to the Greater Parramatta district, the Panel recommended merging Parramatta with Auburn, Holroyd, the western third of Ryde and the North Parramatta area of the Hills. At the end of the submission date (30 June 2015) Parramatta Council agreed to discuss a potential merger arrangement, partly reflecting the above Panel suggestion. However, neither Auburn nor Holroyd councils were interested in that merger arrangement. Both Auburn and Holroyd councils have subsequently been conducting strong campaigns opposing above amalgamation. A statement by Holroyd Council endorsed the State members' opposition to any forced amalgamation of that council, noting:

> At the first 'Hands off Holroyd' Public Meeting last month, Mr. Rohan, the current Member for Smithfield, said he 'will cross the floor' to vote against his own Government if Holroyd is forced to merge with Parramatta, Auburn, and parts of Ryde and The Hills Councils. Mr. Rohan said: "Nobody wants forced amalgamations. I will stand up and support Holroyd and the community against amalgamations. Holroyd is a good Council. Holroyd is a great Council".

The 12th May 2016 *Local Government (Council Amalgamations) Proclamation 2016* under the State Government declared a number of local council mergers, including: (a) the merger of Auburn and Holroyd councils; and (b) boundary changes expanding the City of Parramatta. Looking beyond Greater Parramatta, the councils in the St George district offered no plans to the State to amalgamate. Rockdale Council did indicate (by deadline of submissions to the State) a willingness to discuss the prospects of amalgamation with Kogarah and Hurstville councils. Neither of these two councils, however, indicated a wish to entertain that prospect amalgamating with Rockdale. The 12th May 2016 *Local Government (Council Amalgamations) Proclamation 2016* noted above declared: (a) Hurstville and Kogarah mergers; and, (b) acknowledging a court case objecting to the State proposed merger of Rockdale and Botany City councils. These above mergers will also be addressed in Chap. 18.

2. Local Government Act Changes 2015

Moving back to the Panel's recommendations on the Local Government Act (*NSW Government Taskforce Report*) (NSW 2014b), these were with the State Government in early 2015 to consider. The State asked the Independent Pricing and Regulatory Tribunal (IPART) to make recommendations to the State by the end of 2015 on the 'capacity of each council to deal with a number of future challenges likely to face local government councils' (SMH 12 May 2015). In the article the State noted "that of the forty-three (43) councils there were twenty three (23) that had populations less than 100,000 people. Within the Sydney Inner West for example, Ashfield and Strathfield councils had populations less than 50,000 people." The SMH continued noting that 'the State pushes ahead with Fit for the Future program to reduce councils from 43 to 15. The paper then notes 'in an earlier report by the Office of Local Government it was recommended that 250,000 be the minimum size for a council.' Writing further, the paper notes that 'joint organisations' as strategic alliances could be considered by some councils. Shared services would include, for example, child care, road making and waste collection.

Under current NSW legislation the amalgamations and boundary changes (voluntary or otherwise) were referred to the Boundaries Commission for examination via a public inquiry. While the Minister for Local Government could have sought amendments to the act to give the Minister power to approve amalgamation, the State decided in Dec 2015 to refer any amalgamation proposals to the Boundaries Commission. The Boundaries Commission conducted inquiries into amalgamation proposals (including public consultation) and reported back to the State with recommendations. Overall, by the closing on submissions only a handful of Sydney's 41 councils had put forward proposals for amalgamation, hence most indicating a want to continue as is. Finally, the SMH editorial (29 June 2015) encouraged Greater Sydney councils to consider the Auckland City model. The paper indicated with Auckland's population of 1.4m, had one council (commenced in 2010 replacing seven city and district councils). The SMH continued noting 'with the one city administration there were also created 21 local boards with 149 elected members. These members make local decisions and take them to Auckland Council for funding. The local boards also identify the views of their communities, develop their own local plans, co-ordinate with community organization and special interest groups and identify by-laws for their local government area.‘ The SMH noted ’that this Auckland model suggested amalgamation of councils did not mean that local governments become more detached from their communities. The SMH concluded ‘the amalgamation process is necessary, as 41 local governments for one metropolis is simply too many’. Many other writers argue that strong local government councils contribute in total to an effective metropolis.

3. Local Government Update 2016

In the interim, the Minister for Local Government had commenced the process of forcibly amalgamating Holroyd and Auburn councils. Council also noted the Minister said he expected the new council (with adjusted boundaries) to be ‘proclaimed’ on 1 July 2016 (as noted above the Proclamation on mergers was 12th May 2016). The results of the review of proposed boundary adjustments and merger (Holroyd and Auburn councils) was also referred to the Chief Executive of the Office of Local Government for examination and report under the Local Government Act (the Act). In turn, the Chief Executive had appointed a Delegate who would examine the proposals, conduct a public inquiry review, call for written submissions, and then report back to the Minister.

It can be concluded from the examination of documents to date (Information Box 2.1) that there is a wide range of views on local government reform expressed by the State and councils. The overall voice of councils is that they welcome helpful reforms and improved partnerships with the State, but oppose forced amalgamations (vs voluntary amalgamations). The State report on amalgamations as released in Dec 2015, reflecting the State’s intent to see the number of Greater Sydney councils reduced from 41 to 23 (note the Proclamation above). It is opportune to look at the amalgamation questions by examining a range of options to improve the effectiveness of councils (i.e. in the area of urban planning and infrastructure provisions). The State created in early 2016 a web site (www.councilboundaryreview.nsw.gov.au) on the amalgamation process expected to be followed by the Boundaries Commission (submissions closed 28 Feb 2016). The documents that accommodated this announcement are summarized below and addressed further in Chap. 18 (Directions for Planning Cities in Global Transition).

4. Broader Local Government Reforms

In addition to the subject of amalgamations, the State is examining reforms of local government as noted above. The NSW Department of Planning and Environment’s (NSW 2015) *A Plan for Sydney Growth* (referred to above) reflects aspects of local government reform the State already has in place in urban planning and infrastructure provisions. This plan contains an underlying structural basis of local councils working together to meet the achievements of the State's urban planning agenda (including a Sydney population increase of 1.6m by 2036). Under this plan, there appears a range of administrative arrangements for councils to work together with the State (given also the State local government reform reports noted above). The area of 'urban planning and infrastructure provision' is one of the most significant areas of councils and the State interacting (i.e. planning policies and funding of infrastructure). While there are other issues related to 'efficiency of local government' (i.e. finances as argued by councils above in the amalgamation debate) it can be concluded that the area of sustainable urban planning (SUP) practices is the key to councils showing they can either (under State options): (a) stand alone; (b) work in partnership; or, (c) work within a voluntary amalgamated arrangement. These arrangements will be referred to further in Chap. 18 (Directions for Planning Sustainable Cities).

2.5 Summary

In this chapter a methodology for evaluating the extent of sustainable urban planning (SUP) practices being applied at municipal and district (groups of municipalities) levels was outlined. The chapter commenced with an introduction to the planning of Greater Sydney. Three case study districts were then nominated, being: Sydney Inner West; Greater Parramatta; and, St George. A methodology was adopted for assessing the extent each municipality within the three districts is applying SUP practices (as outlined in Chap. 1). Key planning principles were then adopted and an outline given on how these principles would be applied to case study districts (and municipalities within). These principles are: (1) Developing Sustainable Urban Centers; (2) Developing Sustainable Transport; and, (3) Protecting the Natural Environment. The process of State initiated local government reform and council amalgamations (acknowledging connections to planning Greater Sydney) were also reviewed. Finally, the chapter noted that conclusions reached in examining the selected districts (and municipalities within) would be used in the final chapter (Chap. 18). It was also noted that Chap. 18 will consider Greater Sydney planning reforms that are needed for municipal and district planning (ensuring a sustainable Greater Sydney in the future).

2.6 Addendum 2.1

Information Box 2.1 NSW Government and Council Reports on Local Government Reform

Date + #	Document	Summary
2012 #1	*Destination 2036* NSW 2012b	A State Government strategic plan and delivery program for local government
2	*Strengthening Your Community* NSW 2012c	State Government suggestions on actions that could be taken by Local Government Councils to strengthen their performance.
3	*Leichhardt Council's Feedback to the Strengthening Your Community* Leichhardt 2012a	Council feedback to the *State and the community.*
4	*The Case for Sustainable Change* NSW 2012d	State Government suggestions on reforms for local councils.
5	*Leichhardt Council's Feedback to the Case for Sustainable Change Paper* Leichhardt 2012b	A paper by Leichhardt Council commenting on the State's *The Case for Sustainable Change*
6	*Destination 2036 Outcomes Report* NSW 2012e	State Independent Steering Committee (ISC) report on local government reform to 2036.
7	*Draft Vision for NSW Government* NSW 2012f	State projections on its agenda to 2036.
8	*Leichhardt Council's Outcomes Report Submission* Leichhardt 2012c	Submission to the *Destination 2036 Outcomes Report*
9	*Destination 2036 Draft Action Plan* NSW 2012g	Actions for local government reform to 2036.
2014 10	*Revitalizing Local Government* NSW 2014b	Local Government Review Panel's State Government suggestions on revitalizing local government
11	Taskforce Report on a *New Local Government Act* NSW 2014c	State review of the Local Government Act
12	*Fit for the Future* NSW 2014d	State asked all councils to undertake a self-assessment and prepare a road map on possible amalgamations (with incentive funding for those Councils that merge).
13	*Local Govt. Act Task Force Report* NSW 2014e	Panel's recommendations and final report.
14	*Future Directions for NSW Local Government* NSW 2014f	State Report recommended Leichhardt, City of Sydney, Woollahra, Waverly, Randwick, Botany and Marrickville Councils be amalgamated. The Report also recommended Auburn, Holroyd, the western third of Ryde and the North Parramatta area of the Hills be amalgamated.
15	*Leichhardt Council's Submission on the Future Directions Paper* Leichhardt 2014a	Leichhardt Council's submission objecting to the proposed amalgamation (above).

(continued)

Information Box 2.1 (continued)

Date+#	Document	Summary
16	*Revitalizing Local Government Final Report* NSW 2014g	Local Government Review Panel's Report contained 65 recommendations around local government reform. One recommendation was that Leichhardt, Ashfield, Burwood, Canada Bay, Marrickville and Strathfield amalgamate.
17	*Leichhardt Council's Submission to the Revitalizing Local Government Final Report* Leichhardt 2014b	Leichhardt Council's stated position on the 65 recommendations in the above report, with Council opposing any merger with another council.
18	12th May 2016 *Local Government (Council Amalgamations) Proclamation 2016*	This Proclamation affected a number of councils, including those in: Sydney Inner West; Greater Parramatta; and, St George Districts

Sources noted in column two and listed under references at end of chapter

References

Australian Bureau of Statistics (ABS) (2011) Australian census review. ABS, Canberra
Leichhardt Council (2012a) Leichhardt council's feedback to the strengthening your community. NSW, Leichhardt
Leichhardt Council (2012b) Feedback to the case for sustainable change paper. NSW, Leichhardt
Leichhardt (2012c) Leichhardt council's outcomes report submission. NSW, Leichhardt
Leichhardt (2014a) Leichhardt council's submission on the future directions paper. NSW, Leichhardt
Leichhardt (2014b) Leichhardt council's submission to the revitalizing local government report. NSW, Leichhardt
NSW (1979) Environmental Planning and Assessment Act 1979. N.S.W. Department of Environment and Planning, Sydney
NSW (2009) Local government amendment (Planning and Reporting) bill. N.S.W. : Department of Planning & Infrastructure, Sydney.
NSW (2012a) Draft metropolitan strategy for Sydney. NSW, Sydney
NSW (2012b) Destination 2036. NSW, Sydney.
NSW (2012c) Strengthening your community. NSW, Sydney.
NSW (2012d) Case for sustainable change. NSW, Sydney.
NSW (2012e) Destination 2036 outcomes report. NSW, Sydney.
NSW (2012f) Draft vision for NSW government. NSW, Sydney.
NSW (2012g) Destination 2036 draft action plan. NSW, Sydney.
NSW (2014a) A plan for growing Sydney. NSW Government Planning & Environment, Sydney
NSW (2014b) Revitalizing local government. NSW, Sydney.
NSW (2014c) Taskforce report on a New Local Government Act. NSW, Sydney.
NSW (2014d) Fit for the future. NSW, Sydney.
NSW (2014e) Local Government Act taskforce report (NSW 2014d)
NSW (2014f) Future directions for NSW local government. NSW, Sydney.
NSW (2014g) Revitalizing local government final report. NSW, Sydney.
NSW (2015) Urban green cover technical guidelines. NSW, Sydney.
NSW (2016) Local government (Council Amalgamations) proclamation 2016. NSW, Sydney.
Rauscher and Momtaz (2015) Sustainable neighbourhoods in Australia: city of Sydney urban planning. NSW, Sydney.
Spearritt (1988) Sydney functional plan on Cumberland Plain. Department of Planning, NSW, Sydney.

Part II
Planning Sydney Inner West District

Having been introduced to planning global cities and sustainable urban planning (SUP) (Chap. 1) and provided with a methodology of examining the extent of SUP practices being applied at municipal and district levels (Chap. 2), Part II looks at the Sydney Inner West District. This district consists of the municipalities of Marrickville, Ashfield, Leichhardt, Burwood, Strathfield and Canada Bay. The basis of examining the extent of SUP practices is the selection and application of three key SUP principles (herein called planning principles). These planning principles are: Developing Sustainable Urban Centers; Developing Sustainable Transport; and, Protecting the Natural Environment. With these planning principles in mind, three questions are posed for each municipality (in separate Chaps. 3, 4, 5, 6, 7 and 8): (1) what is the pattern of urban transition the municipality is experiencing; (2) what is the extent SUP practices are being applied by that municipal council; and, (3) what is the impact of Greater Sydney planning on the development of the municipality, including proposed council amalgamations. With conclusions reached on SUP practices at the municipal level, the three key planning principles are applied to the Sydney Inner West district as a whole unit (Chap. 9). Results from these municipal and district assessments (and results of Greater Parramatta and St George assessments in later chapters) will contribute to addressing directions for planning sustainable districts over the whole of Greater Sydney and in any city worldwide (Chap. 18).

Chapter 3
Planning Marrickville, Greater Sydney

Dedication *This chapter is dedicated to Mary Reibey, an enterprising businesswoman who contributed to assisting the early Sydney community. Dictionary of Australian Biography She resided for years in Marrickville and appears on the Australian twenty dollar notes (Plate 3.1).*

Plate 3.1 Mary Reibey

A snippet on Mary Reibey reveals her legacy:

> Mary Reibey (1777–1855) was an Englishwoman who was transported to Australia as a convict, but went on to become a successful businesswoman in Sydney. An enterprising and determined person of strong personality, during her lifetime Mary Reibey earned a reputation as an astute and successful businesswoman in the colony of New South Wales.

R.C. Rauscher, S. Momtaz, *Cities in Global Transition*,
DOI 10.1007/978-3-319-39865-5_3

Abstract The chapter addresses the extent Marrickville Council (herein referred to as Marrickville) is applying sustainable urban planning (SUP) practices (see Chap. 1) in planning the municipal area. The chapter acknowledges the *Local Government (Council Amalgamations) Proclamation 2016* merging Marrickville with Leichhardt and Ashfield councils noted in Chap. 2). Marrickville (in Sydney Inner West District) is one of twelve municipalities within three Greater Sydney Districts (Sydney Inner West, Greater Parramatta and St George), all considered to be in global transition (see Chap. 2). Examining SUP practices is based on the application of three key SUP principles (as adopted in Chap. 2), being: Developing Sustainable Urban Centers; Developing Sustainable Transport; and, Protecting the Natural Environment.

Firstly, a background to Marrickville development and population changes (e.g. global movement of population) is outlined. To examine the pattern of urban development, four phases of Marrickville development are examined: Early Marrickville; Traditional Marrickville; New Marrickville; and, Future Marrickville. Given conclusions reached here, key Marrickville Council planning documents are then critiqued to determine the extent Council applies SUP practices. In addition, given State planning of Greater Sydney affects Marrickville, a second question is posed: what is the impact of Greater Sydney planning and development on Marrickville? In examining this question, examples of State plans (including State council amalgamations) likely to impact on Marrickville are addressed. By chapter's end, conclusions are drawn on Council planning and Greater Sydney planning affecting Marrickville. The chapter notes that these conclusions will be addressed further in: Chap. 9 (examining the extent of SUP practices being applied over the whole of Sydney Inner West District); and, Chap. 18 (examining directions for planning a sustainable Greater Sydney or a sustainable city anywhere).

3.1 Background and Population Change

Marrickville Municipality (herein call Marrickville and under the Proclamation noted above merged with Leichhardt and Ashfield councils) in the context of Sydney Inner West (as introduced in Chap. 2) is adjacent to the City of Sydney. The name Marrickville comes from the 'Marrick' estate of Thomas Chalder. This estate of 60 acres was subdivided and named after the native village Marrick, North Yorkshire, England. One of the more notable people in Marrickville's history is Thomas Holt (1811–1888). Holt was a Sydney businessman who built in 1857 the mansion named 'The Warren' in Marrickville South. A snippet on Holt provides an insight on his legacy:

> The Warren was a landmark in the district for some decades (and the still operating Warren View Hotel in Enmore). Holt was a member of the New South Wales Legislative Council between 1868 and 1883. He was also a member of the New South Wales Legislative Assembly for two periods between 1856 and 1857 and again between 1861 and 1864. Holt was the first Colonial Treasurer in New South Wales.

A brief summary of Marrickville's history reveals:

> Based on artefacts found near the Cooks River and Alexandra Canal, it is believed that the area has been occupied for at least 7,000 years. The area was originally occupied by the Cadigal clan of the Dharug people who spoke the Eora language. Their name for most of the present day local government area was Bullanaming. European settlement of the area began very soon after the arrival of the First Fleet in 1788. The first land grant was made in 1789, and by 1809 all the land had been granted. In the 1830s, the district consisted of five large estates. At this point, the area was still quite rural in nature. Following the subdivision of these estates, municipalities were formed in Marrickville (1861), Camperdown (1861), Newtown (1862), St Peters (1871) and Petersham (1872). The area lost its rural character as Sydney grew, and by the 1920s had become an urbanized industrial area. The population of these municipalities increased greatly in that time, peaking at roughly 113,000 in 1948.

Marrickville became a municipality in 1861 and by 1948 it had merged with neighboring municipalities of St Peters and Petersham to form Marrickville Municipal Council. The suburbs and localities within the Marrickville Council area include (a–z): Camperdown; (partly located the City of Sydney); Dulwich Hill; Enmore; Lewisham; Marrickville; Newtown (partly located within the City of Sydney); Petersham; St Peters; Stanmore; Sydenham; and, Tempe. At the 2011 Census, there were 76,500 people in the Marrickville local government area.

With this background introduction to Marrickville, the extent of Marrickville's development phases and population change are examined next.

The degree that population change is impacting on Marrickville (i.e. global movements of people) is addressed next. The Australian Bureau of Statistics (ABS 2011) reports on Marrickville ancestry and languages spoken at home:

> While the area's background has traditionally been working-class, several waves of immigration and a continuing trend of gentrification have substantially influenced the demographics and character of the area. While migration has always been an important part of Marrickville, this was especially true from the 1950s to the 1970s, as large groups of immigrants arrived from Southern Europe, the Middle East and then Asia. The area is today one of the most culturally diverse in the country. Compared to the national average, at the 2011 Census, Marrickville Council area had a high proportion of households (34.1 %) where two or more languages are spoken (national average was 20.4 %); and a low proportion (62.3 %) where English only was spoken at home (national average was 76.8 %). Just over half (53 %) of Marrickville residents were born in Australia, compared to the national average of 70 %; the next most common countries of birth were Vietnam 6.2 %, Greece 5.1 %, England 2.8 %, New Zealand 2.0 % and Lebanon 1.8 %. (*Source* ABS 2011)

Finally, in author's (Rauscher) qualitative research (2014/2015), there are several poems written to reflect the cultural diversity of Marrickville (end of chapter under Notation 3.1). These poems, include: *Grains of Memories in Marrickville* (# 1); *Newtown's Global Realm* (# 2); and, *White Cockatoos and Marrickville* (# 3).

3.2 Phases of Development

An examination of the development phases of any locality provides a basis of understanding current trends and likely future directions of growth of that locality. Following a look at the background of Marrickville four phases of Marrickville development will be examined. These phases include: Early Marrickville; Traditional Marrickville; New Marrickville; and Future Marrickville. To assist in addressing these phases, author (Rauscher) qualitative research field work was completed in Marrickville from 2014–2015. Early Marrickville (Plate 3.2) centred on Illawarra Rd (top left). Marrickville was urbanized with substantial commercial centers in the late 1800s at Newtown and Marrickville center (top center). Transport was soon arranged between Marrickville suburbs and the City of Sydney and surrounding districts, such as Marrickville to Leichhardt (top right). Marrickville municipality was large enough to gain the Lewisham Hospital (now residential) (bottom left). Important civic ceremonies were usually held at the war memorial plaza outside the Marrickville Town Hall (bottom center). Finally, transport soon improved from horse drawn to electric trams (bottom right).

Traditional Marrickville (Plate 3.3) sees renewal occurring throughout most of the municipality's suburbs. Looking at Marrickville homes, many are detached or terraced Victorian houses (late nineteenth century). Other homes (early twentieth

Plate 3.2 Early Marrickville. *Left to right*: Illawarra Rd; commercial area; transport; Lewisham hospital; War memorial outside council; and, tram service

Plate 3.3 Traditional Marrickville. *Left to right*: restored cottage: detail of dormer and chimney; early estate; streetscape; corner commercial restoration; and, residential street

century) are Federation style. Restored cottages often reflect this Federation era, including multi-coloured roofs (top left). Attention to architectural detail was a priority with early tradesmen (top center). Estates were scattered throughout Marrickville, including spacious gardens (top right). Most of the larger estates have been subdivided, though a few remain. Streetscapes often reflect uniformity of fencing, setbacks and vegetation (bottom left). Older corner shops with residential above are prime for renewal, with heritage features retained (bottom center). Finally, streetscapes can reflect attention to plantings that provide shade for protection from heat of summer (bottom right).

Continuing with the research and field work, the projected New Marrickville (Plate 3.4) reflects the building boom, especially for residential properties throughout the municipality. The conversion of a flour mill (Waratah Mills, Dulwich Hill) shows the premium placed on buildings that are centrally located and capable of restoration (top left). A renewal site can also accommodate new buildings as shown here at the Waratah Mills site (top center). The Council has also designed a new community HUB (top right). The HUB will contain a new library as a central recreation and meeting place for downtown Marrickville (bottom left). Large industrial sites, such as this Cooperage site in Dulwich Hill, provide scope for residential complexes (bottom center). Finally, infill sites throughout Marrickville provides for imaginative designs (bottom right).

Plate 3.4 New Marrickville. *Left to right*: flour mill silos conversion to apartments; New residential apartments to 'Silo' site; proposed new residential on HUB site; proposed community area on HUB site; cooperate residential development; and, tram service new infill residential

Finally, Future Marrickville (Plate 3.5) reflects the continue building boom in residential development at increasing densities throughout the municipality. The quality of buildings show marked increase in design (top-left). The 16 storey residential development (Station St.) was referred to the State Government 'gateway' process as Council considered the height in excess of Council's Local Environment Plan. A split (6 and 8 storey) high rise is illustrated here in Marrickville Rd (top-middle). The Lewisham Towers (Old Canterbury Rd.) is a full residential complex and varying building heights and open space system (top-right). The sketches of the new residential complex Council has designed for the Old Marrickville Hospital (bottom-left) are shown here. The increasingly popular night life in Marrickville Rd is likely to increase in the future (bottom-middle). Finally, the high rise building was not approved by the Council as it was considered excess in height (bottom-right).

Plate 3.5 Future Marrickville. *Left to right*: developments and concept proposals

3.2.1 Conclusion

Given the review of development phases of Marrickville (above) and changes in population demographics, it can be concluded that Marrickville has been experiencing accelerated global transition (especially in the 2000s). The planning of Marrickville (merged with Leichhardt and Ashfield as noted above under the Proclamation) in meeting development demands and applying sustainable urban planning (SUP) practices is examined next.

3.3 Marrickville and Sustainable Urban Planning

Planning of Marrickville Municipality has been taking place since it became a local government body in 1861. A review of planning in Marrickville will pay particular attention to applications of sustainable urban planning (SUP) (as outlined in Chap. 1) practices (including use of indicators of sustainability) (Chap. 2). Marrickville Council key planning documents (a–z with date and web reference) are listed in Information Box 3.1. These key planning documents are reviewed with a goal to

Information Box 3.1 Marrickville municipality planning

Climate Change Action (2001)	www.marrickville.nsw.gov.au
Guideline to Biodiversity Protection (called the 'guideline') (Marrickville 2011c)	www.marrickville.nsw.gov.au
Marrickville Biodiversity Action Plan 2011–2015 (2011b)	www.marrickville.nsw.gov.au
Marrickville Biodiversity Strategy (2011a)	www.marrickville.nsw.gov.au
Marrickville Development Control Plan (2011f)	www.marrickville.nsw.gov.au
Marrickville Local Action 21 (1997)	www.marrickville.nsw.gov.au
Marrickville Local Environment Plan (2011e)	www.marrickville.nsw.gov.au
Marrickville State of Environment Report 2010–2011 (2012)	www.marrickville.nsw.gov.au
Marrickville Urban Strategy (MUS) (Marrickville 2011d)	www.marrickville.nsw.gov.au
Oil Depletion Protocol Marrickville Agreement (2007)	www.marrickville.nsw.gov.au
Strategy for a Water Sensitive Community (2013)	www.marrickville.nsw.gov.au

Source Webs as noted March 2016

assess the extent Council's planning reflects SUP practices. A summary of these documents and a brief critique of each follows (documents can be sourced under the web link as noted in the right column).

3.3.1 Climate Change Action

Starting the review with climate change planning documents, Council has developed a strategy to address greenhouse gas emissions and climate change adaptation. Under climate change action, Marrickville Council established targets in 2001 (Marrickville Council 2001) to reduce greenhouse gas emissions by 20 % on 1997 levels by 2010. A number of projects focused on energy and transport were, and continue to be, implemented to reduce Marrickville Council's greenhouse gas emissions. Council reported in 2011 on its greenhouse gas emissions assessment reporting a 24 % reduction on business emissions. Related to climate change, under Council's *Target Sustainability @ Marrickville* (Marrickville Council 2010) (www.marrickville.nsw.gov.au) council develops a series of environmental programs (Plate 3.6). Examining water sustainability, for example, Council created a water cycle template (top left) to show the current way water is used in the local government area (LGA). The template measures water sources, water consumed and water being drained out of the area. Council concluded the current water used in Marrickville was unsustainable. The Council also reported reducing its carbon emissions and saving on electricity costs after installing 80 new solar panels on the roof of its Administration Building (top right) and an aquatic centre (bottom left). In addition, the Council adopted the Environmental Management System (EMS)

Plate 3.6 Target sustainability @ Marrickville. *Left to right*: water sustainability; Council's administration building; Annette Kellerman Aquatic Centre; and, target sustainability program symbol

(Marrickville 2003) to minimize and manage risk of Council's activities: (a) impacting on the environment; or, (b). contributing to factors likely to affect climate change. This environmental program is illustrated in the widely distributed logo (bottom right).

Moving to energy conservation, Marrickville Council has adopted a position of opposing all new coal-fired power stations in NSW on the grounds that it is incompatible with mitigating climate change. Council supports investment in energy efficiency measures and renewable power sources. Council is also reducing greenhouse gas emissions through energy efficiency measures and investigating sources of renewable energy. All of Council's major projects currently (2015) aim to minimize greenhouse gas emissions, with a result of reducing Council's energy consumption each year for the last 4 years (2011–2015). Generating energy locally from a renewal source, wind power, also assists Council in reducing greenhouse gas emissions. To date (commencing in 2009) Council has investigated Tempe land as a potential site for an urban wind farm. This is seen by Council as an opportunity for the Marrickville community to take action to generate energy locally, though no project has been finalized to date (2015).

3.3.2 Biodiversity Protection

In November 2011 Council adopted its first ever biodiversity strategy. The strategy is divided into three reports: The *Marrickville Biodiversity Strategy 2011–2021* (called 'the Strategy) (Marrickville Council 2011a); The *Marrickville Biodiversity Action Plan 2011–2015* (called'the action plan) (Marrickville Council 2011b); and, *Guideline to Biodiversity Protection* (called the 'guideline') (Marrickville Council 2011c).

3.3.3 Local Environment Plan and Development Control Plan

In planning for the built environment, Marrickville Council, through its *Local Environmental Plan* (LEP) (Marrickville Council 2011e) and Development Control Plans (DCPs) sets controls on new developments to ensure that environmental best practice features are applied in order to preserve and improve the environment. Council adopted in 2011 its main development control plan, *Marrickville Development Control Plan* (2011f).

3.3.4 Local Action 21 Plan

Under its Environment Report goals, Marrickville Council signed up in 1997 to Local Action (LA) 21 under the United Nations Agenda 21 program (Chap. 1). LA 21 identifies what Marrickville Council can do in addressing the protection of the natural environment and related matters. Marrickville Council developed its first LA21 Plan in 1997 (Marrickville Council 1997), located at www.marrickville.nsw.gov.au/Templates/Advanced/Bridge/TrimDownload.aspx?TrimDocNum=32349.03.

The Plan outlines the Council's sustainability vision, goals and objectives and the actions Council will take to implement components of the Plan. The Plan addresses matters such as environmental education, energy efficiency, biodiversity conservation and waste minimization. The Council outlines several objectives, including "protecting environmentally sensitive areas in Marrickville such as the Cooks River; promoting ecologically sustainable development in the area; and, for minimizing Council's operational impacts on the environment."

3.3.5 Marrickville State of Environment Report

Marrickville Council prepares an annual State of the Environment Report (Marrickville Council 2012) (www.marrickville.nsw.gov.au). The Report is structured around the themes of Land; Air; Water; Energy; Biodiversity; Waste; Livability (incorporating for example: noise, Aboriginal and Torres Strait Islander heritage and non-Aboriginal heritage); and, Learning for Sustainability. The report summarizes any new or updated information relating to the environmental indicators and activities undertaken during the reporting period to protect and enhance the local environment. Finally, the report highlights any emerging issues or threats to the environment. Key issues include: climate change; energy conservation; sustainable transport; water conservation; heritage preservation; and, environmental protection. Under conservation, for example, the Council notes that Marrickville flora and fauna communities have been extensively modified as a result of urban development.

3.3.6 Oil Depletion and Transport

Realizing transport is the second biggest contributor to greenhouse gases in NSW, Marrickville Council has demonstrated its commitment to reducing transport related greenhouse gas emissions by signing the *Oil Depletion Protocol* (Marrickville Council 2007) and establishing an internal working group to reduce fuel consumption. Council report on this protocol and the progress to 2010 as follows:

> The Oil Depletion Protocol (ODP) was established to encourage a worldwide reduction in the consumption of, and dependence upon, oil. The protocol is an international agreement designed to deliver these reductions in a collaborative way. Individuals, businesses, non-government organizations and governments have signed up to the ODP. In August 2007 Marrickville Council became the first local Council in Australia to sign the Oil Depletion Protocol. As a signatory to the Oil Depletion Protocol, Council has committed to reducing its oil consumption by 3 % per year.

3.3.7 Water Sensitive Community

Looking further at water (from comments under 'climate change' above), Council states its goal of becoming a water sensitive community in *Strategy for a Water Sensitive Community 2012–2021* (Marrickville Council 2013). Council reported on their goals in this area in 2015:

> A water sensitive community will be achieved through the following strategies: a. reduce the use of potable mains water in homes, businesses, Council facilities and public spaces; b. manage the storm water system and its impacts on the urban environment; c. support regional projects to improve the health of the Cooks River, Botany Bay, Lower Parramatta River, Sydney Harbour and their catchments; d. implement sustainable urban water management (SUWM); and, e. transformation of urban systems from a focus on water supply and wastewater disposal to more complex, flexible systems is needed.

This strategy aims to supply water from within the catchment and provides a range of services through green infrastructure systems. Finally Council developed a Water Saving Action Plan, identifying water efficiency projects and instituting a Sustainable Irrigation Plan (SIP) for use on sports fields (thus substituting potable water use). In addition, Council has increased its sub-catchment planning, including water conservation sub-catchment plans to the year 2050.

3.3.8 Conclusion on Council Planning and SUP Practices

Having reviewed Marrickville Council planning documents and their applications, it can be concluded that Council (acknowledging the 12th May 2016 *Local Government (Council Amalgamations) Proclamation 2016* noted in Chap. 2) was moving in the direction of applying sustainable urban planning (SUP) practices (Question 2 above). The State's Greater Sydney planning (including infrastructure projects) and State council amalgamation proposals likely to affect Marrickville (Question 3) are reviewed next.

3.4 Greater Sydney Planning and Marrickville

There are plans and proposed developments of Greater Sydney likely to impact on Marrickville. Greater Sydney planning requirements were set by the NSW State in 2006, with Marrickville Council adopting subsequently in 2011 the *Marrickville Urban Strategy* (MUS) (Marrickville 2011d). Council's plans had to be consistent with the *Sydney Metropolitan Strategy* (NSW 2011). Council's complimentary town planning documents were the *Marrickville Local Environment Plan* (2011e) and the *Marrickville Development Control Plan* (2011f), noted above. Further, the State adopted its strategy for planning Greater Sydney to 2031 within *A Plan for Growing Sydney* (NSW 2014a). The Plan forecasts a population growth of 1.6 m to be added to the Greater Sydney population of 4.84 m (at June 2014). Marrickville is nominated by the State to be within the Central Subregion within this Plan. Local government areas within Central include: Ashfield, Botany, Burwood, Canada Bay, Leichhardt, Marrickville, Randwick, Strathfield, Sydney, Waverley, and Woollahra. In examining this Plan and its impact on Marrickville, three key planning principles as adopted in Chap. 2 are addressed: (1) Developing Sustainable Urban Centres; (2) Developing Sustainable Transport; and, (3) Protecting the Natural Environment. The planning principle *Developing* Sustainable Urban Centres is examined first. The State is working with Marrickville Council (as well as Canterbury and Bankstown councils) and the local community to investigate opportunities to revitalise communities around rail stations. Within Marrickville these areas include Sydenham and Dulwich Hill.

The State states:

> The first step in the process is to determine if renewal is feasible, and identify what kinds of infrastructure and services would be required to revitalise each area to support the delivery of new homes. The focus area is within 800 m from each station, and opportunities for new homes and neighbourhood renewal will be informed by: traffic and transport analysis; economic feasibility; environmental protection considerations; open green space needs; the local infrastructure needed; and, assessing the community's needs and feedback. A preliminary precinct plan for each station area will soon be prepared as part of a Land Use and Infrastructure Strategy for the Sydenham to Bankstown corridor.

The draft strategy provides an overview of opportunities for more homes and jobs that are walking distance from Bankstown, Punchbowl, Wiley Park, Lakemba, Belmore, Campsie, Canterbury, Hurlstone Park, Dulwich Hill, Marrickville and Sydenham railway stations. The planning principle Developing Sustainable Transport is examined next. Under the Plan, the State asked Marrickville Council (along with other Sydney Inner West councils) in late 2014 to comment on the proposed *Parramatta Rd Renewal Project* (NSW 2014b). The State's WestConnex roads project, also affecting Marrickville Council, is closely tied to the Parramatta Rd renewal project. Both these projects will be examined further in Chap. 9.

In late 2015 the State released the *draft Parramatta Rd Urban Transformation Strategy* (NSW 2015) (www.newparramattard.com.au). The Strategy identifies eight Precincts, specifically targeted for renewal under Precinct Plans. The State explains:

> Precincts were chosen due to their access to jobs, infrastructure, public transport, or how well they can support new housing types. The eight Precincts are places with unique character – diverse places that attract visitors and new residents and that have potential to be revitalized. Each Precinct is subject to a Precinct Implementation Plan, detailed on the following pages. These Plans include: a vision for the Precinct, why it has been identified as a growth Precinct, and what it will be like in the future – including population, housing and jobs projections three plans that illustrate how the vision will be delivered: A. Structure Plan that outlines the way land uses will be distributed and how the Precinct will change over time. B. Built Form Plan will show: the potential height of buildings; and C. Access and Movement Plan that details how people will move in and around the Precinct; and, how the vision can be delivered. (*Source* NSW 2014b)

There are a number of precinct renewal areas under the project, including Precinct 8, the Camperdown area of Marrickville Council (Plate 3.7).

The State spells out more detail for the Camperdown:

> The area is home to narrow disconnected streets and parks, shaped by the subdivision of historical landholdings and disrupted by creek lines that often force pedestrians and cyclists onto Parramatta Road and Pyrmont Bridge Road. The Precinct has a strong industrial history, with rows of Victorian and Federation houses interspersed with larger-scale buildings of both institutional and industrial origin. The vision for the Camperdown Precinct Camperdown Precinct will be home to high-quality housing and workplaces right on the edge of the CBD, well connected to the surrounding city, parklands, health and education facilities and focused on a busy and active local center. What it will be like to live and work there Camperdown Precinct will to evolve into an attractive, highly urbanized neighborhood marked by taller residential and business buildings, with potential for an innovative business and research hub to reflect the area's synergy of health, education, technology and reinvention. New residential development will be adjacent to the central core, complement-

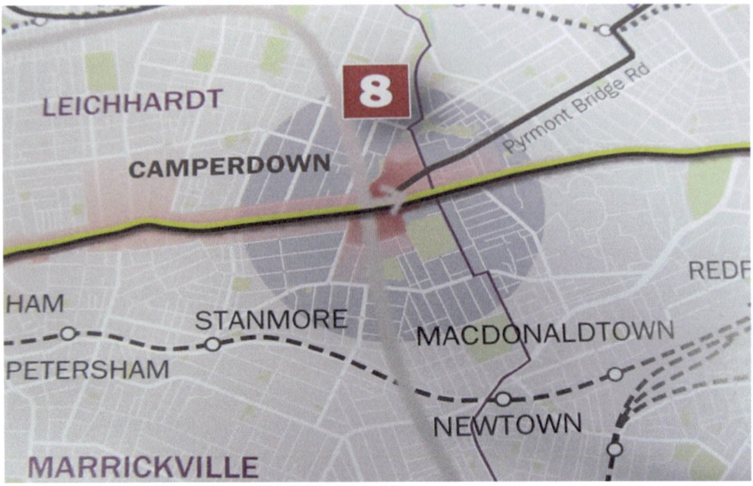

Plate 3.7 Draft Parramatta Rd urban transformation strategy – precinct 8 Camperdown (Source NSW 2015)

> ing the Precinct's industrial heritage and the adjacent heritage conservation area's lower density housing. New open spaces and road connections will provide a network of footpaths to support a fully walkable and revitalized center. The vision can be realized by: connecting new developments to the surrounding neighborhoods and carefully transitioning new, higher-density development to existing conservation areas; adapting and retaining the character of existing industrial heritage buildings; increasing the potential for student housing; addressing the constraints of the north-south street blocks and limited east-west connections; recognizing that the University and Royal Prince Alfred Hospital may expand into their existing land holdings, which may limit the potential to improve connections. (*Source* NSW 2014b)

Finally, the third planning principle Protecting the Natural Environment is examined next. The Plan commits the State to 'protect the health of the subregions waterways and aquatic habitats'. This goal, for example, would apply to the prime conservation challenges in Marrickville. This would include the protection of the Cooks River and its foreshores. Marrickville Council, along with neighbouring councils bordering the Cooks River, has played a major role in restoring the River to improved health. The efforts of these councils in undertaking this work will be outlined in more detail in Chap. 9.

3.4.1 Conclusion on Marrickville and Greater Sydney Planning

Having reviewed above the State's Greater Sydney planning (including infrastructure projects) it can be concluded the implementation of these plans and developments is likely to have a significant impact on Marrickville (Question 3). The extent of this impact will be better understood when these State plans are further examined in the context of the whole of the Sydney Inner West District (including Marrickville) in Chap. 9. Finally, in examining State proposed council amalgamations, options for councils (including Marrickville) working together and in partnership with the State to meet local and district needs will be addressed in Chap. 18 (Planning Reforms and Directions for Planning Sustainable Cities). The subject to amalgamations of councils and how this affects Marrickville is examined next.

3.5 Amalgamations of Councils and Marrickville

The subject of amalgamations of Greater Sydney councils (given implications for planning at Municipal and Greater Sydney levels) was reviewed in Chap. 2. NSW Premier Mike Baird announced in Dec 2015 the State's preference for Marrickville Council to merge with merge with Ashfield and Leichhardt councils. Within the Inner West District these three councils compose the eastern end (right) of the district. The Marrickville Council stated a clear preference to be a stand-alone council. Within a Council survey 75 % of Marrickville residents said they do not want to merge. Council also remained opposed to forced amalgamations. Council also notified the State that, if the State Government forced amalgamations, Council's preference was to merge with Leichhardt and Ashfield councils. Council further resolved in Nov 2015 that Council:

1. Reinforce to its community, staff and the NSW State Government that its preferred option is to stand alone and is fundamentally opposed to forced amalgamations;
2. Given the legal predicament the Council faced and the ultimatum issued by the NSW State Government, submit a Merger Preference for Marrickville, Leichhardt and Ashfield Council by 18 November 2015;
3. Council signal strongly to the State Government it will: a withdraw its Merger Preference if the State Government fails to proceed with forced amalgamations; and, b. not support an Inner West Council amalgamation proposition comprising Marrickville, Leichhardt, Ashfield, Burwood, Strathfield and Canada Bay Councils under any circumstances. (Source: Marrickville Council 10 November 2015)

The Mayors of Marrickville, Ashfield and Leichhardt councils urged all residents to have their say on the issue of forced amalgamations of councils. The Mayors noted that submissions to the new Council Boundary Review (www.councilboundaryreview.nsw.gov.au). In the interim, the Council noted, the Minister for Local Government had commenced the process of forcibly amalgamating Marrickville, Ashfield and Leichhardt councils. Council also noted the Minister said he expected the new council to be 'proclaimed' on 1 July 2016 (note the 12th May 2016 *Local Government (Council Amalgamations) Proclamation 2016* noted in Chap. 2). This came about despite all three councils repeatedly stating their preference to remain stand-alone, and a majority of residents also saying they prefer the current local government boundaries. The results of the review of merger proposals (which includes a new inner west council of Marrickville-Ashfield-Leichhardt) was referred to the Chief Executive of the Office of Local Government for examination and report under the Local Government Act (the Act). In turn, the Chief Executive had appointed a Delegate who was to examine the proposals, conduct a public inquiry, call for written submissions, and then report back to the Minister.

3.5.1 Conclusion on Amalgamations

Marrickville Council made a clear case, based on research and engaging its community, to stand alone and not be amalgamated with Ashfield and Leichhardt councils. At the end of the State review process (Council Boundary Review) the State determined what amalgamations would proceed or not proceed. The 12th May 2016 *Local Government (Council Amalgamations) Proclamation 2016* incorporating the merger of Marrickville, Leichhardt and Ashfield councils was noted. This question will be further addressed in Chap. 18.

3.6 Overall Conclusions

Having completed an assessment of the development and planning of Marrickville, there were a number of conclusions reached. Firstly, Marrickville has been experiencing accelerated development and changes to the composition of its population. Given these changes, and examining Marrickville Council's response, it was concluded Council is moving in the right direction in increasing its sustainable urban planning (SUP) practices. It was also concluded, however, that there would be significant future impacts on Marrickville from Greater Sydney growth and development. In reviewing potential Council and State responses to this future scenario, it was concluded that new partnerships between the Council and the State would be needed. A major initiative from the State (2009 onwards) was to investigate local government reform, including a potential amalgamation of councils (e.g. Marrickville, Ashfield and Leichhardt). Marrickville Council had to date, however, expressed an intent to stand alone and only amalgamate as a second choice (if the State forced amalgamations). The 12th May 2016 *Local Government (Council Amalgamations) Proclamation 2016* that merged Marrickville, Leichhardt and Ashfield councils was acknowledged. These overall chapter conclusions will be examined against results of assessing planning and development over all six council areas within Sydney Inner West District (Chaps. 3, 4, 5, 6, 7 and 8) and the district as a whole in Chap. 9. Results from the district assessment (and assessment of Parramatta and St George districts in later chapters) will contribute assessing how to plan for a sustainable Greater Sydney and a sustainable city anywhere (Chap. 18).

3.7 Summary

The chapter addressed the extent Marrickville Council is applying sustainable urban planning (SUP) practices (see Chap. 1) in planning the municipal area. It was noted that Marrickville, in Sydney Inner West District, was merged with Leichhardt and Ashfield councils under the 12th May 2016 *Local Government (Council Amalgamations) Proclamation 2016* (Chap. 2). It was established that Marrickville, one of twelve municipalities within three Greater Sydney Districts (Sydney Inner West, Greater Parramatta and St George) is considered to be in global transition. SUP practices (based on the application of three key SUP principles) were introduced, being: Developing Sustainable Urban Centers; Developing Sustainable Transport; and, Protecting the Natural Environment. A review of Marrickville background and population changes (i.e. global movement of population) was outlined. In examining the pattern of urban development, four phases of Marrickville development are examined: Early Marrickville; Traditional Marrickville; New Marrickville; and, Future Marrickville. With conclusions drawn, key Marrickville Council planning documents were then critiqued to determine the extent Council applies SUP practices. In addition, given State planning of Greater Sydney affects Marrickville, a second question was posed: what is the impact of Greater Sydney planning and development on Marrickville?. In examining this question, examples of State plans (including State council amalgamations) likely to impact on Marrickville were addressed. By chapter's end, conclusions were drawn on Council planning and impact of Greater Sydney planning on Marrickville. The chapter noted that these conclusions would be addressed further in: Chap. 9 (examining the extent of SUP practices being applied over the whole of Sydney Inner West District); and, Chap. 18 (examining directions for planning a sustainable Greater Sydney and a sustainable city anywhere).

Notation 3.1 Cultural Diversity Poems – Marrickville
Poems (Rauscher 2015 unpublished) reflecting the diversity of Marrickville:

Grains of Memory in Marrickville **(# 1)**

Kanus was an old man whose memories moved like grains of washing sand
To walk streets of Sydney's Marrickville is stepping into memory spaces
Corner store, bake house, shoe repairs, hair salon and haberdashery
The buildings' stucco of green and burgundy, yellow patches and red roofs

Now the footpath sports ceramic tiles preserving the legends once etched deep
Lines of chimneys stacked precariously still lean northward braced
Playgrounds filled with street kids once, now only memories of moms fetching children
You grasp the images and capture the grains of memory in Marrickville

White Cockatoos and Marrickville **(# 2)**

Moved you'll be by Marrickville colors like white cockatoos and rainbow parrots
Paint splashed by new citizens of Sydney, new spirit and new outlooks
Come they did to strings of suburbs laid in a mighty labyrinth
Time of 1960s to 2010s, a relentless pursuit of this idyllic urban place

Cottages, terraces, duplexes, gun barrels or semi's in a military row
Now houses stripped to skeleton, walls crumbled for more light and space
Backyards with fountains, renewed courtyards in terraces of specked brick
Rooms opened, mortgages secured, Marrickville is vision grasped and spirit fed more

Newtown's Global Realm **(# 3)**

Bengali, Sri Lankan, Moroccan, Indian, Pakistani assembled in Newtown
The sweet Turkish delight, Thai curry, Vietnamese vegetarian spices
Nestled in the global Sydney community in brushed up 1890s shops
The accepting Australia, the curious land down under, so willing to learn

The treasures of the world fill these colorful corners and side lanes
Peruvian hats, carpeted rugs, medieval ceramic pots and Lebanese breads
Street front jesters of broad smiles, of new found freedoms and markets to develop
Newtown presents the open society blended as a world model of community goodwill

References

Australian Bureau of Statistics (ABS) (2011) Australian census review. ABS, Canberra
Marrickville Council (1997) Marrickville local action 21. NSW, Marrickville
Marrickville Council (2001) Climate change action. NSW, Marrickville
Marrickville Council (2007) Oil depletion protocol Marrickville agreement, NSW, Marrickville
Marrickville Council (2010) Target sustainability (2010). NSW, Marrickville
Marrickville Council (2011a) Marrickville biodiversity strategy 2011–2021. NSW, Marrickville
Marrickville Council (2011b) Marrickville biodiversity action plan 2011–2015. NSW, Marrickville
Marrickville Council (2011c) Guideline to biodiversity protection Marrickville Urban Strategy (MUS). NSW, Marrickville
Marrickville Council (2011d) Marrickville Urban Strategy (MUS). NSW, Marrickville
Marrickville Council (2011e) Marrickville local environment plan. NSW, Marrickville
Marrickville Council (2011f) Marrickville development control plan. NSW, Marrickville
Marrickville Council (2012) State of environment report 2010–2011. NSW, Marrickville
Marrickville Council (2013) Strategy for a water sensitive community. NSW, Marrickville
New South Wales (2011) Sydney metropolitan strategy. NSW, Sydney
New South Wales (2014a) A plan for growing Sydney. NSW, Sydney
New South Wales (2014b) Parramatta Rd renewal project. NSW, Sydney
New South Wales (2015) Draft Parramatta Rd urban transformation strategy. NSW, Sydney

Chapter 4
Planning Ashfield, Greater Sydney

Dedication *This chapter is dedicated to Pamela Travers (1899–1996), novelist, actress and journalist, who lived in Ashfield. She is best known as the author of Mary Poppins (Plate 4.1).*

Plate 4.1 Pamela travers

R.C. Rauscher, S. Momtaz, *Cities in Global Transition*,
DOI 10.1007/978-3-319-39865-5_4

A snippet on Travers follows:

> Pamela Lyndon Travers, OBE (born Helen Lyndon Goff; was an Australian-British novelist, actress and journalist. In 1924, she immigrated to England where she wrote under the pen name P. L. Travers. In 1933, she began writing her series of children's novels about the mystical and magical English nanny Mary Poppins. For services to literature, Travers was made an Officer of the Order of the British Empire by Her Majesty the Queen in 1977. Her popular books have been adapted many times, including the 1964 film starring Julie Andrews and the Broadway musical originally produced in London's West End.

Abstract The chapter addresses the extent Ashfield Council (herein referred to as Ashfield) is applying sustainable urban planning (SUP) practices (see Chap. 1) in planning the municipal area. The chapter acknowledges the *Local Government (Council Amalgamations) Proclamation 2016* merging Ashfield, Marrickville with Leichhardt councils noted in Chap. 2). Ashfield (in Sydney Inner West District) is one of twelve municipalities within three Greater Sydney Districts (Sydney Inner West, Greater Parramatta and St George), all considered to be in global transition (see Chap. 2). Examining SUP practices is based on the application of three key SUP principles (as adopted in Chap. 2), being: Developing Sustainable Urban Centers; Developing Sustainable Transport; and, Protecting the Natural Environment.

Firstly, a background to Ashfield development and population changes (e.g. global movement of population) is outlined. To examine the pattern of urban development, four phases of Ashfield development are examined: Early Ashfield; Traditional Ashfield; New Ashfield; and, Future Ashfield. Given conclusions reached here, key Ashfield Council planning documents are then critiqued to determine the extent Council applies SUP practices. In addition, given State planning of Greater Sydney affects Ashfield, a second question is posed: what is the impact of Greater Sydney planning and development on Ashfield? In examining this question, examples of State plans (including council amalgamations under the Proclamation noted above) likely to affect Ashfield are addressed. By chapter's end, conclusions are drawn on Council planning and impact of Greater Sydney planning on Ashfield. The chapter notes that these conclusions will be addressed further in: Chap. 9 (examining the extent of SUP practices being applied over the whole of Sydney Inner West District); and, Chap. 18 (examining directions for planning a sustainable Greater Sydney or a sustainable city anywhere).

4.1 Background and Population Change

Ashfield Municipality (herein called Ashfield) in the context of Sydney Inner West (as introduced in Chap. 2) is adjacent to Marrickville to the west of the City of Sydney. Under the *Local Government (Council Amalgamations) Proclamation*

2016 Ashfield was merged with Marrickville and Leichhardt Ashfield councils as noted in Chap. 2. The area now known as Ashfield was home to the Wangal people, inhabiting Ashfield and beyond (Chap. 2). The municipality comprises of the suburbs (a–z): Ashfield; Ashbury (part of); Croydon (part of); Hurlstone Park (part of); Croydon Park (part of): Dobroyd Point; Haberfield; and, Summer Hill. Ashfield was (like most of Sydney Inner West) an industrial municipality. British settlement of the Ashfield district is described as follows:

> By 1790, a rough track had been built between the colony's two settlements at Sydney Cove and Parramatta. This route later became the main artery of the expanding Greater Sydney and, as the northern boundary of what is now Ashfield, dictated early British settlement in the area. The first land grant in the area was made to Rev Richard Johnson in 1793 and all of it had been granted by 1810. By the 1820s, all the grants had been amalgamated into two large estates: Ashfield Park (the northern half between Liverpool Rd and Parramatta Rd) and Canterbury Estate (the area south of Liverpool Rd). Ashfield Park was named by Robert Campbell, whose father was the laird of Ashfield in Scotland. In 1838, Elizabeth Underwood, then owner of Ashfield Park, subdivided part of her land to form the village of Ashfield between Liverpool Rd and Alt St. Part of the subdivision was the building of St John's Church in Alt St in 1841. This is the oldest surviving building in Ashfield. By 1855, the village had about 70 houses and 200 residents. However, the opening of the Sydney-Parramatta railway line that year, with Ashfield as one of its six original stations, led to a population explosion. In 1872, there were enough residents for the area to be granted a municipal council. By 1890, the population had grown to 11,000. While never a noted industrial suburb, Ashfield has had a couple of significant industries. On Parramatta Rd near Frederick St was the Australian Six motor car factory which opened in 1920. The site later became an AWA factory producing radio valves and other components. The site has since been turned into a commercial and residential development.

The degree that population change is impacting on Ashfield (i.e. global movements of people) is addressed next. The Australian Bureau of Statistics (ABS 2011) reports on Ashfield ancestry and languages spoken at home:

> At the 2011 Census, the proportion of residents in Ashfield local government area who stated their ancestry as Chinese was in excess of four times the state and national averages; and the proportion of households where an Asian language was spoken at home was about six times higher than the national average. In addition, there were 41,214 people in the Ashfield local government area. Aboriginal and Torres Strait Islander people made up 0.6 % of the population. The proportion of residents in Ashfield local government area who stated their ancestry as Chinese was in excess of four times the state and national averages; and the proportion of households where an Asian language was spoken at home was about six times higher than the national average. One area where Ashfield differed markedly from the national figures was in its ethnic mix. Australian born residents are a minority with only 41 % of Ashfield residents being Australian-born. Foreigner birth were from 15 % born in China and 6 % born in India. A fifth of the population spoke a Chinese language at home (Mandarin 14 % and Cantonese 6 %). Finally, within Ashfield housing, of the 8,215 occupied private dwellings counted, 63 % were flats (compared to the national figure of just 14 %), 27 % were detached houses, while 9 % were semi-detached or attached houses. The high number of flats contributed to a higher than average number of people renting (46 %) compared to houses owned outright (24 %) or being purchased (26 %). (*Source* ABS 2011)

Finally, in author's (Rauscher) qualitative research (2014/2015), there are two poems written to reflect the cultural diversity of Ashfield (end of chapter under Notation 4.1). These poems, include: *Ashfield's National Blends* (#4) and *Return to Your Summer Hill* (#5). Phases of Ashfield development are examined next.

4.2 Phases of Development

An examination of the development phases of any locality provides a basis of understanding current trends and likely future directions of growth of that locality. Following a look at the background of Ashfield four phases of Ashfield development will be examined. These phases include: Early Ashfield; Traditional Ashfield; New Ashfield; and, Future Ashfield. To assist in addressing this question, the author (Rauscher) completed qualitative research field work in Ashfield Municipality (2014–2015). Early Ashfield (Plate 4.2) is seen here (top left), with staff and community are outside the Ashfield Hotel. One of the most famous citizens of Ashfield was Quong Tart (top center), past business person and civic leader of Ashfield. Ashfield station became a central hub of the municipality, with commercial shops such as the corner 'Charlotte Building' (top right). The churches were always prominent in Ashfield as illustrated here looking at the presbytery of the Ashfield Catholic Church. Ashfield contains (like Marrickville and Leichhardt) a large number of Victorian (late 1800s) and Federation (early 1900s) buildings. The municipality also has a significant number of post-war low-rise flats normally built as apartment blocks. Ashfield provided an 'escape' from the inner areas of the City of Sydney, hence there are many cottages reflecting the owners' desire for more space (bottom

Plate 4.2 Early Ashfield. *Left to right* hotel staff in Hume highway; explorers park, Ashfield; Charlotte building (cafe now) opposite Ashfield Stn.; atholic church presbytery; restored cottage; and gadigal aboriginal

Plate 4.3 Traditional Ashfield. *Left to right* streetscape; corner restoration; federation cottage; cottage restoration; early apartment block; and, town house

center). Ashfield was the home of the Gadigal people, as illustrated here in this sketch (bottom right). As in other parts of Sydney the Aboriginal first inhabitants were soon forced to live further away from the urbanizing areas.

Traditional Ashfield (Plate 4.3) sees renewal occurring (2015) throughout most of the municipality's suburbs. The trim rows of cottages (top left) reflect an owner desire for front garden spaces and individual house allotments (as against shared boundaries with town houses of the inner city areas). The Federation era is prevalent throughout the suburb of Haberfield (Crowe), including those designed for corner blocks (top center). Most of the cottages were substantial and fronted significant front yards (top center). Attention to architectural detail is illustrated here with bull nose veranda and semi-circular scallops trim. Medium density living was accommodated in the early 1900s onwards within Ashfield, here a Summerhill block (bottom center). Finally, the occasional town house (bottom right) can be found in Ashfield, a reminder of the popularity and versatility of this housing form in the inner city areas.

The New Ashfield (Plate 4.4) reflects the dramatic changes the municipality has undergone since the 1970s, especially in the Ashfield commercial areas. A number of medium density complexes (top left) have been added to the landscape of suburbs such as Summerhill and Ashfield. Restoration of buildings have occasionally been done with inclusions of street wall art, as here in Summerhill (top center). Modern office complexes (top right) have changed the character of Ashfield into more of a metropolis reflecting the inner city areas of Sydney. Some newly established public

Plate 4.4 New Ashfield. *Left to right* medium density residential complex; restored with street wall art; commercial complex, Hume highway; new public square with fountain; new community gardens and open space; and, building restoration with street wall art (Sources Raymond Rauscher and Street Wall Artists 2015)

squares have incorporated public art, as in this instance of a square at Summerhill (bottom left). The wish to create garden suburb surrounds in Ashfield resulted in many new open space embellishments (bottom centre). The desire of the Ashfield Council to replicate art reflecting the municipality's natural areas is depicted here in this street wall art (bottom right). Finally, the major feature of Ashfield population today is the multi-cultural nature of its population. A couple of poems at chapter's end (Notation express the contrasts of Ashfield as follows: *Ashfield's National Blends* (Notation 1 #4) and *Return to Your Summer Hill* (Appendix 1 #5) describes the recent physical changes in the Ashfield suburb of Summer Hill.

Finally, Future Ashfield (Plate 4.5) reflects the continue building boom in residential development at increasing densities throughout the municipality (left to right). The quality of buildings planned or being under development show marked increase in design styles (top left). An eight storey residential corner property, with shops below (top-middle). A six storey building here creates a new streetscape of cafes in Liverpool Rd (top-right). Many residential apartment buildings have been selling 'off the plan'. With demands outstripping supply (2015) selling off the plan suits the development industry and some buyers. Finally, Future Ashfield reflects the continuing building boom in residential development at increasing densities

Plate 4.5 Future Ashfield

throughout the municipality. Factors of design, such as height, density, housing affordability, overshadowing and open space provisions, are likely to remain issues of development importance and contention within communities at the local government level into the future.

4.2.1 Conclusion

Given the review of development phases of Ashfield (above) and changes in population demographics, it can be concluded that Ashfield has been experiencing accelerated global transition (2000s) (Question 1). It was noted that under the *Local Government (Council Amalgamations) Proclamation 2016* Ashfield *was* merged with Marrickville and Leichhardt councils. The planning of Ashfield in meeting development demands now and in the future is examined next.

4.3 Ashfield and Sustainable Urban Planning

Planning of Ashfield Municipality has been taking place since it became a local government body in 1861. A review of planning in Ashfield will pay particular attention to applications of sustainable urban planning (SUP) (as outlined in Chap. 1) practices (including use of indicators of sustainability) (Chap. 2). A review of planning in Ashfield will focus on applications of sustainable urban planning (SUP) practices (as outlined in Chap. 1) and planning of a municipality (Chap. 2). Ashfield Council key planning documents (a–z with date and web reference) are listed in Information Box 4.1. A summary of these documents and a brief critique assist with this review of Ashfield planning (documents can be sourced under the web link as noted in the right column).

Information Box 4.1 Ashfield municipality planning

Ashfield Integrated Water Management Plan 2010–2014 (2010a)	www.ashfield.nsw.gov.au
Ashfield Urban Strategy Plan (2010b)	www.ashfield.nsw.gov.au
Ashfield Development Control Plan (2007)	www.ashfield.nsw.gov.au
Ashfield Local Environment Plan (2013)	www.ashfield.nsw.gov.au
State of Environment Report 2010–2011 (2012)	www.ashfield.nsw.gov.au
WWF Footprint Calculator (2015)	http://www.wwf.org.au

Source Webs as noted February 2016

4.3.1 Biodiversity

Within the State of Environment (SoE) reporting, Council addresses its biodiversity protection programs. In terms of catchment, rivers and waterways planning Ashfield Council points out that its local government area lies within two catchments, the Parramatta River Catchment (97 %) and the Cooks River Catchment (3 %). In addressing vegetation conservation and native animal protection under biodiversity, Council states:

> Most of the native vegetation in our local area has been cleared since European settlement and only a few small areas of remnant vegetation remain. There are two endangered ecological communities still represented in the area, including remnant species from the Sydney Turpentine Ironbark Forest community found at Ashfield Park, and a small community or Sydney Coastal Estuary Swamp Forest Complex that is now confined to a few remaining swamp mahogany trees (*Eucalyptus Robusta*) in Robson Park. Ashfield Council is working to maintain the integrity of these remnant specimens. The Ashfield Council area is home to many native animals, including birds, reptiles and mammals. An endangered population of Long-nosed Bandicoots lives in the Inner West, with recent sightings recorded in 2011.

4.3.2 Catchment Planning

The Ashfield Local Government Area lies within two catchments, the Parramatta River Catchment (97 %) and the Cooks River Catchment (3 %). The Council is engaged on local waterway projects in cooperation with other councils and the State Government. To widen the Council's actions it is a member of the Cooks River Alliance and the Parramatta River Catchment Group (comprised of representatives from local councils and from State Government agencies). The purpose of these groups is to facilitate and coordinate improvements to these rivers and their catchments. In addition, the Hawthorne Canal, along with Iron Cove Creek, is a significant tributary partly within Ashfield. The canal was originally a natural waterway, was straightened and given artificial banks. A further background on the Cooks River and the Cooks River Greenway is found in Chap. 9 (Planning Sydney Inner West District, Greater Sydney).

4.3.3 Ecological Footprint Measurements

Ashfield Council has taken a series of actions to move towards adopting and implementing a sustainable urban planning program. Overall Council encourages residents and businesses to use Ecological Footprint Measurements (Chap. 1) in considering activities and lifestyles. Council encourages use of the WWF Footprint Calculator (http://www.wwf.org.au). (World Wildlife Fund 2015) to find out ecological footprints, and ways to reduce these footprints.

4.3.4 Ashfield Local Environment Plan

In planning for the built environment, Ashfield Council, through its *Local Environmental Plan* (LEP) (Ashfield 2013) and *Development Control Plan* (Ashfield 2007) sets controls on new developments to ensure that environmental best practice features are applied in order to preserve and improve the environment.

4.3.5 State of the Environment (SoE)

Ashfield Council is committed to the State of the Environment (SoE) reporting (Ashfield 2012) as noted above, and states:

> State of Environment (SoE) reporting is a continuous process that provides the community with a summary of the physical condition of the environment within the Ashfield Local Government Area (LGA). It also informs the community and Council on the impacts we are having on our environment. The SoE also highlights the activities of Council, the community and businesses have undertaken to protect and enhance our environment. Council uses the information contained within the SoE Reports to develop environmental initiatives and targets for both corporate and community action.

4.3.6 Sustainability Planning

Council has embraced sustainability planning under actions within: state of environment reporting; biodiversity and catchments protection; and, vegetation and native animals' protection. Under Council's *Integrated Water Management Plan* (Ashfield 2010a) Council has implemented a range of water saving programs. In addition, a major planning project has been within the Greenway Sustainability Project. Here Council participates (a partnership between Ashfield, City of Canterbury, Leichhardt and Marrickville Councils) in working to complete the vision within the *Cooks River to Iron Cove GreenWay* (Marrickville 2009), defining the urban green corridor for wildlife, community and walking and cycling to cover

Sydney's Inner West (as outlined Chap. 3). The GreenWay Sustainability Project ran from 2009 to 2012. The Project was funded by the NSW Environmental Trust through its Urban Sustainability Program. The four partner Councils are still working together to achieve the GreenWay vision. The Ashfield Council notes some of the GreenWay projects (see Chap. 9):

> The GreenWay Project has implemented a range of projects which have focused on strategic partnerships and planning for sustainability, as well as worked with the community and on the ground to improve environmental conditions in the GreenWay. Activities included developing and piloting a GreenWay sustainability program with local primary schools, including Kegworth, Summer Hill, Ashbury and Petersham Public Schools; developing an Active Transport strategy for the GreenWay; and, forward planning for bushcare activities in the GreenWay. (*Source* Ashfield Council 2015)

4.3.7 Conclusion on Council Planning and SUP Practices

It can be concluded, having reviewed Ashfield Council planning documents and their applications, it can be concluded that Council was moving in the direction of applying sustainable urban planning (SUP) practices (Question 2 above). The State's Greater Sydney planning (including infrastructure projects) and State council amalgamation proposals likely to affect Ashfield (Question 3) are reviewed next.

4.4 Greater Sydney Planning and Ashfield

There are planning and developments of Greater Sydney likely to impact on Ashfield. In many instances Ashfield Council had the task of providing planning input on these plans. Greater Sydney planning requirements were set by the NSW State in 2006, with Ashfield Council subsequently adopting in 2010 the Ashfield Urban Strategy Plan (Ashfield 2010a, b). Council's plans had to be consistent with the *Sydney Metropolitan Strategy* (NSW 2011). Council's complimentary town planning documents were the *Ashfield Local Environment Plan* (2013) and the *Ashfield Development Control Plan* (2007). The State then adopted its strategy for planning Greater Sydney to 2031 (*A Plan for Growing Sydney* (NSW 2014). The Plan forecasts a population growth of 1.6 m (to be added to the Greater Sydney population of 4.84 m at June 2014).). In examining this Plan and its impact on Ashfield, three key planning principles as adopted in Chap. 2 are addressed: (1) Developing Sustainable Urban Centres; (2) Developing Sustainable Transport; and, (3) Protecting the Natural Environment.

The Plan's forecast that the district centre of Burwood will continue to expand in residential, office and services, Ashfield (given common boundary with Burwood) will likely continue to accommodate new residential development and services. The current Ashfield has a number of major office buildings and high density around the

Ashfield rail station. Looking at *Developing Sustainable Transport*, Ashfield Council, along with other Sydney Inner West and Greater Parramatta councils, were asked in late 2014 to comment on the proposed *Parramatta Rd Renewal Project* (NSW 2014). In late 2015 the State released *The Draft Parramatta Rd Urban Transformation Strategy* (NSW 2015) (www.newparramattard.com.au). The Strategy identifies eight Precincts, specifically targeted for renewal under Precinct Plans. The State explains:

> Precincts were chosen due to their access to jobs, infrastructure, public transport, or how well they can support new housing types. The eight Precincts are places with unique character – diverse places that attract visitors and new residents and that have potential to be revitalized. Each Precinct is subject to a Precinct Implementation Plan, detailed on the following pages. These Plans include: a vision for the Precinct, why it has been identified as a growth Precinct, and what it will be like in the future – including population, housing and jobs projections three plans that illustrate how the vision will be delivered: A. Structure Plan that outlines the way land uses will be distributed and how the Precinct will change over time. B. Built Form Plan will show: the potential height of buildings; and C. Access and Movement Plan that details how people will move in and around the Precinct; and, how the vision can be delivered.

Under 'precinct renewal' areas the State has nominated a Precinct 6 (Plate 4.6), partly within the Summerhill suburb of Ashfield.

The State notes:

> Taverners Hill will be an urban village with many links to public transport and the adjoining GreenWay. The Precinct will enjoy access to neighborhood parks, squares and leafy streets just off Parramatta Road. Tebbutt Street will be the Precinct's main street with residential development, including taller buildings at Parramatta Road, close to the Light Rail stops. New buildings will respond to heritage elements and the larger commercial buildings. New east-west pedestrian and cycling connections will make it easier to get around. Taverners Hill enjoys a quirky and diverse range of building types, from low-rise early period homes to small-scale industrial buildings. Its intimate neighborhoods and places have excellent access to transport, including heavy and light rail and bus services to the Sydney CBD, as well as the leafy GreenWay corridor, which follows the Hawthorne Canal. Taverners Hill will be an urban village with walking and cycling links via the GreenWay, access to many

Plate 4.6 Parramatta Rd urban transformation strategy – precinct 6 Summer Hill/Lewisham

public transport modes and many neighborhood parks, squares and leafy streets The GreenWay itself will be enhanced with better links under Parramatta Road and Longport Street. New taller buildings will mark the Precinct at Parramatta Road and will also be located near the light rail stops. Residents will enjoy easy access to neighborhood parks, squares and leafy streets just off Parramatta Road. New east-west connections will make it easier to walk around and access public transport. The vision can be realized by: providing better crossings for Parramatta Road and Longport Street and across the light rail corridor; transitioning new, higher-density development to existing areas and conservation areas; providing more open space areas and make them easier to get to; designing for the impact of major through-traffic roads; addressing aircraft noise; focusing services within residential areas rather than on Parramatta Road; addressing small residential lots and fragmented ownership, especially south of Parramatta Road; and, using the right mechanisms to fund public infrastructure, including high quality public places.

Moving to transport, the State's WestConnex roads project affects Ashfield. The motorway, is closely tied to the Parramatta Rd renewal project. Both the Westconnex and the Parramatta Rd Renewal projects, as they affect Ashfield and the whole of Sydney Inner West, will be addressed further in Chap. 9. Finally, under '3. *Protecting the Natural Environment*' the Plan commits the State to 'protect the health of the subregions waterways and aquatic habitats'. This goal, for example, would apply to the prime conservation challenges in Ashfield. This would include the protection of the Cooks River and its foreshores. Ashfield Council, along with neighbouring councils bordering the Cooks River, has played a major role in restoring the River to improved health. The efforts of these councils in undertaking this work will be outlined in more detail in Chap. 9.

4.4.1 Conclusion on Ashfield and Greater Sydney Planning

Having reviewed above the State's Greater Sydney planning (including infrastructure projects) it can be concluded the implementation of these plans and developments is likely to have a significant impact on Ashfield (Question 3). The extent of this impact will be better understood when these State plans are further examined in the context of the whole of the Sydney Inner West District (including Marrickville) in Chap. 9. Finally, in examining State proposed council amalgamations, options for councils (including Ashfield) working together and in partnership with the State to meet local and district needs will be addressed in Chap. 18 (Planning Reforms and Directions for Planning Sustainable Cities). The subject to amalgamations of councils and how this affects Ashfield is examined next.

4.5 Amalgamations of Councils and Ashfield

The subject of amalgamations of Greater Sydney councils (given implications for planning at municipal and Greater Sydney levels) was reviewed in Chap. 2. NSW Premier Mike Baird announced in Dec 2015 that Ashfield Council will be forced to

Plate 4.7 Sydney inner west councils – Marrickville, Ashfield, and Leichhardt

merge with Marrickville and Leichhardt councils (Plate 4.7). These three councils compose the eastern end of Sydney Inner West District.

Ashfield Council had stated a clear preference to be a stand-alone council. Within a Council survey the majority of residents said they do not want to merge. Council also remained opposed to forced amalgamations. Council notified that if the State Government forced amalgamations, Council's preference was to merge with Marrickville and Leichhardt councils. The Mayors of Marrickville, Ashfield and Leichhardt councils were urging all residents in early 2016 to have their say on the issue of forced amalgamations of councils. The Mayors noted that submissions to the new Council Boundary Review were open till the end of February www.councilboundaryreview.nsw.gov.au In the interim, the Council noted, the Minister for Local Government had commenced the process of forcibly amalgamating Ashfield, Marrickville, and Leichhardt councils. Council also noted the Minister said he expected the new council to be 'proclaimed' on 1 July 2016 (took place 12th May 2016 under the *Local Government (Council Amalgamations) Proclamation 2016* noted in Chap. 2). This came about despite all three councils repeatedly stating their preference to remain stand-alone, and a majority of residents also saying they prefer the current local government boundaries. The results of the review of merger proposals (which includes a new inner west council of Marrickville-Ashfield-Leichhardt) were referred to the Chief Executive of the Office of Local Government for examination and report under the Local Government Act (the Act). In turn, the Chief Executive was appointed a Delegate to examine the proposals, conduct a public inquiry, call for written submissions, and then report back to the Minister.

4.5.1 Conclusion on Amalgamations

Ashfield Council made a clear case, based on research and engaging its community, to stand alone. Council also notified that State that, if the State Government forced amalgamations, Council's preference was to merge with Marrickville and Leichhardt councils. At the end of the State review process (Council Boundary Review) the State determined what amalgamations as proposed would proceed or not proceed. The 12th May 2016 *Local Government (Council Amalgamations) Proclamation 2016* merging of Leichhardt, Marrickville and Ashfield councils was noted. This question will be further addressed in Chap. 18.

4.6 Overall Conclusions

Having completed an assessment of the development and planning of Ashfield, there are a number of overall conclusions reached. Firstly, Ashfield has been experiencing accelerated development and changes to the composition of its population. Given these changes, and examining Ashfield Council's response, it was also concluded Council is moving in the right direction in increasing its sustainable urban planning (SUP) practices. It was also concluded, however, that there would be significant future impacts on Ashfield from Greater Sydney growth and development. In reviewing potential Council and State responses to this future scenario, it was concluded that new partnerships between the Council and the State would be needed. A major initiative from the State (2009 onwards) was to investigate local government reform, including a potential amalgamation of councils (e.g. Ashfield, Marrickville, and Leichhardt). The Council had to date, however, expressed an intent to stand alone and only amalgamate as a second choice (if the State forced amalgamations). The 12th May 2016 *Local Government (Council Amalgamations) Proclamation 2016* that merged Ashfield, Marrickville, and Leichhardt councils was acknowledged. These overall chapter conclusions will be examined against results of assessing planning and development over all six council areas within Sydney Inner West District (Chaps. 3, 4, 5, 6, 7 and 8) and the district as a whole in Chap. 9. Results from the district assessment (and assessment of Parramatta and St George districts in later chapters) will contribute assessing how to plan for a sustainable Greater Sydney and a sustainable city anywhere (Chap. 18).

4.7 Summary

The chapter addressed the extent Ashfield Council was applying sustainable urban planning (SUP) practices (see Chap. 1) in planning the municipal area. It was noted that Ashfield, in Sydney Inner West District, was merged with Marrickville and Leichhardt councils under the 12th May 2016 *Local Government (Council*

Amalgamations) Proclamation 2016 (Chap. 2). It was concluded that Ashfield, one of twelve municipalities within three Greater Sydney Districts (Sydney Inner West, Greater Parramatta and St George) is considered to be in global transition. SUP practices (based on the application of three key SUP principles) were introduced, being: Developing Sustainable Urban Centers; Developing Sustainable Transport; and, Protecting the Natural Environment. A review of Ashfield background and population changes (i.e. global movement of population) was outlined. In examining the pattern of urban development, four phases of Ashfield development are examined: Early Ashfield; Traditional Ashfield; New Ashfield; and, Future Ashfield. With conclusions drawn, key Ashfield Council planning documents were then critiqued to determine the extent Council applies SUP practices. In addition, given State planning of Greater Sydney affects Ashfield, a second question was posed: what is the impact of Greater Sydney planning and development on Ashfield?. In examining this question, examples of State plans (including State council amalgamations) likely to impact on Ashfield were addressed. By chapter's end, conclusions were drawn on Council planning and impact of Greater Sydney planning on Ashfield. The chapter noted that these conclusions would be addressed further in: Chap. 9 (examining the extent of SUP practices being applied over the whole of Sydney Inner West District); and, Chap. 18 (examining directions for planning a sustainable Greater Sydney and a sustainable city anywhere).

Notation 4.1 Cultural Diversity Poems – Ashfield

Poems (Rauscher unpublished 2015) reflecting the diversity of Ashfield:

Ashfield's National Blends **(#4)**

Brahma Spiritual, St Johns Anglican, and St Vincent's anchor Ashfield Park
A multi-cultural 2015 society, statutes from Quong Tart to Mary Poppins
Street wall art paintings reflecting from streets where nature once dominated
New complex of civic heart and plaza, architectural balances for new urban scene

Proud Ashfield of ribbon rail and Liverpool Rd, all nations here now roam
Hotel wall depicts citizens Alt, Holden, and Pratten tall standing proud
Australia blends the world's nations on foundations of rugged past and new hopes
All welcome, sleeves of commerce rolled up, the Ashfield of international blends rises

Return to Your Summer Hill **(#5)**

Years may lapse, decades eclipse but return you will to Summer Hill
Your cottage is where those awakening days of youth took place
When days were work, learning the rules and ways, glimpsing at lives
When growing up was unquestioned, with young ambitions fueled here

Glimpsing a past in decades skipping spaces, Summer Hill memories slip
Return in time as grown wiser, a heart fuller, a life more ffulfilled
These streets are now paved, old is new, colour everywhere, people of all nations
Capture you will those years so fleeting to reclaim the Summer Hill once knew

References

Ashfield Council (2007) Ashfield development control plan. NSW, Ashfield
Ashfield Council (2010a) Ashfield integrated water management plan 2010–2014. NSW, Ashfield
Ashfield Council (2010b) Ashfield urban strategy plan. NSW, Ashfield
Ashfield Council (2012) Ashfield state of environment report 2010–2011. NSW, Ashfield
Ashfield Council (2013) Ashfield local environment plan. NSW, Ashfield
Australian Bureau of Statistics (ABS) (2011) Australian census review. ABS, Canberra
Marrickville Council (2009) Cooks river to iron cove greenway. NSW, Marrickville
New South Wales (2015) Draft Parramatta Rd urban transformation strategy. NSW, Sydney
NSW (2014) A plan for growing Sydney. NSW Government Planning & Environment, Sydney.
NSW Government (2011) Sydney metropolitan strategy. NSW, Sydney
NSW Government (2014) Parramatta Rd renewal project. NSW, Sydney
World Wildlife Fund (2015) WWF footprint calculator. NSW, Sydney

Chapter 5
Planning Leichhardt, Greater Sydney

Dedication *This chapter is dedicated to Nick Origlass (1906–1996), former mayor of Leichhardt and defender of the right of the public to be heard (Plate 5.1).*

Plate 5.1 Nick Origlass (*Source* Australian Biography 2016)

A snippet on Origlass follows:

> Nick Origlass joined the Workers Party in 1934 and by 1937 was its leading figure. Origlass formed the Labor Socialist Group. In 1952, he ceased publication of the newspaper, *The Socialist*, in order to join the ALP. In 1958 he was elected to Leichhardt Council for the ALP. A decade later he was expelled from the ALP but re-elected to the council on an "Independent Labor" ticket.' (*Source* Wikipedia link under Origlass 2015).

R.C. Rauscher, S. Momtaz, *Cities in Global Transition*,
DOI 10.1007/978-3-319-39865-5_5

Issy Wyner, close working colleague of Origlass, added these words at the Origlass funeral:

> In his lifetime Nick achieved something that is not commemorated in bronze or stone. Rather, it resides in the knowledge and memory of the many generations who have witnessed and learned of the essence of democracy and fighting for the worthwhile things in life. This is demonstrated through our involvement in a variety of struggles over more than 60 years: for egalitarianism in every sphere of life; for protection of the environment and ecology; and, for people's right to participate in decision-making that affects their life, liberty and pursuit of happiness. (*Source* Wikipedia link under Origlass 2015)

Abstract The chapter addresses the extent Leichhardt Council (herein referred to as Leichhardt) is applying sustainable urban planning (SUP) practices (see Chap. 1) in planning the municipal area The chapter acknowledges the *Local Government (Council Amalgamations) Proclamation 2016* merging Leichhardt, Marrickville and Ashfield councils noted in Chap. 2. Leichhardt (in Sydney Inner West District) is one of twelve municipalities within three Greater Sydney Districts (Sydney Inner West, Greater Parramatta and St George), all considered to be in global transition (see Chap. 2). Examining SUP practices is based on the application of three key SUP principles (as adopted in Chap. 2), being: Developing Sustainable Urban Centers; Developing Sustainable Transport; and, Protecting the Natural Environment.

Firstly, a background to Leichhardt development and population changes (e.g. global movement of population) is outlined. To examine the pattern of urban development, four phases of Leichhardt development are examined: Early Leichhardt; Traditional Leichhardt; New Leichhardt; and, Future Leichhardt. Given conclusions reached here, key Leichhardt Council planning documents are then critiqued to determine the extent Council applies SUP practices. In addition, given State planning of Greater Sydney affects Leichhardt, a second question is posed: what is the impact of Greater Sydney planning and development on Leichhardt? In examining this question, examples of State plans (including State proposed council amalgamations) likely to impact on Leichhardt are addressed. By chapter's end, conclusions are drawn on Council planning and Greater Sydney planning. The chapter notes that these conclusions will be addressed further in: Chap. 9 (examining the extent of SUP practices being applied over the whole of Sydney Inner West District); and, Chap. 18 (examining directions for planning a sustainable Greater Sydney or a sustainable city anywhere).

5.1 Background and Population Change

Leichhardt Municipality (herein called Leichhardt) (part of Sydney Inner West as introduced in Chap. 2) is adjacent and north of Marrickville and adjacent and to the west of the City of Sydney. Under the *Local Government (Council Amalgamations)*

Proclamation 2016 Leichhardt was merged with Marrickville and Ashfield councils as noted in Chap. 2. The area now known as Leichhardt was home to the Wangal people (Chap. 2). Leichhardt is named after the Prussian explorer Ludwig Leichhardt, who in the 1840s was feted for his 4800 km (3000 mi) expedition in search of an overland route from southern Queensland to Port Essington, a British settlement on the far northern coast of Australia. He later vanished without trace on his attempt to cross the continent from the Darling Downs to the Swan River Colony on the Western Australia coast.

Leichhardt was proclaimed a municipality in 1871. In 1949, it was merged with the municipalities of Annandale and Balmain. Suburbs of Leichhardt include: Annandale; Balmain; Balmain East; Birchgrove; Cockatoo Island; Leichhardt; Lilyfield; and, Rozelle. At the 2011 Census, there were 52,198 people in the Leichhardt local government area. Aboriginal and Torres Strait Islander people made up 1.0 % of the population. As at the 2011 Census, the Municipality of Leichhardt had approximately double the national and state average of people who spoke Italian at home; and just under double the New South Wales average of people who claimed Italian ancestry. With this background introduction to Leichhardt Municipality (herein referred to as Leichhardt), the extent of Leichhardt's global transition can be addressed. The degree that population change is impacting on Marrickville (i.e. global movements of people) is addressed next. The Australian Bureau of Statistics (ABS 2011) reports on Leichhardt ancestry and languages spoken at home:

> Since the middle of the twentieth century, Leichhardt has been a center for the Italian community, reflected in the many Italian-owned businesses in the area. Although it has become less distinctively Italian with the increasing gentrification of the suburb and movement of families to suburbs with larger blocks of land, its Italian character is still palpable, particularly in the popular restaurants and cafés of its main street, Norton Street. Many residents moved to suburbs further away from Sydney as the city has expanded and larger housing blocks became available. As at the 2011 Census, the Municipality of Leichhardt had approximately double the national and state average of people who spoke Italian at home; and just under double the New South Wales average of people who claimed Italian ancestry. Other common national groups were – England 4.8 %, Italy 4.4 %, New Zealand 3.1 %, Greece 0.7 % and Ireland 0.6 %. English was stated as the only language spoken at home by 72.6 % of persons usually resident in Leichhardt. In terms of housing, approximately one third of dwellings were separate density housing houses with 22 % apartments and 36 % terraces, semis or townhouses. 63.7 % of persons usually resident in Leichhardt stated they were born in Australia.

Finally, in author's (Rauscher) qualitative research (2014/2015), there are several poems written to reflect the cultural diversity of Leichhardt (end of chapter under Notation 5.1). These poems, include: *Annandale Crimson* (# 6); *Balmain Remnants* (# 7); and, *Annandale Crimson* (# 8); *Heartland Leichhardt Revives* (# 9); and, *Taverners Hill Lights Up* (# 10). With this background to Leichhardt Municipality, the extent of Leichhardt's development phases examined next.

5.2 Phases of Development

Having reviewed a background on Leichhardt, four phases of Leichhardt development and impact of that development are now examined. These phases are: Early Leichhardt; Traditional Leichhardt; New Leichhardt; and, Future Leichhardt Directions. To assist in addressing this question, author (Rauscher) qualitative research field work was completed in Leichhardt Municipality in 2014–2015. Early Leichhardt (Plate 5.2) illustrates the intense urbanization that took place in the late 1800s along Parramatta Rd (top left). Industry often was dependent on waterways, hence these smoke stacks (top center) reflect heavy industry operations along Leichhardt foreshores. This industry included paint factories and other polluting operations as shown here at Balmain (top right). The old Glebe Island Bridge as the main connector of Leichhardt to the city center (bottom left). Substantial residential buildings, often with shops below, were built along thoroughfares (bottom center). Finally, the artist sketch here of the water pipeline viaduct (bottom right) that runs through Annandale reflects the extent of infrastructure required to service early Leichhardt municipality.

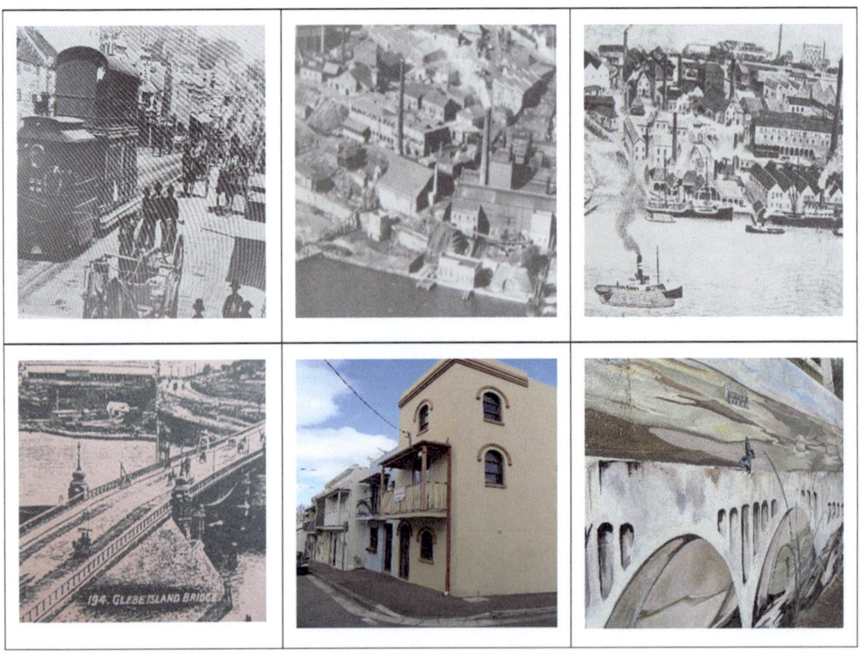

Plate 5.2 Early Leichhardt Municipality. *Left to right* Transport along Parramatta Rd; Industries along foreshores; Factories at Balmain; Old Glebe Island Bridge; Early residential buildings; and, Water main viaduct as street wall art (*Sources* Leichhardt Council Archives, Raymond Rauscher, and Street Wall Artist 2015)

Plate 5.3 Traditional Leichhardt. *Left to right* Streetscape; Decorative cottages; Modest cottages; Landscaped streets; Row housing with uniform coloured fencing; and stand-alone terrace

Traditional Leichhardt (Plate 5.3) is experiencing renewal (2016) throughout most of the municipality's suburbs. The homes in Leichhardt range from Victorian to modern homes (mostly in fill) and medium density developments. The trim rows of cottages (top left) reflect a variety of early architecture. The Federation touch (top center) is shown in this truncated a-frame cottage. Modest cottages are scattered throughout the municipality. The landscaping of streets (bottom left) contrasted with the tighter inner city streets. The uniformity of row housing was broken with brighter fencing and roofing (bottom center). Finally, throughout Leichhardt can be found stand-alone (bottom right) or rows of terraces.

The New Leichhardt (Plate 5.4) reflects an even spread of changes throughout the municipality since the 1970s, especially in the commercial areas. The 'Forum' (top left) was an early innovative residential and commercial square created within the Leichhardt suburb shopping center. A range of residential infill projects (top center) occur throughout Factory buildings have also been converted to residential use (top right). New architecture has added a new urban dimension to Leichhardt, as reflected in this bus depot building (bottom left). Within the Orphan School Creek (bottom center) area of Annandale a number of new residential buildings, including this one with vertical vegetation, have formed a residential precinct. The scale of this precinct is reflected in these high rise residential buildings (bottom right) built on a square.

Plate 5.4 New Leichhardt. *Left to right* The Forum; Infill residential; Factory conversion; Bus depot; Orphan School Creek complex; Residential around square

Finally, Future Leichhardt (Plate 5.5) reflects the continuing building boom in residential development at increasing densities throughout the municipality. Factors of design, such as height, density, housing affordability, overshadowing and open space provisions, are likely to remain issues of development importance and contention within communities at the local government level into the future.

Current and future developments where major public space allocations and recreation uses were considered critical by Leichhardt Council included: Old Harold Park residential site; Callan Park (once a mental health hospital); and, the Bays Project. The Bays Project is reviewed on a Sydney Inner West District level in Chap. 9. A glimpse at Leichhardt of old and on the streets today is captured in several poems. A picture of the renewed Annandale is described in *Annandale Crimson* (Appendix 1 #6). The importance of retaining suburb history (Balmain in this instance) is reviewed in *Balmain Remnants* (Appendix 1 #7). Moving to the Norton St end of Leichhardt, the multi-cultural centre of Norton St is described in *Growing City Cultures in Norton St* (Appendix 1 #8). The overall recent urban revival of Leichhardt is outlined in *Heartland Leichhardt Revives* (Appendix 1 #9). Finally, the history and changes in the furthest western end of Leichhardt, Taverners Hill, is described in *Taverners Hill Lights On* (Appendix 1 #10).

In examining the development phases of Leichhardt it can be concluded that Leichhardt has recently (2000s) shown a major increase in investment in renewal

Plate 5.5 Future Leichhardt

projects and new developments. The degree that this increase in development is being influenced by Leichhardt experiencing accelerated global transition.

5.2.1 Conclusion

Given the review of development phases of Leichhardt (above) and changes in population demographics, it can be concluded that Leichhardt has been experiencing accelerated global transition (2000s) (Question 1). The planning of Leichhardt (merged with Leichhardt and Ashfield as noted above under the Proclamation) in meeting development demands now and in the future is examined next. Particular attention is paid to Question 2 above, whether Leichhardt Council planning is moving in the direction of sustainable urban planning (SUP) practices (Chaps. 1 and 2).

5.3 Leichhardt and Sustainable Urban Planning

Planning of Leichhardt Municipality has been taking place since it became a local government body in 1871. A review of planning in Leichhardt will focus on applications of sustainable urban planning (SUP) practices (as outlined in Chap. 1) and planning of a municipality (Chap. 2). Leichhardt Council key planning documents

Information Box 5.1 Leichhardt municipality planning

Climate Change Plan (2013a)	www.leichhardt.nsw.gov.au
Environmental Sustainability Strategy 2010–2014 (2010)	www.leichhardt.nsw.gov.au
Leichhardt 2025 + Strategic Plan (2013e)	www.leichhardt.nsw.gov.au
Leichhardt Development Control Plan (2013c)	www.leichhardt.nsw.gov.au
Leichhardt Integrated Transport Plan (2012)	www.leichhardt.nsw.gov.au
Leichhardt Local Environment Plan (2013b)	www.leichhardt.nsw.gov.au
Public Disclosure Summary (2013e)	www.leichhardt.nsw.gov.au
State of Environment Report 2012–2013 (2013d)	www.leichhardt.nsw.gov.au

Source Webs as noted February 2016

(a–z with date and web reference) are listed in Information Box 5.1. A summary of these documents and a brief critique assist with this review of Leichhardt planning (documents can be sourced under the web link as noted in the right column).

5.3.1 Biodiversity Protection

Moving from climate change and energy to biodiversity protection, Council stresses importance of preserving biodiversity in its *Leichhardt Local Environment Plan* (Leichhardt 2013c), Council notes here that Leichhardt has diverse areas of flora and fauna even though the original native flora and fauna communities within the area have been extensively modified as a result of urban development.

5.3.2 Climate Change Protection

Under climate change Leichhardt Council adopted the *Climate Change Plan* (Leichhardt 2013a) (www.leichhardt.nsw.gov.au) in May 2013. The Plan is Council's principal policy framework for corporate decision-making in response to climate change. The Plan includes objectives, strategies, and actions to improve energy efficiency, increase the amount of renewable energy Council produces through a Renewable Energy Master Plan, and make climate change a key consideration within all of our Council systems. Examining potential impacts of climate change, the Sydney Coast Councils Group has provided Leichhardt Council with potential inundation maps over its local government area. The object here is assist councils like Leichhardt to better plan programs of public works and to educate residents and authorities in response to this potential hazard. Leichhardt Council has also adopted, under its climate change initiatives, a carbon neutrality plan. This neutrality is when the net carbon emissions associated with an organization are equal to zero through the acquisition and cancellation of carbon offsets that meet stringent criteria (Chap. 1). In 2009, Leichhardt Council formed a Climate Change

Taskforce to investigate the policies and actions required for the Council to become carbon neutral by 2012.

5.3.3 Energy Planning

Moving to energy planning, Council is looking at a better approach to accessing energy for the municipality. The Council has examined renewable energy policies in its participation in the Southern Sydney Region Organisation of Councils (SSROC). The SSROC is an association of Councils in the southern area of Sydney, with Marrickville, Ashfield and Leichhardt participating. SSROC (in 2011) resolved to adopt regional renewable energy targets for 2030. These targets are summarized as follows: (1) More than 20 % of energy consumed in the region should be derived from renewable sources; and, (2) SSROC member Councils aim to be using a minimum of 30 % of energy from renewable sources. The SSROC has set a path to meet the target under a SSROC *Renewable Energy Master Plan* (2013). All renewable energy options to achieve the target and recommend the most cost effective and feasible solutions including analysis of governance and regulatory aspects have been addressed.

Council has installed a large number of solar panel units on different service facilities (i.e. neighborhood centres and swimming pool complex) under a program called Energy Icon Project. Council also uses photo voltaic systems and have fitted real-time monitoring system on most of these. This system allows Council staff and the community to view energy performance and generation of data online via a website called the Sunny Portal (http://www.sunnyportal.com/templates/start.aspx?ReturnUrl=/the).

Leichhardt Council has completed a carbon reduction project, the cogeneration plant at Leichhardt Park Aquatic Centre (Plate 5.6).

Plate 5.6 Cogeneration system

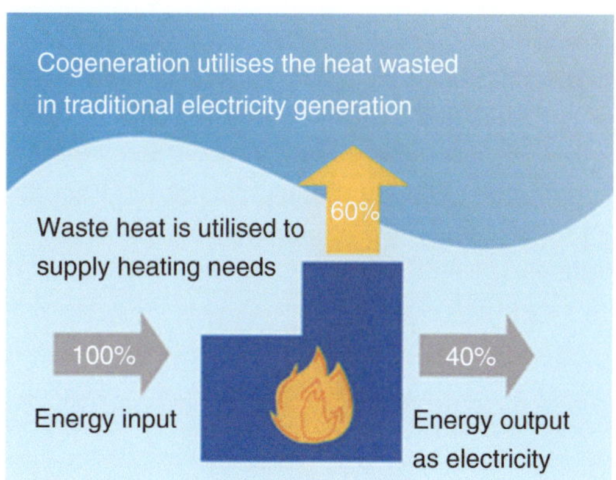

5.3.4 Leichhardt Environment Plan

In planning for the built environment, Leichhardt Council, through its *Local Environmental Plan* (LEP) (Leichhardt 2013b) and Development Control Plan (DCP) (Leichhardt 2013c) sets controls on new developments to ensure that environmental best practice features are applied in order to preserve and improve the environment.

5.3.5 Environmental Sustainability Plan

Council is currently (2015) developing a longer-term strategy: the *10-year Environmental Sustainability Plan* (www.leichhardt.nsw.gov.au). The new Plan will: (1) Build on the objectives of the previous *Environmental Sustainability Strategy 2010–2014*, the *Leichhardt 2025 + Strategic Plan*, and other Council strategic plans; (2) Articulate Council's strategic direction and priorities for Environmental Sustainability to 2025; (3) Inform all aspects of Council's business so that environmental sustainability objectives are integrated within plans, policies and activities; (4) Guide Council to meet legislative requirements and responsibilities; (5) and, respond to both current and emerging issues. Leichhardt Council is finding urban consolidation is increasing the density of residential areas within the municipality. There are requirements for Leichhardt to deliver 2000 additional dwellings by 2031 as indicated by the NSW government's Inner West Subregional Strategy (established in 2005).

5.3.6 Leichhardt 2025 + Strategic Plan

A significant background to urban planning in Leichhardt is contained in Council's *Leichhardt 2025 + Strategic Plan* (Leichhardt 2013e) (www.leichhardt.nsw.gov.au) (details of the Plan to be examined later in this Chapter). Council adopted this plan to establish a clear direction and framework to achieve a sustainable and liveable community. The main objectives of the Plan are translated into measurable indicators under headings (or 'themes') as follows: land; climate change and air quality; transport; water; biodiversity; and, waste and consumption.

5.3.7 Leichhardt Local Environmental Plan

Council's two most important planning instruments – the *Leichhardt Local Environmental* Plan (Leichhardt 2013b) and an accompanying *Development Control Plan* (Leichhardt 2013c) were adopted in 2013. Planning and design outcomes that Council is pursuing include sustainable and contemporary building and place design, transit oriented development, heritage conservation, active streets and laneways and a diversity of land use that supports economic, environmental and social sustainability. For the first time, explicit recognition of the growing prominence of food security as an issue is addressed within the DCP, encouraging initiatives such as community gardens and green roofs. An Emerging Technologies Clause gives Council scope to approve new and innovative technologies as they emerge without having to fully amend the DCP.

5.3.8 State of Environment Report

Council produces a SoE Report each year. The SoE provides information about the local environment and about progress towards meeting environmental objectives as set out in the *Environmental Sustainability Strategy*. Within Council's SoE reporting specific attention is given, for example, to addressing: land uses; air and climate change; water conservation; biodiversity protection; waste reduction; noise abatement; heritage protection; and flora and fauna protection.

5.3.9 Sustainable Neighbourhoods

Finally, addressing sustainable neighbourhoods Council has established the *Sustainable Neighbourhoods Program* (www.leichhardt.nsw.gov.au). Under this program Council implements a sustainable education program and has piloted a street in Annandale (Taylor St). The program aims to improve the environmental and social sustainability of Leichhardt neighbourhoods by building capacity within the community to identify, plan for, and implement sustainability and climate change actions on a neighbourhood-by-neighbourhood scale.

5.3.10 Transport

Moving from biodiversity to sustainable transport, *Leichhardt Integrated Transport Plan* (Leichhardt Council 2012) comprises of the 10 Year Strategic Plan (and 4 Year Service Delivery Plan). The Plan, identifies nine objectives for accessibility, environmental improvement, equity, access and accessibility, social inclusion, cultural engagement and community wellbeing. Leichhardt Council is creating via the *Leichhardt Bicycle Strategy* (Leichhardt 2010) (www.leichhardt.nsw.gov.au) a network of bicycle routes and making streets more bike friendly. Council notes that it maintains more than 55,000 m of bicycle routes. These routes vary in types, including: on-road; off-road' shared (cycle/pedestrian) paths; and, mixed traffic (where signage is provided to indicate to drivers and cyclists that cycling in the traffic stream is the expected).

5.3.11 Conclusion on Council Planning and SUP Practices

It can be concluded, having reviewed Leichhardt Council planning documents and their applications, it can be concluded that Council is moving in the direction of applying sustainable urban planning (SUP) practices (Question 2 above). The State's Greater Sydney planning (including infrastructure projects) and State council amalgamation proposals likely to affect Leichhardt (Question 3) are reviewed next.

5.4 Greater Sydney Planning and Leichhardt

Planning of Leichhardt has been taking place since it became a local government body in 1871. A review of planning in Leichhardt will pay particular attention to applications of sustainable urban planning (SUP) (as outlined in Chap. 1) practices (including use of indicators of sustainability) (Chap. 2). Greater Sydney planning requirements were set by the NSW State in 2006, with Leichhardt Council adopting in 2013 the *Leichhardt 2025 + Strategic Plan* (Leichhardt 2013e). Council's plans had to be consistent with the *Sydney Metropolitan Strategy* (NSW 2011). During 2012, the NSW Government started to develop new plans for the long-term future of the State linked to *NSW 2021*. The *Metropolitan Strategy* is a 20-year plan to build livable places across Sydney. It guides future planning and investment decisions covering housing, economic development and jobs, open space and the transport we need to connect homes, jobs and education and recreation facilities. The Metropolitan Plan integrates land use, urban and funded–transport planning to provide a framework for sustainable growth and development across the city to 2036. It will also meet the targets in the *NSW 2021,* notably in integrated transport and land use planning. The Plan also sets the boundaries for future urban development and identifies the strategic transport corridors and major centres best placed to focus sustainable commercial and residential growth in the future. It provides strategic direction for a sustainable city by limiting urban sprawl, protecting Sydney's valuable resource lands, encouraging more development near public transport and reducing the need for car travel. It means more jobs located near transport; more homes located near transport; and more jobs located near homes.

Key elements of the Draft Strategy include: (1) linking land use and transport planning together; (2) promoting the future of Parramatta Road as a strategic issue for metropolitan Sydney; (3) highlighting the importance of housing and employment in the future of the metropolitan area; and, (4) increasing the focus of affordable housing. These fundamental strategic planning principles have informed the preparation of Leichhardt Municipal Council's community strategic plan and LEP 2013. The State Government projects a low growth rate in Leichhardt with a population of 57,800 by 2036. The population profile gains a stronger balance of young people as more of the 25–39 family age group have children and stay in the area. Less rental accommodation, more owner purchasers, high enrollments in childcare and schools, and new higher birth rates signal a trend to a more balanced profile.

Further, the State adopted its strategy for planning Greater Sydney to 2031 within *A Plan for Growing Sydney* (NSW 2014a). The Plan forecasts a population growth of 1.6 m to be added to the Greater Sydney population of 4.84 m (at June 2014). Marrickville is nominated by the State to be within the 'Central Subregion' within this Plan. Local government areas with 'Central' include: Ashfield, Botany, Burwood, Canada Bay, Leichhardt, Marrickville, Randwick, Strathfield, Sydney, Waverley, and Woollahra. In examining this Plan and its impact on Leichhardt, three key planning principles as adopted in Chap. 2 are addressed: (1) Developing Sustainable Urban Centres; (2) Developing Sustainable Transport; and, (3) Protecting the Natural Environment. With these principles in mind, several Greater Sydney State planning proposals currently affecting Leichhardt are examined next. These proposals, for example, include: WestConnex (www.westconnex.com.au) motorway proposal; *Parramatta Rd Renewal Project* (NSW 2014b); Bays Precinct development; Callan Park restoration; aircraft noise monitoring; and, assessing the State's *Revitalization of Local Government Report* (NSW 2014c).

Leichhardt Council, along with other Sydney Inner West and Greater Parramatta councils, were asked in late 2014 to comment on the proposed Westconnex and Parramatta Rd projects in 2014. The section of the WestConnex that passes through Leichhardt is a tunnel under Parramatta Road, from Hawthorne Canal to Camperdown. There are a number of 'precinct renewal' areas under the project, including Precinct 6 which is partly within Leichhardt Municipality as well as Ashfield and Marrickville, as addressed Chap. 4. The State's WestConnex roads project, is closely tied to the Parramatta Rd renewal project. Both projects, as they affect Sydney Inner West, will be addressed further in Chap. 9.

Moving to another Greater Sydney planning proposal, the State is progressing the urban consolidation project entitled The Bays Precinct development (partly within Leichhardt Council area). The project generally comprises the areas known as White Bay, Glebe Island, Rozelle Bay and Blackwattle Bay and the former rail yards west of Glebe Island. The area falls within the Local Government Boundaries of both the City of Sydney and Leichhardt Municipal Council. Significant areas of the site are controlled by Sydney Ports and the Sydney Harbour Foreshore Authority.

The current Bays Precinct Taskforce Report is available at: http://www.shfa.nsw.gov.au/sydney-Our_places_and_projects-Our_projects-Bays_Precinct.htm.

Moving to another State project within Leichhardt, Callan Park planning is being progressed. Callan Park is a large parkland in the Leichhardt area. At over 60 hectares and equivalent to the size of the Sydney Botanical Gardens, Callan Park is one of the largest and most regionally significant public facilities in Sydney's Inner West. Council comments on Callan Park include:

> Built as an enlightened hospital for the insane in 1885, it has a number of exceptional heritage buildings. Callan Park is owned and managed by the NSW State Government. Master planning of the site is a state government responsibility however, Leichhardt Council took on this challenge as a result of many previous unsuccessful state government attempts. With the transfer of mental health beds away from Callan Park in 1983 after the Richmond Report was initiated, the last twenty years have seen a number of successful community led campaigns to oppose state government redevelopment plans. These campaigns led to the passage of the Callan Park (Special Provisions) Act 2002 to: preserve the public ownership of Callan Park; to protect its current features; and, to restrict its future use.

Another ongoing State controlled project affecting Leichhardt (as well other local government areas in Sydney Inner West) is expansion of Mascot Airport. Though the airport is outside Leichhardt, the aircraft noise affects most suburbs of Leichhardt. Council notes:

> Aircraft noise is an issue for the Leichhardt local community because when the new third runway opened at Sydney Airport in 1995 it resulted in an increase in aircraft movements over the area. There are ongoing issues with the levels of aircraft movements increasing. In 1995, the 3rd runway of Sydney Airport was officially opened for use resulting in a massive increase in aircraft movements over the suburbs to the north of the Airport, particularly the suburbs of Annandale, Lilyfield and Leichhardt. As the Leichhardt Local Government Area is located close to Sydney Airport, the residents of the Leichhardt Municipality have suffered ever increasing aircraft noise as a result of a significant increase in aircraft passenger numbers and traffic since 1995 which is projected to get worse over the next two decades.

On Greater Sydney's housing provisions, Leichhardt Council has expressed concerns about the State Government moves to sell long standing public housing. For example, the Council notes, the State Government announced in 2013 plans to sell off 293 Housing NSW properties in Millers Point (as was implemented in 2014). Leichhardt Council is concerned that this sets a dangerous precedent. Housing NSW owns a large amount of housing stock in Leichhardt Municipality and neighboring municipalities and this may be targeted next. Having people from different social and economic backgrounds living together in the same local community is what makes Leichhardt neighborhoods great, the Council argues with a call to protect public housing in the Leichhardt Municipality.

The State then adopted its strategy for planning Greater Sydney to 2031 (*A Plan for Growing Sydney* (NSW 2014a). The Plan forecasts a population growth of 1.6 m (to be added to the Greater Sydney population of 4.84 m at June 2014). In examining this Plan and its impact on Leichhardt, three key Planning Principles (Chap. 2) are addressed: (1) Developing Sustainable Urban Centres; (2) Developing Sustainable Transport; and, (3) Protecting the Natural Environment.

Examining the principle Developing Sustainable Urban Centres, the Plan's forecast is that the district centre of Burwood will continue to expand in residential, office and services, Leichhardt (given proximity to Burwood) will likely continue to accommodate new residential development and services. The current Leichhardt has a number of medium density areas developing close to foreshores and around shopping centres served with bus routes or light rail. Looking at Developing Sustainable Transport, Leichhardt Council, along with other Sydney Inner West and Greater Parramatta councils, were asked in late 2014 to comment on the proposed *Parramatta Rd Renewal Project* (NSW 2014b). The State's WestConnex roads project, also affecting Leichhardt Council, is closely tied to the Parramatta Rd renewal project.

In late 2015 the State released *The Draft Parramatta Rd Urban Transformation Strategy* (NSW 2015) (www.newparramattard.com.au). The Strategy identifies eight Precincts, specifically targeted for renewal under Precinct Plans as outlined in earlier chapters. There are a number of P Renewal' areas under the project, including Precinct 7, the Norton St area of Leichhardt (Plate 5.7).

Finally, examining he principle Protecting the Natural Environment, the Plan commits the State to 'protect the health of the subregions waterways and aquatic habitats'. This goal, for example, would apply to the prime conservation challenges in Leichhardt. This would include the protection of bushlands, waterways and fore-

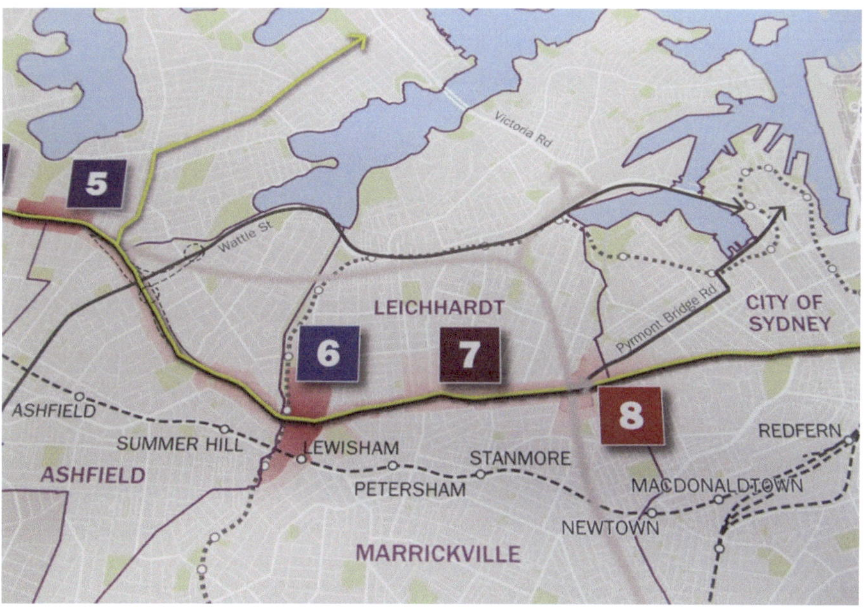

Plate 5.7 Parramatta Rd Transformation Project – Precinct 7 Leichhardt (*Source* NSW 2014b)

shores within Leichhardt. Leichhardt Council, along with neighbouring councils developing the Inner West Open Space Corridor, has played a major role in this project. The efforts of these councils in undertaking this work will be outlined in more detail in Chap. 9.

5.4.1 Conclusion on Leichhardt and Greater Sydney Planning

Having reviewed above the State's Greater Sydney planning (including infrastructure projects) it can be concluded the implementation of these plans and developments is likely to have a significant impact on Leichhardt (Question 3). The extent of this impact will be better understood when these State plans are further examined in the context of the whole of the Sydney Inner West District (including Leichhardt) in Chap. 9. Finally, in examining State proposed council amalgamations, options for councils (including Leichhardt) working together and in partnership with the State to meet local and district needs will be addressed in Chap. 18 (Planning Reforms and Directions for Planning Sustainable Cities). The subject to amalgamations of councils and how this affects Leichhardt is examined next.

5.5 Amalgamations of Councils and Leichhardt

The subject of amalgamations of Greater Sydney councils (given implications for planning at municipal and Greater Sydney levels) was reviewed in Chap. 2. The planning of Leichhardt (merged with Marrickville and Ashfield councils as noted above under the 12th May 2016 Proclamation) in meeting development demands and applying sustainable urban planning (SUP) practices is examined next. NSW Premier Mike Baird had announced in Dec 2015 that Leichhardt Council will be forced to merge with Marrickville and Ashfield councils (Plate 5.8). Looking at the Inner West District these three councils compose the eastern end (right) of the district.

Leichhardt Council stated a clear preference to be a stand-alone council. Within a Council survey the majority of Leichhardt residents said they do not want to merge. Council also remained opposed to forced amalgamations. Council also notified the State that, if the State Government forced amalgamations, Council's preference was to merge with Marrickville and Ashfield councils. The Mayors of Leichhardt, Marrickville, and Ashfield councils urged all residents to have their say on the issue of forced amalgamations of councils. The Mayors noted that submissions to the new Council Boundary Review were open till the end of February 2016 at www.councilboundaryreview.nsw.gov.au. In the interim, the Council noted, the Minister for Local Government had commenced the process of forcibly amalgamating Marrickville, Ashfield and Leichhardt councils. Council also noted the Minister said he expected the new council to be 'proclaimed' on 1 July 2016. This came

Plate 5.8 Sydney Inner West Councils – Marrickville, Ashfield, and Leichhardt

about despite all three councils repeatedly stating their preference to remain standalone, and a majority of residents also saying they prefer the current local government boundaries. The results of the review of merger proposals (which includes a new inner west council of Marrickville-Ashfield-Leichhardt) would be referred to the Chief Executive of the Office of Local Government for examination and report under the Local Government Act (the Act). In turn, the Chief Executive had appointed a Delegate who would examine the proposals, conduct a public inquiry, call for written submissions, and then report back to the Minister. The Council on 19 Jan 2016 revolved to conduct a public information campaign on the merger proposal, encouraging full participation of citizens. This included Council conducting a statistically valid phone survey to poll the opinion of local residents on: a. whether or not the Government's merger proposal should proceed; b. citizen concerns and expectations for what the merger would mean for citizens; and, citizen priorities for their local council's future policy direction.

5.5.1 Conclusion on Amalgamations

Leichhardt Council made a clear case, based on research and engaging its community, to stand alone and not be amalgamated with Marrickville and Ashfield councils. Council also notified the State that, if the State Government forced amalgamations, Council's preference was to merge with Marrickville and Ashfield councils. The 12th May 2016 *Local Government (Council Amalgamations) Proclamation 2016* merging Leichhardt, Marrickville and Ashfield councils was noted. This question will be further addressed in Chap. 18. At the end of the State review process (Council Boundary Review) the State would determine what amalgamations as proposed would proceed or not proceed. This question will be further addressed in Chap. 18.

5.6 Overall Conclusions

Having completed an assessment of the development and planning of Leichhardt, there are a number of overall conclusions reached. Firstly, Leichhardt has been experiencing accelerated development and changes to the composition of its population. Given these changes, and examining Leichhardt Council's response, it was also concluded Council is moving in the right direction in increasing its sustainable urban planning (SUP) practices. It was also concluded, however, that there would be significant future impacts on Leichhardt from Greater Sydney growth and development. In reviewing potential Council and State responses to this future scenario, it was concluded that new partnerships between the Council and the State would be needed. A major initiative from the State (2009 onwards) was to investigate local government reform, including a potential amalgamation of councils (e.g. Leichhardt, Ashfield and Leichhardt). The Council had to date, however, expressed an intent to stand alone and only amalgamate as a second choice (if the State forced amalgamations). The 12th May 2016 *Local Government (Council Amalgamations) Proclamation 2016* that merged Leichhardt, Marrickville and Ashfield councils was acknowledged. These overall chapter conclusions will be examined against results of assessing planning and development over all six council areas within Sydney Inner West District (Chaps. 3, 4, 5, 6, 7 and 8) and the district as a whole in Chap. 9. Results from the district assessment (and assessment of Parramatta and St George districts in later chapters) will contribute to assessing how to plan for a sustainable Greater Sydney and a sustainable city anywhere (Chap. 18).

5.6.1 Summary

The chapter addressed the extent Leichhardt Council was applying sustainable urban planning (SUP) practices (see Chap. 1) in planning the municipal area. It was noted that Leichhardt, in Sydney Inner West District, was merged with Marrickville and Ashfield councils under the 12th May 2016 *Local Government (Council Amalgamations) Proclamation 2016* (Chap. 2). It was established that Leichhardt, one of twelve municipalities within three Greater Sydney Districts (Sydney Inner West, Greater Parramatta and St George) is considered to be in global transition. SUP practices (based on the application of three key SUP principles) were introduced, being: Developing Sustainable Urban Centers; Developing Sustainable Transport; and, Protecting the Natural Environment. A review of Leichhardt background and population changes (i.e. global movement of population) was outlined. In examining the pattern of urban development, four phases of Leichhardt development are examined: Early Leichhardt; Traditional Leichhardt; New Leichhardt; and, Future Leichhardt. With conclusions drawn, key Leichhardt Council planning documents were then critiqued to determine the extent Council applies SUP practices. In addition, given State planning of Greater Sydney affects Leichhardt, a second question was posed: what is the impact of Greater Sydney planning and development on Leichhardt?. In examining this question, examples of State plans (including State council amalgamations) likely to impact on Leichhardt were addressed. By chapter's end, conclusions were drawn on Council planning and impact of Greater Sydney planning on Leichhardt. The chapter noted that these conclusions would be addressed further in: Chap. 9 (examining the extent of SUP practices being applied over the whole of Sydney Inner West District); and, Chap. 18 (examining directions for planning a sustainable Greater Sydney and a sustainable city anywhere).

Notation 5.1 Cultural Diversity Poems – Leichhardt
Poems reflecting the diversity of Leichhardt:

Annandale Crimson **(# 6)**

Crimson sun beams creasing the sky across Annandale Australia
Southland earthy ochre colors reflect rows of bright cottages
Gargoyles hunch, terraces of basalt stone, stucco plastered whites
Aqua blue paint drips off facades, orange doors, and golden brown lanterns

The new agers discovery of streets Booth and Johnston, of The Abbey overarching all
Plasterers, stone restorers, painters, plumbers and solar installers flown in
The generation securing funding for a new urban renaissance Sydney wide
The leaves will fray and scatter down streets in Annandale, as dreams reborn again

Balmain Remnants **(# 7)**

Gladstone St, Mullens St, Lilyfield Rd, Albert St, all remnants of Balmain past
Rozelle School of sandstone edifice, memories etched by the Balmain boys
Residents touch of new age, now following footsteps in search of family traces
Narrow side streets, bays, and memories of Masonic days and temperance too

The glue of a nation that pioneers worked, sweated and stitched together
Wander to Balmain's wharf, past galleys of shipping, timber and rusted bolts and tin
Colors of cream, pink and yellow with doors of red and crimson, once adored Edwardian
Drifting ebb, revealing morsels of a lifelong time rich, new footsteps, keep discovering

Boys from Balmain don't cry, proud Premier Warn eloquently told Parliament
Glistening peninsula, reflective of colonial Sydney, to Blackwattle Bay long ago
Darling Street string of early institutions, Masonic to Temperance and religious halls
Building was Victorian, Edwardian steeped in European tradition, royal right through

Balmain opens its heart to multicultural and new settler dreams, over two century's long
Bangladesh curries now, beside New Guinea coffees, and fresh breads and candle makers
Town Hall refitted, library world connected and barista's row with delicacies stacked
Gladstone Park as Heye's and Long's impressionisms remind a return to civil life here

Parkes, Uren, Issy Weiner and Nick Origlass all left imprints on Balmain
Round the shore's edges scatters docks and alcoves forged
Governor Darling memorialized in the winding road Rozelle to Balmain
Unity Hotel, Town Hall, and Court House reflect justice to hold high

A new age generation with sanders and plasterers descending
Colossus of tradition here, now highlighted in yellows and royal reds abounding
Preserved are narrow lanes and streetscapes filled with pride of place
Rest now workers who made this metropolis, as others will rebuild yet again

Cultures of Norton St, Leichhardt **(# 8)**

Italian ices, Sicilian pizza, Milan art and Trieste Coffee, Norton St Leichhardt
All ingredients here for cultural basis of growing cities anywhere worldwide
Nations to thrive in the future will be multicultural in smartness and tolerance
Communities languishing could mirror this change seeking global new tastes

For all the new apartments a center sprouts of a generous city's urban success
A people united in heritage bonds over the decades of settlement here
Leichhardt taps the heart of a place, its shared history and ambitious hopes
Italians, Greeks, Asians, and more bring a cultural basis of downtown

Heartland Leichhardt Revives **(# 9)**

Back with deep desire to walk again Leichhardt paths of heartland's call
A mighty Anzac bridge you traverse and plunge into a labyrinth of urban lanes
Lilyfield's rolling streets cast an artist's random brushstrokes
Rozelle tumbles into foreshores and angular juts of stone and mortar

A city spreads bastions of hotels and worker halls of suburbs's pride
New age families wire brush the bricks and stone to reveal new life
Trundling down to Balmain East shores where trams once docked with supplies
The heartland of democracy, Leichhardt spirit welcomes you once again

Taverners Hill Lights Up **(# 10)**

On Taverners Hill, Leichhardt, acclaimed vistas of endless Sydney
The trading vehicles westward bound, drove business to distant grounds
Years passed and window vacancies arose, signing off an era of past trade
An immigrant community arose here and that would change all forever

The Italian and multi-cultural mix settling Australia breathed a new life
Surrounding Taverners Hill Hotel of solid masonry, brick and mortar
Mansions, terraces and humble worker cottages, new paint now splashed
Glory of past, Taverners Hill shines again a beacon across the plains of Sydney

5.7 Summary

The chapter addressed the extent Leichhardt Council is applying sustainable urban planning (SUP) practices (see Chap. 1) in planning the municipal area. Leichhardt (in Sydney Inner West District). It was established that Marrickville, one of twelve municipalities within three Greater Sydney Districts (Sydney Inner West, Greater Parramatta and St George) is considered to be in global transition. SUP practices (based on the application of three key SUP principles) were introduced, being: Developing Sustainable Urban Centers; Developing Sustainable Transport; and, Protecting the Natural Environment. A review of Leichhardt background and population changes (i.e. global movement of population) was outlined. In examining the pattern of urban development, four phases of Leichhardt development are examined: Early Leichhardt; Traditional Leichhardt; New Leichhardt; and, Future Leichhardt. With conclusions drawn, key Leichhardt Council planning documents were then critiqued to determine the extent Council applies SUP practices. In addition, given State planning of Greater Sydney affects Leichhardt, a second question was posed: what is the impact of Greater Sydney planning and development on Leichhardt?. In examining this question, examples of State plans (including State proposed council amalgamations) likely to impact on Leichhardt were addressed. By chapter's end, conclusions were drawn on Council planning and Greater Sydney planning. The chapter noted that these conclusions would be addressed further in: Chap. 9 (examining the extent of SUP practices being applied over the whole of Sydney Inner West District); and, Chap. 18 (examining directions for planning a sustainable Greater Sydney and a sustainable city anywhere).

References

Australian Bureau of Statistics (ABS) (2011) Australian census review. ABS, Canberra
NSW Government (2011) Sydney metropolitan strategy. NSW, Sydney
NSW Government (2014a) A plan for growing sydney. NSW, Sydney
NSW Government (2014b) Parramatta Rd renewal project. NSW, Sydney
NSW Government (2014c) Revitalization of local government report. NSW, Sydney
New South Wales (2015) Draft Parramatta Rd urban transformation strategy. NSW, Sydney
Leichhardt Council (2010) Leichhardt Council environmental sustainability strategy 2010–2014. NSW, Leichhardt
Leichhardt Council (2012) Leichhardt integrated transport plan. NSW, Leichhardt
Leichhardt Council (2013a) Climate change plan. NSW, Leichhardt
Leichhardt Council (2013b) Leichhardt local environment plan. NSW, Leichhardt
Leichhardt Council (2013c) Leichhardt development control plan. NSW, Leichhardt
Leichhardt Council (2013d) State of environment report 2012–2013. NSW, Leichhardt
Leichhardt Council (2013e) Leichhardt 2025 + strategic plan. NSW, Leichhardt

Chapter 6
Planning Burwood, Greater Sydney

Dedication *This chapter is dedicated to Sir Charles Lloyd Jones (1878–1958) (Plate 6.1), a businessman and patron of the arts born in Burwood.*

Plate 6.1 Sir Charles Lloyd Jones

A profile on Sir Jones provides an insight into his life:

> Sir Charles Lloyd Jones (1878–1958) was an Australian businessman and patron of the arts, serving as Chairman of David Jones Limited from 1920 to his death in 1958. He was also publisher of Art in Australia. He was the inaugural Chairman of the Australian Broadcasting Commission 1932–34. He was made a Knight Bachelor in the 1951 New Year's Honors.

R.C. Rauscher, S. Momtaz, *Cities in Global Transition*,
DOI 10.1007/978-3-319-39865-5_6

Abstract The chapter addresses the extent Burwood Council (herein referred to as Burwood) is applying sustainable urban planning (SUP) practices (see Chap. 1) in planning the municipal area. The chapter acknowledges the *Local Government (Council Amalgamations) Proclamation 2016* merging Burwood with Strathfield and Canada Bay councils noted in Chap. 2. Burwood (in Sydney Inner West District) is one of twelve municipalities within three Greater Sydney Districts (Sydney Inner West, Greater Parramatta and St George), all considered to be in global transition (see Chap. 2). Examining SUP practices is based on the application of three key SUP principles (as adopted in Chap. 2), being: Developing Sustainable Urban Centers; Developing Sustainable Transport; and, Protecting the Natural Environment.

Firstly, a background to Burwood development and population changes (e.g. global movement of population) is outlined. To examine the pattern of urban development, four phases of Burwood development are examined: Early Burwood; Traditional Burwood; New Burwood; and, Future Burwood. Given conclusions reached here, key Leichhardt Council planning documents are then critiqued to determine the extent Council applies SUP practices. In addition, given State planning of Greater Sydney affects Burwood, a second question is posed: what is the impact of Greater Sydney planning and development on Burwood? In examining this question, examples of State plans (including council amalgamations under the Proclamation noted above) likely to affect Bufrwood are addressed. By chapter's end, conclusions are drawn on Council planning and impact of Greater Sydney planning on Burwood. The chapter notes that these conclusions will be addressed further in: Chap. 9 (examining the extent of SUP practices being applied over the whole of Sydney Inner West District); and, Chap. 18 (examining directions for planning a sustainable Greater Sydney or a sustainable city anywhere).

6.1 Background and Population Change

Burwood Municipality (herein called Burwood) in the context of Sydney Inner West (as introduced in Chap. 2) is adjacent to Ashfield to the west of the City of Sydney. Under the *Local Government (Council Amalgamations) Proclamation 2016* Burwood *was* merged with Strathfield and Canada Bay councils as noted in Chap. 2. The area now known as Burwood Municipality was home to the Wangal people, inhabiting a major portion of Sydney Inner West. Burwood derived its name from a grant of 250 acres made by Governor Hunter in 1799 to Captain Thomas Rowley of the New South Wales Corps, who named the land after the Burwood Farm on which he had lived in his native Cornwell, England. Burwood was once the area inhabited by the Wangal Aboriginal people, as was neighbouring local government areas The Municipality of Burwood was incorporated by proclamation in the Government Gazette in 1874. A historical note addresses the changing population of Burwood 1900–1930s as follows:

> The population of Burwood rose from 7400 in 1900 to over 20, 000 by 1930. In 1912, the old steam trams were replaced by electric trams. The steam tram from Ashfield to Enfield was extended to Burwood and Mortlake in 1899 and to Cabarita in 1907. In 1948, the trams were taken off the road and replaced with the diesel buses you see today. The motor bus service was begun by F.H. Stewart from Chullora (Enfield) to Sydney in 1915. Burwood's streets were lit by gas until the change was made to electricity in 1921. The 1920s were a great period for brick houses in Burwood. Two huge brick pits supplied the building materials for the site. Subdivisions in the Burwood area in the 1830s propelled the growth of a village and by 1855, when the railway line opened, Burwood was one of the initial six stops on the Sydney-to-Parramatta route. The railway led to a huge growth in population. In 1874, the area became a municipality. Burwood has a mixture of residential, commercial, and light industrial developments. The main shopping strip runs along Burwood Road, beside Burwood railway station. Westfield Burwood is a large regional shopping center, north of the railway line, on Burwood Road opposite Burwood Park. Burwood Plaza is a smaller shopping center on Railway Parade, south of the railway line. High rise residential and commercial buildings are also found in surrounding streets and along the railway line. Commercial and light industrial developments are located along Parramatta Road.

Burwood Local Government Area (LGA) is the heart of Sydney Inner West. It covers an area of 7.26 km^2, strategically located midway between the major centres of Sydney City and Parramatta. Burwood LGA comprises the following suburbs – Burwood Heights, Croydon and parts of Croydon Park, Enfield and Strathfield. Burwood has been identified as a Major Regional Centre by the NSW Government. The State notes that Burwood is set to grow significantly in the next 25 years (2015–2040), taking advantage of the area's strong local economy and excellent transport infrastructure. The municipality has a population of 32,423 people and is predominantly urban in character, with a strong commercial centre. The suburb of Burwood (about 1/3rd the population of the municipality) records 69.6 % of its population born outside Australia. Two of the most common other countries of Burwood citizen birth were China and India. Other important population characteristics include:

> The Burwood Local Government Area (LGA) population is expected to increase by 49 % over the next 20 years. This population is mainly made up of young people and adults. The predominant age range with the highest number of people is 25–29 years followed closely by 20–24 years. However the Burwood LGA has an aging population with the highest proportional population growth occurring in the 80–100+ age bracket. (ABS 2011 Census).

The Australian Bureau of Statistics (ABS 2011) comments on Burwood Municipality ancestry and languages spoken at home:

> Burwood's LGA population is culturally diverse with about 58 % of the population being born overseas. The five most commonly spoken languages other than English are (in rank order) Mandarin, Cantonese, Italian, Korean and Arabic. A small proportion (0.4 %) of the Burwood LGA is Aboriginal or Torres Strait Islander. At the 2011 Census, the proportion of residents in Burwood local government area who stated their ancestry as Chinese was in excess of seven times the state and national averages; and the proportion of households where an Asian language was spoken at home was approximately eight times higher than the national average. Finally, in terms of housing, the proportion of people who own their own home is 34 %.

Finally, in author's (Rauscher) qualitative research (2014/2015), there is a poem written to reflect the cultural diversity of Burwood (end of chapter under Notation 6.1: *Burwood Finds Metropolis's Heart)*. With this background introduction to Burwood, the extent of Burwood's development phases are examined next.

6.2 Phases of Development

An examination of the development phases of any locality provides a basis of understanding current trends and likely future directions of growth of that locality. Following a look at the background of Burwood four phases of Burwood development will be examined. These phases include: Early Burwood; Traditional Burwood; New Burwood; and Future Burwood. To assist in addressing this question, author (Rauscher) qualitative research field work was completed in Burwood Municipality (2014–2015). Plate 6.2 illustrates the early development of Burwood, from the1800s. This decorative late 1800's house (top-left) is reflective of the investments in residential properties away from the inner City of Sydney. Burwood sports many examples of Victorian and Federation styles. The strength of Burwood as a municipality

Plate 6.2 Early Burwood. *Left to right* Residential estate; Municipality insignia; Masonic temple; Presbyterian Church; Historic corner shop with residents; Historic corner home

Plate 6.3 Traditional Burwood. *Left to right* Burwood Post Office; Residential High rise Complex; Commercial Offices Downtown; Heritage Hotel; Mural in CBD; and, High rise with shops below

from 1874 shows the long history of local government in Burwood (top-centre). The strong institutions of Burwood is reflective in this Masonic Temple (top-right) (currently the subject of a major development project). The religious groups are well represented in Burwood as noted in this well designed Presbyterian Church (bottom-left). In the 1800s shopping locally was popular, hence corner shops (bottom-center) with residential above were seen on street corners of Burwood. Finally, this historic residential timber home (bottom-right) reflects country style living that existed alongside the urban higher densities of Burwood in the 1800s.

Renewal (2016) is occurring in Traditional Burwood (Plate 6.3) throughout most of the municipality's suburbs. A prominent restoration in Burwood has been the Burwood Post Office (top-left). A residential high rise complex of several buildings captures the streetscape of downtown Burwood (top-centre). Commercial offices have found Burwood to be an alternative to Sydney inner city locations (top-right). The heritage listed Burwood Rd hotel has been converted into a club (bottom-left). The community spirit of Burwood is reflected in this mural expressing the multicultural community. Finally, high rise residential buildings are seen increasingly in Burwood (bottom-right), with a number of further projects underway in 2015.

A picture of the New Burwood is illustrated in Plate 6.4. Burwood Downtown is seen here from a distance (top-left). The entry to the Burwood CBD presents a high density character (top-middle). Burwood has a number of business parks within the

Plate 6.4 New Burwood

Plate 6.5 Future Burwood

CBD (top-right). An early high rise cluster (bottom-left) was built adjacent to the railway line. Commercial premises below high rise residential is popular in Burwood (bottom-middle). Finally, a number of new style high rise buildings can be seen in Burwood today (bottom-right).

Finally, Future Burwood (Plate 6.5) reflects the continuing building boom in residential development at increasing densities throughout the municipality (left to right). More dramatic designs are being trialled (top-left). Some sites on main roads will afford a greater attention to vegetation needs (middle). Finally, the stand alone residential high rise with commercial below will continue to be popular and meet the demand for apartment living close to services (right). Finally, Future Burwood reflects the continued building boom in residential development at increasing densities throughout the municipality. Factors of design, such as height, density, housing affordability, overshadowing and open space provisions, are likely to remain issues of development importance and contention within communities at the local government level into the future.

6.2.1 Conclusion

Given the review of development phases of Burwood (above) and changes in population demographics, it can be concluded that Burwood has been experiencing accelerated global transition (2000s) (Question 1). It was noted that under the *Local Government (Council Amalgamations) Proclamation 2016* Burwood *was* merged with Strathfield and Canada Bay councils. The planning of Burwood in meeting development demands now and in the future is examined next. Particular attention is paid to Question 2 above, whether Burwood Council planning is moving in the direction of sustainable urban planning (SUP) practices (Chaps. 1 and 2).

6.3 Burwood and Sustainable Urban Planning

Planning of Marrickville Municipality has been taking place since it became a local government body in 1874. A review of planning in Marrickville will pay particular attention to applications of sustainable urban planning (SUP) (as outlined in Chap. 1) practices (including use of indicators of sustainability) (Chap. 2). Burwood Council key planning documents (a–z with date and web reference) are listed in Information Box 6.1. A summary of these documents and a brief critique assist with this review of Burwood planning (documents can be sourced under the web link as noted in the right column).

Information Box 6.1 Burwood municipality planning

BURWOOD2030 (2013a)	www.burwood.nsw.gov.au
Community Strategic Plan (2012a)	www.burwood.nsw.gov.au
Development Control Plan (2013b)	www.burwood.nsw.gov.au
Green Action Plan 2012 (2012b)	www.burwood.nsw.gov.au
Local Environmental Plan (LEP) (2012c)	www.burwood.nsw.gov.au
Sustainability Action Plan (2012d)	www.burwood.nsw.gov.au

Source Webs as noted February 2016

6.3.1 Burwood 2030

Burwood has adopted a strategic planning process under its *BURWOOD2030* (Burwood 2013a) document. This process aims to take Burwood to sustainability by 2030.

6.3.2 Community Strategic Plan

Following an extensive community engagement program with community and stakeholders, the Council developed the 20 year *Community Strategic Plan* (Burwood 2012a). The Plan is set to become the blueprint for any Council activity in the future. A number of strategic goals included in this plan are connected to the NSW State Government's *NSW 2021* (NSW 2012). The NSW State Government identified Burwood as a major centre in its Metropolitan Strategy and projected that Burwood would have significant employment growth (36.5 %) between 2001 and 2031. The Metropolitan Strategy projects the Inner West will have an additional 30,000 homes by 2031 and many of these homes are planned for Burwood.

6.3.3 Green Action Plan 2012 and Sustainability Action Plan (SAP)

Complementing the Strategic Plan noted above, measures to protect the environment are contained in the Council's Green Action Plan 2012 (Burwood 2012b). Council has also established a community based Environment and Sustainability Committee in November of 2008 with a primary function to provide advice to the Council on the implementation of Action Plan 2012 and producing the *Sustainability Action Plan* (SAP) (Burwood 2012d). Council comments:

> The Committee has worked with the Council in developing a series of individual action plans and programs aimed at reducing the Council's carbon footprint and addressing sustainability issues across all areas of the environment. The plans are described in the Green Action Plan Council previously prepared an annual State of the Environment Report to provide a report card on Council's environmental actions. This reporting method has now been changed by legislation with environmental actions and the Green Action Plan now incorporated into the delivery plan and quarterly operational reports.

Under greenhouse and climate change policies, Burwood Council joined the Cities for Climate Protection Australia Program in 2006 so that Council could work towards reducing greenhouse gas emissions and implementing environmental sustainability initiatives. As a member of this program Council is working towards achieving five milestones, and take a leading role in the reduction of greenhouse gas emissions at a local community level. The five milestones that comprise the program are outlined below:

Milestone 1: Conduct an inventory and forecast for Community and Corporate (Council) greenhouse gas emissions.
Milestone 2: Establish an emissions reduction goal.
Milestone 3: Develop and adopt a Local Action Plan.
Milestone 4: Implement the Local Action Plan.
Milestone 5: Monitor and report on achievements

In 2006, Burwood Council established a greenhouse gas emissions inventory and forecast (Milestone 1). Council also set Corporate and Community emissions reduction goals (Milestone 2). Burwood Council has completed (Milestone 3) adopting a Local Action Plan (LAP) outlining the strategy that will be undertaken to reduce greenhouse emissions in the Local Government Area.

6.3.4 Local Environmental Plan (LEP)

In planning for the built environment, Burwood Council, through its *Local Environmental Plan* (LEP) (Burwood 2012c) and Development Control Plan (Burwood 2013b) sets controls on new developments to ensure that environmental best practice features are applied in order to preserve and improve the environment. The Council notes that it's LEP and DCPs place an emphasis on environmental management and ecologically sustainable development (ESD) principles to be integrated into the development.

Council notes that it is committed to improving the facilities and services to the community to ensure that Burwood responds to being the Sydney Inner West District center. A brief summary of several Council major projects include (Plate 6.6):

Council notes these projects as Railway Square (left); Burwood Park upgrade (center); and, Wangal Park improvements (right). The Railway Square is to become an open space, including heritage and upgraded new spaces in the heart of the Town

Plate 6.6 Burwood Council facility projects. *Left to right* Railway Square; Burwood Park; and, Wangal Park

Centre. Council hopes that this project will improve pedestrian traffic flow around Burwood Station. The Burwood Park upgrade will see a new cafe and eco-friendly amenities block. Finally, Wangal Park will be a recreational and education space in the middle of Burwood. This will include: a wetland, surrounded by natural bush; a boardwalk surrounding the wetland; and, an observation deck; a kiosk featuring a terrace overlooking the wetland; a full fitness jogging/cycle circuit; a new playground for children; and, solar lighting throughout the park.

6.3.5 Conclusion on Council Planning and SUP Practices

It can be concluded, having reviewed Burwood Council planning documents and their applications, it can be concluded that Council is moving in the direction of applying sustainable urban planning (SUP) practices (Question 2 above). The State's Greater Sydney planning (including infrastructure projects) and State council amalgamation proposals likely to affect Burwood (Question 3) are reviewed next.

6.4 Greater Sydney Planning and Burwood

There are planning and developments of Greater Sydney likely to impact on Burwood. In many instances Burwood Council has had the task of providing planning input on these plans. This includes, for example, input on: making and reviewing public submissions; implementing State planning policies at the local level; joint partnership on service or infrastructure projects with State or private companies; and, monitoring State or private development projects affecting Burwood. Other Council planning responsibilities under State legislation that also faces the Council includes: engaging the municipality community (e.g. residents, business, services, development interests, and institutions) on State planning policies and development proposals; statutory plan making (e.g. adopting local environment plans (LEPs)); and, reporting back to the State on a range of planning matters (i.e. State proposed roads, state of environment reporting and infrastructure projects).

Greater Sydney planning requirements were set by the NSW State in 2006; with Burwood Council adopting in 2013 the *Burwood 2030* (Burwood 2013a). Council's plans had to be consistent with the *Sydney Metropolitan Strategy* (NSW 2011). The State then adopted its strategy for planning Greater Sydney to 2031 (*A Plan for Growing Sydney* (NSW 2014)). The Plan forecasts a population growth of 1.6 m (to be added to the Greater Sydney population of 4.84 m at June 2014). *A Plan for Growing Sydney* (NSW 2014). Burwood is nominated by the State to be within Subregion 'Central' within this Plan. Local government areas with 'Central' include: Ashfield, Botany, Burwood, Canada Bay, Leichhardt, Marrickville, Randwick, Strathfield, Sydney, Waverley, and Woollahra. In examining this Plan and its impact

Plate 6.7 Parramatta Rd urban transformation strategy – Precinct 4 Burwood (*Source* NSW 2015)

on Burwood, three key planning principles as adopted in Chap. 2 are addressed: (1) Developing Sustainable Urban Centres; (2) Developing Sustainable Transport; and, (3) Protecting the Natural Environment. In Developing Sustainable Urban Centres the Plan notes the State, with council, is to provide capacity for additional mixed-use development in Burwood including offices, retail, services and housing. Moving to Developing Sustainable Transport, under the Plan, the State asked Burwood Council (along with other Sydney Inner West councils) asked in late 2014 to comment on the proposed *Parramatta Rd Renewal Project*. The *Draft Parramatta Rd Urban Transformation Strategy* (NSW 2015) (www.newparramattard.com.au) followed in 2015 and identifies eight Precincts, specifically targeted for renewal under Precinct Plans. Precincts were selected due to their access to jobs, infrastructure, public transport, or how well they can support new housing types. Precinct 4 is the Burwood centre around the Burwood CBD (Plate 6.7). Burwood Council is closely tied to the Parramatta Rd renewal project.

Looking at Burwood the Strategy notes:

> The Burwood Precinct is a northern extension of the Burwood Town Centre, which is a major hub with shopping center, main street shops, a transport interchange, higher density housing and a beautiful city park. Its many transport options provide great access to Sydney Olympic Park, Parramatta CBD and Sydney CBD. In other parts of the Precinct, heritage houses, schools and corner stores show how the area has changed over time. Large areas of parkland are scattered throughout the Precinct's north. The vision for the Burwood Precinct will be a commercial gateway to Burwood Town Centre based around the enlivened spine of Burwood Road building upon existing amenity for new residents. Burwood Precinct will be built around the proposed rapid transit from Burwood to Sydney CBD and an improved focus for Burwood Road as a lively main street. Both Burwood and Parramatta Roads will be marked by taller residential and mixed use buildings.

Looking further at transport in Burwood, the State is also to investigate a potential light rail corridor from Parramatta to Sydney CBD via Parramatta Road, including the implications for Burwood. This planning of light rail and both the Westconnex and the Parramatta Rd Renewal projects, as they affect Sydney Inner West, will be addressed further in Chap. 9. Finally, looking at the third planning principle, Protecting the Natural Environment, the Plan commits the State to 'protect the health of the subregions waterways and aquatic habitats'. This goal, for example, would apply to the prime conservation challenges in Burwood. This would include the protection of bushlands, waterways and foreshores within Burwood. Burwood Council, along with neighbouring councils developing the Inner West Open Space Corridor and upgrading the Cooks River, has played a role in this project. The efforts of these councils in undertaking this work will be outlined in more detail in Chap. 9.

6.4.1 Conclusion on Burwood and Greater Sydney Planning

Having reviewed above the State's Greater Sydney planning (including infrastructure projects) it can be concluded the implementation of these plans and developments is likely to have a significant impact on Burwood (Question 3). The extent of this impact will be better understood when these State plans are further examined in the context of the whole of the Sydney Inner West District (including Burwood) in Chap. 9. Finally, in examining State proposed council amalgamations, options for councils (including Burwood) working together and in partnership with the State to meet local and district needs will be addressed in Chap. 18 (Planning Reforms and Directions for Planning Sustainable Cities).

6.5 Amalgamations of Councils and Burwood

The subject of amalgamations of Greater Sydney councils (given implications for planning at municipal and Greater Sydney levels) was reviewed in Chap. 2. NSW Premier Mike Baird announced in Dec 2015 the State's preference for Burwood Council to merge with Strathfield and Canada Bay councils. Within the Inner West District, these three councils compose the western (left) end of the district. Burwood Council stated a preference to merge with Auburn and Canada Bay councils. Council also remained opposed to forced amalgamations. The Mayor of Burwood urged all residents to have their say on the issue of amalgamations of councils. Council noted that submissions to the new Council Boundary Review were open till the end of February www.councilboundaryreview.nsw.gov.au In the interim, the

Council noted, the Minister for Local Government had commenced the process of forcibly amalgamating Burwood, Canada Bay and Strathfield councils. Council also noted the Minister said he expected the new council to be 'proclaimed' on 1 July 2016 (note the 12th May 2016 *Local Government (Council Amalgamations) Proclamation 2016* noted in Chap. 2). This came about despite all three councils repeatedly stating their preference to remain stand-alone, and a majority of residents also saying they prefer the current local government boundaries. The results of the review of merger proposals (which includes a new inner west council of Burwood, Canada Bay and Strathfield) would be referred to the Chief Executive of the Office of Local Government for examination and report under the Local Government Act (the Act). In turn, the Chief Executive had appointed a Delegate who would examine the proposals, conduct a public inquiry, call for written submissions, and then report back to the Minister.

6.5.1 Conclusion on Amalgamations

Burwood Council made a clear case, based on research and engaging its community, to merge with Canada Bay and Auburn. The Council had, however (late in the State amalgamation process) expressed an interest to amalgamate with Canada Bay and Strathfield councils. The Minister for Local Government had commenced the process in 2016 (given Strathfield Council desired to stand alone and not amalgamate) to forcibly amalgamate Burwood, Canada Bay and Strathfield councils. At the end of the State review process (Council Boundary Review) the State would determine what amalgamations as proposed would proceed or not proceed. The 12th May 2016 *Local Government (Council Amalgamations) Proclamation 2016* merging Burwood, Strathfield and Canada Bay councils was noted. This question will be further addressed in Chap. 18.

6.6 Overall Conclusions

Having completed an assessment of the development and planning of Burwood, there are a number of overall conclusions reached. Firstly, Burwood has been experiencing accelerated development and changes to the composition of its population. Given these changes, and examining Burwood Council's response, it was also concluded Council is moving in the right direction in increasing its sustainable urban planning (SUP) practices. It was also concluded, however, that there would be significant future impacts on Burwood from Greater Sydney growth and development.

In reviewing potential Council and State responses to this future scenario, it was concluded that new partnerships between the Council and the State would be needed. A major initiative from the State (2009 onwards) was to investigate local government reform, including a potential amalgamation of councils (e.g. Burwood, Canada Bay and Strathfield councils). The Council had (late in the State amalgamation process) expressed an interest to amalgamate with these two councils. The 12th May 2016 *Local Government (Council Amalgamations) Proclamation 2016* that merged Burwood, Strathfield and Canada Bay councils was acknowledged. These overall chapter conclusions will be examined against results of assessing planning and development over all six council areas within Sydney Inner West District (Chaps. 3, 4, 5, 6, 7 and 8) and the district as a whole in Chap. 9. Results from the district assessment (and assessment of Parramatta and St George districts in later chapters) will contribute to assessing how to plan for a sustainable Greater Sydney and a sustainable city anywhere (Chap. 18).

6.7 Summary

The chapter addressed the extent Burwood Council was applying sustainable urban planning (SUP) practices (see Chap. 1) in planning the municipal area. It was noted that Burwood, in Sydney Inner West District, was merged with Strathfield and Canada Bay councils under the 12th May 2016 *Local Government (Council Amalgamations) Proclamation 2016* (Chap. 2). It was concluded that Burwood, one of twelve municipalities within three Greater Sydney Districts (Sydney Inner West, Greater Parramatta and St George) is considered to be in global transition. SUP practices (based on the application of three key SUP principles) were introduced, being: Developing Sustainable Urban Centers; Developing Sustainable Transport; Burwood and, Protecting the Natural Environment. A review of Burwood background and population changes (i.e. global movement of population) was outlined. In examining the pattern of urban development, four phases of Burwood development are examined: Early Burwood; Traditional Burwood; New Burwood; and, Future Burwood. With conclusions drawn, key Burwood Council planning documents were then critiqued to determine the extent Council applies SUP practices. In addition, given State planning of Greater Sydney affects Burwood, a second question was posed: what is the impact of Greater Sydney planning and development on Burwood?. In examining this question, examples of State plans (including State proposed council amalgamations) likely to impact on Burwood were addressed. By chapter's end, conclusions were drawn on Council planning and impact of Greater Sydney planning on Burwood. The chapter noted that these conclusions would be addressed further in: Chap. 9 (examining the extent of SUP practices being applied over the whole of Sydney Inner West District); and, Chap. 18 (examining directions for planning a sustainable Greater Sydney and a sustainable city anywhere).

Notation 6.1 Cultural Diversity Poem – Burwood

Poem reflecting the diversity of Burwood:

Burwood Finds Metropolis's Heart (#11)

Burwood for most is a memory of one long stretch of haberdashery and convenient shopping
The Burwood today is a metropolis, district heart, multi-cultural new start venue
The centre of office buildings, chic cafes, and high rise living, international poised
Contributing ethnic groups, Greek and Italian foundations, now Middle East and Asia too

They all now call Burwood home, days of bustle and evening's entertainment
The international flair runs alongside steep traditions of pioneers' legacies
Now Burwood looks to open more plaza spaces and pedestrian squares
The streets of Georgian and Federation homes to be brushed up for metropolis heart

References

Australian Bureau of Statistics (ABS) (2011) Australian census review. ABS, Canberra

Burwood Council (2012a) Community strategic plan. NSW, Burwood

Burwood Council (2012b) Green action plan 2012. NSW, Burwood

Burwood Council (2012c) Local environmental plan. NSW, Burwood

Burwood Council (2012d) Sustainability action plan. NSW, Burwood

Burwood Council (2013a) Burwood 2030. NSW, Burwood

Burwood Council (2013b) Development control plan. NSW, Burwood

New South Wales (2011) Sydney metropolitan strategy. NSW, Sydney

New South Wales (2012) NSW 2021. NSW, Sydney

New South Wales (2014) A plan for growing Sydney. NSW, Sydney

New South Wales (2015) Parramatta Rd transformation project. NSW, Sydney

Chapter 7
Planning Strathfield, Greater Sydney

Dedication *This chapter is dedicated to Miriam Hyde (1913–2005), who was a musician and a poet contributing to Strathfield's cultural development (Plate 7.1).*

Plate 7.1 Miriam Hyde

A snippet on Miriam Hyde gives an insight into her life:

> Miriam's special awards included the OBE (1981), AO (1991), International Woman of the Year (1991–92) for service to music, bestowed by the International Biographical Centre, Cambridge. Miriam played at times with the Strathfield Symphony Orchestra. A gracious compliment was paid her by Macquarie University in the form of an Honorary Doctorate.

R.C. Rauscher, S. Momtaz, *Cities in Global Transition*,
DOI 10.1007/978-3-319-39865-5_7

Abstract The chapter addresses the extent Strathfield Council (herein referred to as Strathfield) is applying sustainable urban planning (SUP) practices (see Chap. 1) in planning the municipal area. The chapter acknowledges the *Local Government (Council Amalgamations) Proclamation 2016* merging Strathfield with Burwood and Canada Bay councils noted in Chap. 2). Strathfield (in Sydney Inner West District) is one of twelve municipalities within three Greater Sydney Districts (Sydney Inner West, Greater Parramatta and St George), all considered to be in global transition (see Chap. 2). Examining SUP practices is based on the application of three key SUP principles (as adopted in Chap. 2), being: Developing Sustainable Urban Centers; Developing Sustainable Transport; and, Protecting the Natural Environment.

Firstly, a background to Strathfield development and population changes (e.g. global movement of population) is outlined. To examine the pattern of urban development, four phases of Strathfield development are examined: Early Strathfield; Traditional Strathfield; New Strathfield; and, Strathfield Burwood. Given conclusions reached here, key Strathfield Council planning documents are then critiqued to determine the extent Council applies SUP practices. In addition, given State planning of Greater Sydney affects Strathfield, a second question is posed: what is the impact of Greater Sydney planning and development on Strathfield? In examining this question, examples of State plans (including council amalgamations under the Proclamation noted above) likely to affect Strathfield are addressed. By chapter's end, conclusions are drawn on Council planning and impact of Greater Sydney planning on Strathfield. The chapter notes that these conclusions will be addressed further in: Chap. 9 (examining the extent of SUP practices being applied over the whole of Sydney Inner West District); and, Chap. 18 (examining directions for planning a sustainable Greater Sydney or a sustainable city anywhere).

7.1 Background and Population Change

Strathfield Municipality (herein called Strathfield) in the context of Sydney Inner West (as introduced in Chap. 2) is adjacent to Burwood to the west of the City of Sydney. Under the *Local Government (Council Amalgamations) Proclamation 2016* Strathfield was merged with Burwood and Canada Bay councils as noted in Chap. 2). Strathfield was once the area inhabited by the Wangal Aboriginal people (Chap. 2). Strathfield was proclaimed a municipality in 1871. The municipality comprises of the suburbs (a–z): Strathfield; Strathfield South; Homebush; Homebush West; Belfield (shared with City of Canterbury); Greenacre (shared with City of Bankstown); and, Flemington. With this background introduction to Strathfield, the extent of Strathfield's development phases and changing demographics are examined next.

The Australian Bureau of Statistics (ABS) comments on Strathfield Municipality ancestry and languages spoken at home:

> The population of Strathfield is made up of a number of ethnic groups, and about 48 % of the population born overseas. As a result there are a number of services for newly arrived immigrants and overseas students who live in the area. Two such organizations are the Russian Ethnic Community Council of NSW Inc. (RECNSW), and the Sydney Tamil Resource Centre Inc. (STRC). In the 1990s, a large number of South Korean migrants settled in Strathfield. The locality of Strathfield now has a large number of shops selling Korean food and cultural items. According to the 2011 census, Strathfield had a total population of 23,639 people. It has become one of the most culturally diverse suburbs in Sydney, particularly so in the high-density housing regions around Strathfield railway station. Overall 62.6 % of people were born overseas, with migrants born in China (9.8 %), South Korea (9.0 %), India (8.1 %), Nepal (3.8 %) and Sri Lanka (2.4 %). Only 16 % of residents had Australian parents. Overall, common stated ancestries include Chinese (18.9 %), Korean (9.3 %), Australian (8.6 %), Indian (8.2 %) and English (7.3 %). (*Source* ABS 2011)

Finally, in author's (Rauscher) qualitative research (2014/2015), there is a poem written to reflect the cultural diversity of Strathfield (end of chapter under Notation 7.1): *Who Took His Strathfield Away*. The phases of Strathfield development are examined next.

7.2 Phases of Development

An examination of the development phases of any locality provides a basis of understanding current trends and likely future directions of growth of that locality. Following a look at the background of Strathfield four phases of Strathfield development will be examined. These phases include: Early Strathfield; Traditional Strathfield; New Strathfield; and, Future Strathfield Directions. To assist in addressing this question, author (Rauscher) qualitative research field work was completed in Strathfield Municipality (2014–2015). Early Strathfield (Plate 7.2) is firstly illustrated in this Federation house and garden setting (top-left). The Municipal Chambers continue to operate from the same building (top-center). The large Russian community is served by the local Russian Orthodox Church (top-right). Another established church is the Uniting Church (bottom-left). Many wealthy families preferred building in Strathfield (bottom-center), instead of the Sydney inner city area. Finally, there was a wide variety of architecture featured in early Strathfield as illustrated in the Presbyterian Church (bottom-right). In contrast the industrial developments were kept mostly in North Strathfield. As the area becomes more popular for residential development, many industrial sites are also being developed as high density residential developments.

Strathfield renewal (Plate 7.3) is occurring (2015) throughout most of the municipality's suburbs, especially in the Strathfield CBD as examined here. Strathfield Outdoor Plaza is a major gathering point beside the railway station (top-left). A number of churches in Strathfield offer multi-lingual services as illustrated here at

Plate 7.2 Early Strathfield. *Left to right* Federation house in garden setting; Municipal Chambers; Russian Orthodox Church; Uniting Church; Residential estate; Presbyterian Church

the Uniting Church (top-right). Commercial outdoor eateries contribute to a cosmopolitan Strathfield (top-center). Retail below and offices above take advantage of the rail hub that Strathfield represents (top-right). Outside dining is popular during the day, as well as night time (bottom-left). The Strathfield Council (bottom-center) operate from the town hall, a short distance from the CBD. Finally, the 'Welcome to Strathfield' square (bottom-right) demonstrates the community spirit of the municipality. The most dramatic change in Strathfield has been the introduction of high rise residential blocks with commercial below (bottom-center), or stand-alone residential (bottom-right).

The New Strathfield (Plate 7.4) reflects the emerging new Strathfield. A number of high rise residential with shops below have been built in Strathfield (top-left). One complex has restaurants linked with an arcade under offices (top-center). Scattered throughout Strathfield are new single residential premises (top-right). A corner 10-story residential high rise (bottom-left) anchors this corner. Notable in Strathfield is this block residential tower (bottom-center). Finally, postmodern experimental architecture (bottom-right) is replacing older residences.

Plate 7.3 Traditional Strathfield. *Left to right* Strathfield Plaza; commercial outdoor eateries; Uniting Church Korean Services; Railway Sq. corner; Municipal Offices; and, Welcome to Strathfield plaza

Future Strathfield (Plate 7.5) reflects the continuing building boom in residential development at increasing densities throughout the municipality. The quality of buildings will show marked increase in design (top left). A number of new high rise projects are planned for Strathfield from Homebush (top-left) through to Strathfield (bottom-right).

7.2.1 Conclusion

Given the review of development phases of Strathfield (above) and changes in population demographics, it can be concluded that Strathfield has been experiencing accelerated global transition (2000s) (Question 1). The planning of Strathfield (merged with Burwood and Canada Bay councils as noted above under the Proclamation) in meeting development demands and applying sustainable urban planning (SUP) practices is examined next.

Plate 7.4 New Strathfield. *Left to right* High rise residential with shops below; restaurants linked with arcade; modern residential architecture; Corner 10 story residential high rise; block residential tower; and, postmodern architecture

Plate 7.5 Future Strathfield

7.3 Strathfield and Sustainable Urban Planning

Planning of Strathfield Municipality has been taking place since it became a local government body in 1861. A review of planning in Strathfield will pay particular attention to applications of sustainable urban planning (SUP) (as outlined in Chap. 1) practices (including use of indicators of sustainability) (Chap. 2). Strathfield Council key planning documents (a–z with date and web reference) are listed in Information Box 7.1. A summary of these documents and a brief critique assist with this review of Strathfield planning (documents can be sourced under the web link as noted in the right column). A summary of these documents (a–z) and a brief critique of each follows.

7.3.1 Strathfield 2025

Commitments have been made by the Council to adopt key principles of ESD which are reflected in Council's community strategic plan entitled *Strathfield 2025* (Strathfield 2012a). Within the plan Council defines its approach to biodiversity (as required by the NSW *Local Government Act 1993)* (NSW 1993), for example, as:

> Biodiversity is defined as the variety of all life forms; the different plants, animals and microorganisms, their genes and the ecosystems of which they are a part. There are a number of small pockets of remnant vegetation throughout the Strathfield Local Government Area and many of these are listed as Endangered Ecological Communities. These endangered ecological communities include Cooks River/Castlereagh Ironbark Forest, Sydney Turpentine Ironbark Forest and Coastal Saltmarsh. The fragmentation of these natural areas leaves them particularly vulnerable, making their management an important priority for Council.

Up to 53 % of Strathfield LGA is located in the Cooks River catchment (mostly upper reaches). The river's catchment area covers about 100 square km's across 13 local council areas. Care and control of the river is a complicated issue, where responsibility is shared between local Councils, the community and the State Government through agencies (further examined in Chap. 9 at a district level). Council states:

> Over recent years, the condition of the river and its foreshores has seen some improvement, due in large part to the efforts of these stakeholders and the local community, however it remains a heavily polluted river with much more to be done. Strathfield Council works with

Information Box 7.1 Strathfield municipality planning

Strathfield 2025 (Strathfield (2012a)	www.strathfield.nsw.gov.au
Strathfield Development Control Plan (2005)	www.strathfield.nsw.gov.au
Strathfield Local Environment Plan (2012b)	www.strathfield.nsw.gov.au
Strathfield State of Environment Report (2008a)	www.strathfield.nsw.gov.au
Strathfield Town Centre – Draft Concept Master Plan (2008b)	www.strathfield.nsw.gov.au

Source Webs as noted February 2016

state agencies, other councils and the community to improve the condition of the river system.

Strathfield Council is a member of the Cooks River Alliance, an eight council partnership focused on addressing river issues (and raised in Chaps. 3, 4, and 5). The Alliance was formed in May 2011. The Alliance evolved from the Cooks River Foreshore Working Group (1998–2011) and the Cooks River Sustainability Initiative (2007–2011).

Moving from the Cooks River to wetlands in Strathfield, Council notes:

> Yarrowee Wetlands, Strathfield, is located within Freshwater Park in the upper reaches of the Cooks River. The wetland was restored in 2010, providing a feeding habitat and breeding site for a variety of native wildlife including frogs, skinks and small birds. The project included the construction of a wetland to treat storm water flowing from several local streets and the planting of over 2000 locally native shrubs, grasses and ground covers to serve as a seed orchard for the Strathfield Council Native Nursery. The project was managed in partnership with the Sydney Metropolitan Catchment Authority.

In looking at specifics of biodiversity (i.e. under the NSW *Threatened Species Conservation Act 1995) (NSW* 1995*),* Council points out:

> A number of endangered species are found in Strathfield including the Green and Golden bell-frog (*Litoria aurea*) and the Downy Wattle (*Acacia pubescens*). Mason Park Wetlands in Homebush also attracts migratory Wader birds from the Northern Hemisphere through spring and summer each year. Council aims to enhance biodiversity across the municipality through forming partnerships with the State and Commonwealth Governments, catchment management groups, bushcare groups, and the local community.

Strathfield Council's *Vertebrate Fauna Study* (Strathfield 2009) of parks and reserves within the Strathfield LGA is a comprehensive study and assessment of the variety of wildlife in the community. The study provides recommendations on how to enhance biodiversity, with Council noting:

> A key finding of the study found that Mason Park Wetlands and its surrounding vegetation is the most species-rich reserve within the Strathfield area. This result was largely due to the presence of migratory shorebirds that arrive during spring and summer. Bushcare is a group of community volunteers that work with Council to assist in the protection and enhancement of natural areas within the Strathfield Local Government Area.

7.3.2 Strathfield Local Environment Plan

In planning for the built environment, Strathfield Council, through its *Local Environmental Plan* (LEP) (Strathfield 2012b) and Development Control Plan (2005) sets controls on new developments to ensure that environmental best practice features are applied in order to preserve and improve the environment.

7.3.3 *Strathfield State of Environment Report*

Other sustainable urban planning initiatives the Council has taken are within the areas of: Climate protection and Strathfield Town Centre Upgrade. These are summarized below.

7.3.4 *Climate Protection*

Strathfield Council, with assistance from the International Council for Local Environmental Initiatives (ICLEI), has made a strong commitment to reduce its greenhouse gas emissions through completing the Cities for Climate Protection (CCP) program. CCP is a 5 milestone process, as noted in Chap. 6 (Burwood) designed to reduce Council and Communities production of Greenhouse Gases. In noting the five milestones, Strathfield Council comments:

> Council has completed Milestone 5 of the CCP program. A key finding from this process was the discovery that Council had successfully reduced its greenhouse gas emissions by 12 % or 671 tons of carbon dioxide during the 2008 financial year. Since then Council has continued to reduce its carbon emissions through its operations and is beginning to see significant changes. Council's monitoring and reporting program is through Planet Footprint.

Looking at climate protection on a regional level, Council notes:

> As well as having global effects on climate, it is predicted that a range of regional specific effects will also be felt in the Sydney region that may include:
>
> (a) An increase in extreme weather events including; heat waves, hail storms, extended droughts, extreme winds, and storms events
> (b) Precipitation levels will become particularly hard to predict, however there will be an overall drop in average rainfall over Sydney's catchments. Currently the average annual rainfall is around 1100 ml which has been subject to a 55 ml drop per decade since the 1950s. The current forecasts are that there will be a further 5 % reduction by 2030 under a moderate change in climate.
> (c) Possible changes in vegetation in the Sydney Basin as climatic changes may favor some species over others. Invasive species may also become more prolific or access parts of the region that have been traditionally too cool.

7.3.5 *Strathfield Town Centre Upgrade*

Moving to urban design, the Strathfield Town Centre is a medium size retail, commercial and residential center and is the most important commercial area in Strathfield LGA. Council is embarking on a bold process to redesign and revitalize the Strathfield Town Centre (Plate 7.6) within the *Strathfield Town Centre – Draft*

Plate 7.6 Plans for Strathfield Town Center (*Source* Strathfield Council 2008b)

Concept Master Plan (Strathfield 2008b). The objective for the Council is to achieve a public domain of exceptional quality to meet the needs of Strathfield and the broader community well into the future, Council noting:

> A key feature of the project process involves Council working closely with the community and all stakeholders in a series of workshops and consultations. This would draw on the collective issues, ideas and visions to facilitate and determine a long-term vision and master plan for the future of the Strathfield Town Centre. Following an extensive design and public consultation process to revitalize and upgrade the Strathfield Town Centre, Council publically exhibited the Strathfield Town Centre – Draft Concept Master Plan.

Council is currently finalizing further design and feasibility investigations, such as detailed traffic modeling. These investigations, in addition to issues raised from the public eexhibition, will guide the next steps in this project. Council summarizes these issues as:

1. Quality public space
2. Community Building with possible offices, residences or seniors living above
3. Bus tunnels linking other streets
4. Plaza development will include a mandatory green roof park with a restaurant/cafe strip
5. Future extension of Town Square to create an even larger public space
6. Link north & south Strathfield via an attractive overhead bridge, east of the Station
7. Develop key sites further around the Station, for higher density development (residential and mixed use with sustainable energy and water efficiencies)

Plate 7.7 Proposed Strathfield Town Center Details. *Left to right* underground bus interchange; seating areas; performance space; open air plaza; linking spaces; and, creative use of light

8. Town Centre to be carbon, energy and water neutral
9. Improve public transport access and pedestrian and public transport access & pedestrian and cycle access

Project design incorporates a number of upgrades, commencing with an underground bus interchange (Plate 7.7) (top-left). New seating areas will be created in an outdoor piazza setting (top-right). Within this setting an outdoor performance space will be included (middle-left). The outdoor plaza will also include meeting nodes (middle-right). New spaces will be linked to increased pedestrian connections (bottom-left). Finally, creative use of light, day and night, will be used to attract activities and create attractive spaces (bottom-right).

7.3.6 Conclusion on Council Planning and SUP Practices

It can be concluded, having reviewed Strathfield Council planning documents and their applications, it can be concluded that Council is moving in the direction of applying sustainable urban planning (SUP) practices (Question 2 above). The State's Greater Sydney planning (including infrastructure projects) and State council amalgamation proposals likely to affect Strathfield (Question 3) are reviewed next.

7.4 Greater Sydney Planning and Strathfield

There are planning and developments of Greater Sydney likely to impact on Strathfield. In many instances Strathfield Council has had the task of providing: planning input on State proposals affecting Strathfield. These include making and reviewing public submissions; implementing State planning policies at the local level; joint partnership on service or infrastructure projects; and, monitoring State or private development projects affecting Strathfield. Other Council planning responsibilities under State legislation that also faces the Council include: engaging the municipality community (e.g. residents, business, services, development interests, and institutions) on State planning policies and development proposals; statutory plan making (e.g. adopting local environment plans (LEPs); and, reporting back to the State on a range of planning matters (i.e. State proposed roads, state of environment reporting and infrastructure projects).

Greater Sydney planning requirements were set by the NSW State in 2006, with Strathfield Council adopting the *Strathfield 2025* (Strathfield 2012a). Council's plans had to be consistent with the *Sydney Metropolitan Strategy* (NSW 2011). The State in 2014 adopted its strategy for planning Greater Sydney to 2031 (*A Plan for Growing Sydney* (NSW 2014a). The Plan forecasts a population growth of 1.6 m to be added to the Greater Sydney population of 4.84 m (at June 2014). Strathfield is nominated by the State to be within Subregion 'Central' within this Plan. Local government areas with 'Central' include: Ashfield, Botany, Burwood, Canada Bay, Leichhardt, Marrickville, Randwick, Strathfield, Sydney, Waverley, and Woollahra. In examining this Plan and its impact on Marrickville, three key planning principles as adopted in Chap. 2 are addressed: (1) Developing Sustainable Urban Centres; (2) Developing Sustainable Transport; and, (3) Protecting the Natural Environment. In Developing Sustainable Urban Centres, and noting Strathfield is adjacent to strategic growth centre of Burwood, it is likely Strathfield will also be providing for additional residential, commercial and service developments. Moving to Developing Sustainable Transport, the State asked Strathfield Council (along with other Sydney Inner West councils) in late 2014 to comment on the proposed *Parramatta Rd Renewal Project* (NSW 2014b). The *Draft Parramatta Rd Urban Transformation Strategy* (NSW 2015) (www.newparramattard.com.au) followed in 2015 and identifies eight Precincts, specifically targeted for renewal under Precinct Plans. Precincts were selected due to their access to jobs, infrastructure, public transport, or how well they can support new housing types. Precinct 3 is the Strathfield centre in the vicinity of Homebush (Plate 7.8). Strathfield Council is closely tied to the Parramatta Rd renewal project.

The State explains:

The State spells out more details for Homebush:

> Sitting between Sydney's two main CBDs, Homebush can be transformed into an active and varied hub, blending higher density housing and a mix of different uses, supported by a network of green links and open spaces with walking access to four train stations. Homebush will be a focus for high density housing, with a hub of activity between Homebush, North Strathfield, Concord West and Strathfield Stations. Both Parramatta

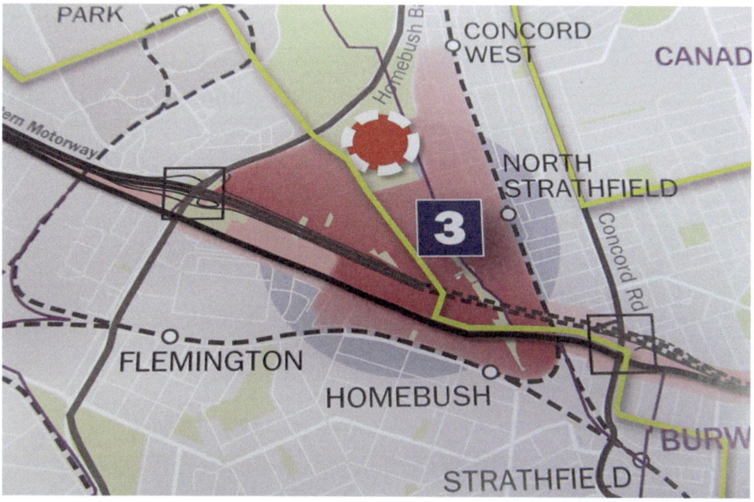

Plate 7.8 Parramatta Rd Urban transformation strategy – Strathfield Precinct 3 (*Source* NSW 2015)

> Road and George Street will form main streets to build on the character of the Bakehouse Quarter and the curve of Parramatta Road. Taller residential buildings will mark the center of activity at the Precinct's core. The network of streets to the north and west from here will be easy and safe to walk through, with medium-density housing and the green corridor of Powell's Creek. The area around Flemington Markets will have a new employment and retail focus.

The State's WestConnex roads project, also affecting Strathfield Council, is closely tied to the Parramatta Rd renewal project. Strathfield Council sees the WestConnex (www.westconnex.com.au) motorway as a major transport project facing the Strathfield LGA (as well as other SIW council areas) (noted in earlier chapters). The section of the WestConnex that passes through Strathfield affects North Strathfield and Homebush. Both the Westconnex and the Parramatta Rd Renewal projects, as they affect Sydney Inner West, will be addressed further in Chap. 9.

7.4.1 Protecting the Natural Environment

Finally, under '3. *Protecting the Natural Environment*' the Plan commits the State to 'protect the health of the subregions waterways and aquatic habitats'. This goal, for example, would apply to the prime conservation challenges in Strathfield. This would include the protection of bushlands, waterways and foreshores within Strathfield. Strathfield Council, along with neighbouring councils developing the Inner West Open Space Corridor and upgrading the Cooks River, has played a role in this project. The efforts of these councils in undertaking this work will be outlined in more detail in Chap. 9.

7.4.2 Conclusion on Strathfield and Greater Sydney Planning

Having reviewed above the State's Greater Sydney planning (including infrastructure projects) it can be concluded the implementation of these plans and developments is likely to have a significant impact on Strathfield (Question 3). The extent of this impact will be better understood when these State plans are further examined in the context of the whole of the Sydney Inner West District (including Strathfield) in Chap. 9. Finally, in examining State proposed council amalgamations, options for councils (including Strathfield) working together and in partnership with the State to meet local and district needs will be addressed in Chap. 18 (Planning Reforms and Directions for Planning Sustainable Cities).

7.5 Amalgamations of Councils and Strathfield

The subject of amalgamations of Greater Sydney councils (given implications for planning at municipal and Greater Sydney levels) was reviewed in Chap. 2. NSW Premier Mike Baird announced in Dec 2015 the State's preference for Strathfield Council to merge with Burwood and Canada Bay councils. Within the Inner West District these three councils compose the western end of the district. Strathfield. Strathfield Council had stated a preference to stand alone and not to be merged with any other council and to oppose any forced amalgamations. The Mayor of Strathfield urged all residents to have their say on the issue of amalgamations of councils. Council noted that submissions to the new Council Boundary Review were open till the end of February (www.councilboundaryreview.nsw.gov.au). In the interim, the Council noted, the Minister for Local Government had commenced the process of forcibly amalgamating Strathfield, Burwood, and Canada Bay councils. Council also noted the Minister said he expected the new council to be 'proclaimed' on 1 July 2016 (note the 12th May 2016 *Local Government (Council Amalgamations) Proclamation 2016* noted in Chap. 2). This came about despite Strathfield Council repeatedly stating its preference to remain stand-alone, and a majority of residents also saying they prefer the current local government boundaries. The results of the review of merger proposals (which includes a new inner west council of Strathfield, Burwood, and Canada Bay) would be referred to the Chief Executive of the Office of Local Government for examination and report under the Local Government Act (the Act). In turn, the Chief Executive had appointed a Delegate who would examine the proposals, conduct a public inquiry, call for written submissions, and then report back to the Minister.

7.5.1 Conclusion on Amalgamations

Strathfield Council made a clear case, based on research and engaging its community, to be a stand-alone council and not merge with any other council. The Minister for Local Government in 2016 had commenced the process, however, of forcibly amalgamating Burwood, Canada Bay and Strathfield councils. At the end of the State review process (Council Boundary Review) the State would determine what amalgamations as proposed would proceed or not proceed. The 12th May 2016 *Local Government (Council Amalgamations) Proclamation 2016* merging Strathfield, Burwood and Canada Bay councils was noted. This question will be further addressed in Chap. 18.

7.6 Overall Conclusions

Having completed an assessment of the development and planning of Strathfield, there are a number of overall conclusions reached. Firstly, Strathfield has been experiencing accelerated development and changes to the composition of its population. Given these changes, and examining Strathfield Council's response, it was also concluded Council is moving in the right direction in increasing its sustainable urban planning (SUP) practices. It was also concluded, however, that there would be significant future impacts on Strathfield from Greater Sydney growth and development. In reviewing potential Council and State responses to this future scenario, it was concluded that new partnerships between the Council and the State would be needed. A major initiative from the State (2009 onwards) was to investigate local government reform, including a potential amalgamation of councils (e.g. Strathfield, Burwood and Canada Bay). The Council had to date, however, expressed an intent to stand alone and only amalgamate as a second choice if the State forced amalgamations. The 12th May 2016 *Local Government (Council Amalgamations) Proclamation 2016* that merged Strathfield, Burwood and Canada Bay councils was acknowledged. These overall chapter conclusions will be examined against results of assessing planning and development over all six council areas within Sydney Inner West District (Chaps. 3, 4, 5, 6, 7, and 8) and the district as a whole in Chap. 9. Results from the district assessment (and assessment of Parramatta and St George districts in later chapters) will contribute to assessing how to plan for a sustainable Greater Sydney and a sustainable city anywhere (Chap. 18).

7.7 Summary

The chapter addressed the extent Strathfield Council was applying sustainable urban planning (SUP) practices (see Chap. 1) in planning the municipal area. It was noted that Strathfield, in Sydney Inner West District, was merged with Burwood and Canada Bay councils under the 12th May 2016 *Local Government (Council Amalgamations) Proclamation 2016* (Chap. 2). It was concluded that Strathfield, one of twelve municipalities within three Greater Sydney Districts (Sydney Inner West, Greater Parramatta and St George) is considered to be in global transition. SUP practices (based on the application of three key SUP principles) were introduced, being: Developing Sustainable Urban Centers; Developing Sustainable Transport; and, Protecting the Natural Environment. A review of Strathfield background and population changes (i.e. global movement of population) was outlined. In examining the pattern of urban development, four phases of Strathfield development are examined: Early Strathfield; Traditional Strathfield; New Strathfield; and, Future Strathfield. With conclusions drawn, key Strathfield Council planning documents were then critiqued to determine the extent Council applies SUP practices. In addition, given State planning of Greater Sydney affects Strathfield, a second question was posed: what is the impact of Greater Sydney planning and development on Strathfield?. In examining this question, examples of State plans (including State proposed council amalgamations) likely to impact on Strathfield were addressed. By chapter's end, conclusions were drawn on Council planning and impact of Greater Sydney planning on Strathfield. The chapter noted that these conclusions would be addressed further in: Chap. 9 (examining the extent of SUP practices being applied over the whole of Sydney Inner West District); and, Chap. 18 (examining directions for planning a sustainable Greater Sydney and a sustainable city anywhere).

Notation 7.1 Cultural Diversity Poem – Strathfield
Poem reflecting the diversity of Strathfield:

Who Took His Strathfield Away? (# 12)

Vincent lives in South Strathfield and feels someone has taken his neighbourhood
Born and raised midst Californian bungalows, Federation and Victorian homes
The global urban shock has dissected his suburb, his city, his heart
The change in the migration numbers that have settled in his neighbourhood upsets

The transformation of place can be a piercing of some people's souls
Familiarity is comfort knowing is stability, welcoming is etched in many streets
A whole new arrangement of buildings rising and new faces in the shops
In planning neighbourhoods, remember the Vincent's, their neighbourhood not to erase

References

Australian Bureau of Statistics (ABS) (2011) Australian census review. ABS, Canberra
Council (2008b) Strathfield Town Centre – draft concept master plan. Strathfield
New South Wales (1993) NSW Local Government Act 1993. Sydney
New South Wales (1995) Threatened Species Conservation Act 1995. Sydney
New South Wales (2011) Sydney metropolitan strategy. Sydney
New South Wales (2014a) A plan for growing Sydney. Sydney
New South Wales (2014b) Parramatta Rd renewal project. Sydney
New South Wales (2015) Draft Parramatta Rd urban transformation strategy. Sydney
Strathfield Council (2005) Strathfield development control plan. Strathfield
Strathfield Council (2008a) Strathfield state of environment report. Strathfield
Strathfield Council (2009) Vertebrate fauna study. Strathfield
Strathfield Council (2012a) Strathfield 2025. Strathfield
Strathfield Council (2012b) Strathfield local environment plan. Strathfield

Chapter 8
Planning Canada Bay, Greater Sydney

Dedication *This chapter is dedicated to Dame Edith Walker, carried out numerous charitable works in Australia with her father, Thomas Walker (Plate 8.1).*

Plate 8.1 Edith Walker (*Source* Wikipedia 2016)

A snippet on Eadith Walker gives an insight into her legacy:

> Dame Edith Campbell Walker, DBE (1861–1937) was an Australian philanthropist. Eadith Campbell Walker was the only child of Australian banker and politician, Thomas Walker. They lived in the family home, Yaralla, an Italianate mansion in Concord West, an inner-western suburb of Sydney. She and her father carried out numerous charitable works in Australia. (*Source* Wikipedia 2015)

R.C. Rauscher, S. Momtaz, *Cities in Global Transition*,
DOI 10.1007/978-3-319-39865-5_8

Abstract The chapter addresses the extent Canada Bay Council (herein referred to as Canada Bay) is applying sustainable urban planning (SUP) practices (see Chap. 1) in planning the municipal area. The chapter acknowledges the *Local Government (Council Amalgamations) Proclamation 2016* merging Canada Bay with Burwood and Strathfield councils noted in Chap. 2. Canada Bay (in Sydney Inner West District) is one of twelve municipalities within three Greater Sydney Districts (Sydney Inner West, Greater Parramatta and St George), all considered to be in global transition (see Chap. 2). Examining SUP practices is based on the application of three key SUP principles (as adopted in Chap. 2), being: Developing Sustainable Urban Centers; Developing Sustainable Transport; and, Protecting the Natural Environment.

Firstly, a background to Canada Bay development and population changes (e.g. global movement of population) is outlined. To examine the pattern of urban development, four phases of Canada Bay development are examined: Early Canada Bay; Traditional Canada Bay; New Canada Bay; and, Future Canada Bay. Given conclusions reached here, key Canada Bay Council planning documents are then critiqued to determine the extent Council applies SUP practices. In addition, given State planning of Greater Sydney affects Canada Bay, a second question is posed: what is the impact of Greater Sydney planning and development on Canada Bay? In examining this question, examples of State plans (including council amalgamations under the Proclamation noted above) likely to affect Canada Bay are addressed. By chapter's end, conclusions are drawn on Council planning and impact of Greater Sydney planning on Canada Bay. The chapter notes that these conclusions will be addressed further in: Chap. 9 (examining the extent of SUP practices being applied over the whole of Sydney Inner West District); and, Chap. 18 (examining directions for planning a sustainable Greater Sydney or a sustainable city anywhere).

8.1 Background and Population Change

Canada Bay City (herein called Canada Bay in the context of Sydney Inner West (as introduced in Chap. 2) is north of Burwood and Strathfield to the west of the City of Sydney. Under the *Local Government (Council Amalgamations) Proclamation 2016* Canada Bay was merged with Burwood and Strathfield councils as noted in Chap. 2). The area now known as Canada Bay was home to the Wangal people, inhabiting a major portion of Sydney Inner West. A background of Aboriginal occupation of the district is contained in Chap. 9 (examining the 'whole of the district'). Canada Bay name is named in remembrance of the 58 French Canadian prisoners deported in 1840, following the Lower Canada Rebellion. Canada Bay, like most of the Sydney Inner West District, was once the area inhabited by the Wangal clan of the Eora Aboriginal people (Chap. 2). The City of Canada Bay encompasses a total area of 20 km^2, including many parks, reserves and foreshores. A brief background on Canada Bay follows:

> The City of Canada Bay is located in the inner-western suburbs of Sydney, about 6–12 kilometers from the Sydney GPO. The City of Canada Bay is bounded by the Parramatta River in the north and east, the Municipality of Ashfield, the Burwood and Strathfield Council areas in the south and the Auburn Council in the west. The City of Canada Bay includes the suburbs of Abbotsford, Breakfast Point, Cabarita, Canada Bay, Chiswick, Concord, Concord West, Drummoyne, Five Dock, Liberty Grove, Mortlake, North Strathfield, Rhodes, Rodd Point, Russell Lea, Strathfield (part) and Wareemba. European settlement dates from 1793, with land used mainly for farming. Residential growth took place during the late 1800s and early 1900s, aided by improved access and industrial growth. Significant development occurred during the post-war years. The population was relatively stable between 1991 and 1996, then increased from nearly 54,000 in 1996 to over 65,000 in 2006. Recent growth has been largely due to the redevelopment of previous industrial sites into residential developments, particularly high density housing. Population growth is expected to continue, especially in the Rhodes and Breakfast Point areas. Major features of the City include the Parramatta River, Birkenhead Point Shopping Centre, Rhodes Shopping Centre, Yaralla Dame Edith Walker Hospital, Concord Repatriation General Hospital, Concord Golf Club and many parks and reserves. The City is served by the Western Motorway, Parramatta Road and the main northern railway line. (*Source* Wikipedia 2015)

The suburbs of Canada Bay include: Abbotsford, Breakfast Point, Cabarita, Canada Bay, Chiswick, Concord, Concord West, Drummoyne, and Five Dock. The Australian Bureau of Statistics (ABS 2011) comments on Canada Bay City ancestry and languages spoken at home in examining the suburb of Canada Bay:

> According to the 2011 census, there were 1,296 residents in Canada Bay. 63.7 % of people were born in Australia. The most common countries of birth were Italy 7.1 %, China 3.7 %, England 2.8 %, New Zealand 1.5 % and Hong Kong 1.2 %. In Canada Bay 57.2 % of people only spoke English at home. Other languages spoken at home included Italian 13.1 %, Cantonese 4.1 %, Greek 3.6 %, Mandarin 3.2 % and Spanish 2.2 %. The most common responses for religion in Canada Bay were Catholic 46.6 %, No Religion 13.8 %, Anglican 12.7 %, Eastern Orthodox 7.0 % and Buddhism 3.4 %.

Finally, in author's (Rauscher) qualitative research (2014/2015), there is a poem written to reflect the cultural diversity of Canada Bay (end of chapter under Notation 8.1 *hodes' New Order Modernism).* The phases of Canada Bay development are examined next.

8.2 Phases of Development

An examination of the development phases of any locality provides a basis of understanding current trends and likely future directions of growth of that locality. Following a look at the background of Canada Bay four phases of Burwood development will be examined. These phases include: Early Canada Bay; Traditional Canada Bay; New Canada Bay; and, Future Canada Bay. To assist in addressing this question, author (Rauscher) qualitative research field work was completed in Canada Bay City (2014–2015). Plate 8.2 illustrates the early suburb of Rhodes, within the City of Canada Bay. Housing commenced in the early 1800s, with this

Plate 8.2 Early Canada Bay. *Left to right* Early Rhodes – Early house 1800s; Rail Station built 1886 with Northern Line opening; Fire station early 1900s; Meadowbank Punt, 1930; Union Carbide 1960s; and Industry 1970

house built by Thomas Walker in 1823 (top-left). The northern rail line (Strathfield to Hornsby) came to Rhodes in 1886 (top-center). In the early 1900s Rhodes was serviced by its own fire station (top-right). To cross the Parramatta River to Rhodes a punt service was available at Meadowbank (bottom-left). The extent of heavy industry operating at Rhodes in the 1960s is illustrated here at Union Carbide (bottom-center). Finally, the continuation of industry in the 1970s (though starting to phase out) is shown here (bottom-right).

Traditional Canada Bay shows renewal (Plate 8.3) is occurring (2014) throughout most of the city's suburbs. A new shopping center (top-left) serves the Canada Bay district. The Historic Bridge (top-center) connects Canada Bay to Meadowbank. Rhodes Public School (top-right) serves the surrounding suburbs. The multi-cultural composition of the population is reflected in this Greek Orthodox Church (bottom-left). The Rhodes Rail Station is a central transport point for Canada Bay City (bottom-center). Finally, a public square with a water sculpture (bottom-right) has created a mini-plaza.

The new Canada Bay is reflected in this scan of Rhodes (Plate 8.4). Most streetscapes in Rhodes have long distance views and are planted (top-left). With the freedom of a master plan, medium density here can follow the curve of the street (top-center). Developed on high ground this six story complex overlooks Homebush Bay (top-right). Rhodes has scattered throughout the suburb high rise residential

Plate 8.3 Traditional Canada Bay. *Left to right* Shopping Center; Historic Bridge to Meadowbank; Rhodes Public School; Greek Orthodox Church; Rhodes Rail Station; and, Public Square with water sculpture

(bottom-left). Architecture in the suburb varies, with a degree of experimentation (bottom-center). Finally, many of the medium density projects have been planned with landscaped surrounds.

Finally, Future Canada Bay (Plate 8.5) reflects the continue building boom in residential development at increasing densities throughout the municipality. The quality of buildings will show marked increase in design (top left). A model layout of Rhodes West is shown here (top-left). A residential development proposal with an inner courtyard consumes a whole block (top-middle). A high rise proposal (top-right) overlooks Homebush Bay.

8.2.1 Conclusion

Given the review of development phases of Canada Bay (above) and changes in population demographics, it can be concluded that Canada Bay has been experiencing accelerated global transition (2000s) (Question 1). The planning of Canada Bay (merged with Burwood and Strathfield councils as noted above under the Proclamation) in meeting development demands and applying sustainable urban planning (SUP) practices is examined next.

Plate 8.4 New Canada Bay. *Left to right* Rhodes – Streetscape; Medium density following street curve; Six-story overlooking Homebush Bay; High rise; New architecture; and, Medium density in landscaped surrounds *(Source* Raymond Rauscher 2015)

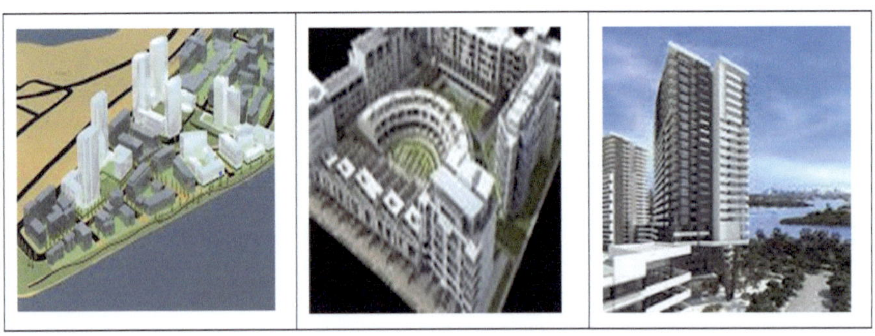

Plate 8.5 Future Canada Bay

8.3 Canada Bay and Sustainable Urban Planning

Planning of Canada Bay City has been taking place since it became a local government body in 1861. A review of planning in Canada Bay will pay particular attention to applications of sustainable urban planning (SUP) (as outlined in Chap. 1) practices (including use of indicators of sustainability) (Chap. 2). A review of

Information Box 8.1 Canada Bay city planning

Canada Bay development control plan (Canada Bay 2013b)	www.canadabay.nsw.gov.au
*Canada Bay local environment plan (*Canada Bay 2013a*)*	www.canadabay.nsw.gov.au
Environmental sustainability strategy (Canada Bay 2014)	www.canadabay.nsw.gov.au
FuturesPlan20 (Canada Bay 2008)	www.canadabay.nsw.gov.au
State of environment report (Canada Bay 2010a)	www.canadabay.nsw.gov.au
Sustaining our environment program (Canada Bay 2010b)	www.canadabay.nsw.gov.au

Source: Webs as noted February 2016

planning in Canada Bay will focus on applications of sustainable urban planning (SUP) practices (as outlined in Chap. 1) and planning of a municipality (Chap. 2). Canada Bay Council key planning documents (a–z with date and web reference) are listed in Information Box 8.1. A summary of these documents and a brief critique assist with this review of Canada Bay planning (documents can be sourced under the web link as noted in the right column).

8.3.1 Canada Bay LEP and Development Control Plan

Council's two most important planning instruments, the *Canada Bay Local Environmental Plan* (Canada Bay 2013a) and an accompanying *Canada Bay Development Control Plan* (Canada Bay 2013b), were adopted in 2013. Planning and design outcomes that Council is pursuing include sustainable and contemporary building and place design, transit oriented development, heritage conservation, active streets and laneways and a diversity of land use that supports economic, environmental and social sustainability. For the first time, explicit recognition of the growing prominence of food security as an issue is addressed within the DCP, encouraging initiatives such as community gardens and green roofs. An Emerging Technologies Clause gives Council scope to approve new and innovative technologies as they emerge without having to fully amend the DCP.

8.3.2 Environmental Sustainability Strategy

Council adopted the *Environmental Sustainability Strategy* (Canada Bay 2014) in 2014. The strategy provides guidelines on how the council can meet sustainability principles in coming years.

8.3.3 FuturesPlan20

Canada Bay Council's future planning is contained in *FuturesPlan20* (Canada Bay 2008). Council explains the Plan and its updating as follows:

> Canada Bay Council's *FuturesPlan20* is a plan outlining what the City of Canada Bay would look like in the future. The Plan is a whole-of-community blueprint for the future of the City of Canada Bay. The Plan was first developed in 2008 based on what the community saw as its vision for the City. It was updated in 2010 and has just been reviewed, revised and readopted in 2013 after extensive community and stakeholder consultation and engagement. *FuturesPlan20*, in addition to development projections, the Plan also has a Delivery Plan and a Resourcing Strategy.

8.3.4 State of Environment (SoE)

Council produces the *State of Environment Report* (Canada Bay 2010a) each year. This report provides information about the local environment and about progress towards meeting environmental objectives as set out in the *Environmental Sustainability Strategy*. Within Council's SoE reporting specific attention is given, for example, to addressing: land uses; air and climate change; water conservation; biodiversity protection; waste reduction; noise abatement; heritage protection; and flora and fauna protection.

8.3.5 Sustaining Our Environment Program

Moving to examine sustainability planning in Canada Bay Municipality the Council works under it's *Sustaining our Environment Program* (Canada Bay 2010b), a program that delivers a number of projects which focus on achieving environmental sustainability outcomes across the City. These projects are funded through the Waste and Sustainability Improvement Payment (WaSIP) program, which is a State Government fund that supports sustainability and waste initiatives delivered by local government. Below (Table 8.1) is a list of Council focused on protecting waterways, saving energy and conserving water.

Commenting on environmental challenges facing the City of Canada, the Council notes:

> Carbon abatement, greenhouse gas emission reduction and the cost of energy are issues facing Council. City of Canada Bay is improving its energy efficiency and has developed an energy management plan, which includes projects to address greenhouse gas emissions. *FuturesPlan20* identifies greenhouse gas emissions and addressing carbon management as a key target area. The Sister City Environmental Project will enable Council's to share expertise and solutions to some of these issues.

In additional to the above planning documents there are other planning projects the Council and other parties are engaged in. Examples of these projects include: Rhodes Peninsula Planning; Wentworth Point Planning;

Table 8.1 Sustaining our environment programs

(a) Energy management plan	This project involves the development and implementation of an Energy Management Plan, which builds on the old Energy Savings Action Plans. The overall target is to reduce both energy consumption and greenhouse gas emissions through efficiency programs and infrastructure upgrades throughout Council's assets.
(b) Water efficiency plan	The Water Efficiency Plan identifies the top ten water consuming sites of Council buildings and sporting facilities, looking at the amount of water used and how the water is used. Water saving opportunities are identified for each site and prioritized according to costs and payback period.
(c) Climate change mitigation and adaptation	This project involves Council adopting the City of Canada Bay Climate Adaptation and Mitigation Plan, which includes a risk assessment. An implementation plan will be developed for adoption, which will provide a staged order of priorities.
(d) Implementing sustainable procurement	The City of Canada Bay adopted a Sustainable Procurement Policy in 2011, and is currently reviewing existing practices to integrate the policy into all areas of business across the organization. This project aims to increase the level of sustainable procurement.
(e) Sustainable fleet and emissions reduction	Through the Sustainable Fleet Program, Council seeks to review the existing fleet policy with a view to reducing fleet emissions. A system for monitoring Council fleet and fleet emissions, and the development of a program and policy which will be applied across the organization will also form part of this strategy.
(f) Sister City environmental project	The City of Canada Bay established a Friendship Agreement with Parkes Shire Council in September 2010. One of the key outcomes of the agreement is a greater understanding of the issues facing each area, and the opportunity to investigate and pursue opportunities for working together.

Source: Canada Bay Council Web February 2016

8.3.6 Rhodes Peninsula Planning

A major growth area in Canada Bay City is the Rhodes Peninsula. The Rhodes Peninsula, which includes the suburbs of Rhodes and Liberty Grove, is currently (2015) in a state of development and change. This development includes high density with buildings four to eight stories, with a projected population over 11,000 when redevelopments completed. One major pubic investment project is the Rhodes Community Precinct (Plate 8.6), which is projected to benefit the entire Rhodes Peninsula and residents living at the western end of Canada Bay.

Following community consultation, the ‘Central Park’ playground design (Plate 8.7) has been developed by Canada Bay Council.

Plate 8.6 Proposed Rhodes Community Center and Residential Living Areas

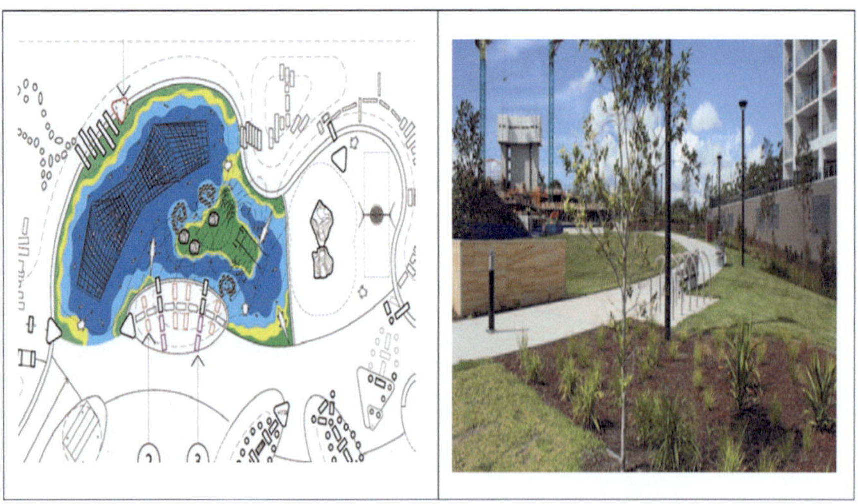

Plate 8.7 Proposed Rhodes Central Park. *Left to right*: Sketch of park (*left*) and perspective of park walkway (*right*)

Plate 8.8 Proposed Bride across Homebush Bay and Promenade (*below*)

8.3.7 *Homebush Bay Bridge*

Moving to another Canada Bay Municipality project, the Homebush Bay Bridge (Plate 8.8) has moved a step closer to becoming a reality in 2015 with construction expected to commence. Homebush Bay Bridge was approved by the State Government in March, 2013. The Bridge will carry buses, cyclists, pedestrians and emergency vehicles. Private Vehicles and cars will not be permitted. The $43 million bridge is being provided by a developer consortium in return for additional development density at Wentworth Point.

8.3.8 *Wentworth Point Planning*

The vision for Wentworth Point is to create a sustainable local community with access to public transport and an active town center with restaurants, community facilities, and a library and retail hub. Plans for Wentworth Point include a waterfront promenade and a bay walk around the foreshore from Wentworth Point to Rhodes. Looking at another residential development in Rhodes, the Wentworth Point Urban Activation Area (Plate 8.9) is an 18-hectare area of land at the northern most end of Wentworth Point which is being planned to accommodate approximately 2300 new dwellings.

In describing the *Wentworth Point Urban Activation Area Project* details, Council notes:

Plate 8.9 Wentworth Point Urban Activation Area

The vision for Wentworth Point is to create a sustainable local community with access to public transport and an active town centre with restaurants, community facilities, and a library and retail hub. Plans for Wentworth Point include a waterfront promenade and a bay walk around the foreshore from Wentworth Point to Rhodes. There will be a mix of building heights, including six, 25-story towers similar to those at Rhodes and Sydney Olympic Park. The proposal includes a 3.9-hectare peninsula park as well as three additional pocket parks; a new primary school with 18 classrooms and playing fields; a continuous pedestrian and cycle path around the foreshores of both Parramatta River and Homebush Bay; and a new maritime plaza lined with shops and cafes.

8.3.9 Renewable Energy Policies

Moving to energy alternatives, Canada Bay has examined renewable energy policies in its participation in the Southern Sydney Region Organisation of Councils (SSROC). The SSROC is an association of Councils in the southern area of Sydney, with Marrickville, Ashfield and Leichhardt participating. SSROC (in 2011) resolved to adopt regional renewable energy targets for 2030. These targets are summarized as follows: (1) More than 20 % of energy consumed in the region should be derived from renewable sources; and, (2) SSROC member Councils aim to be using a minimum of 30 % of energy from renewable sources. The SSROC has set a path to meet the target under a SSROC *Renewable Energy Master Plan* (2013). All renewable energy options to achieve the target and recommend the most cost effective and feasible solutions including analysis of governance and regulatory aspects have been addressed.

Having reviewed the key Canada Bay Council urban planning documents above, it can be concluded the Council appears to be pursuing practices of sustainable urban planning (SUP). The planning and developments of Greater Sydney likely to impact on Canada Bay is reviewed next.

8.3.10 Conclusion on Council Planning and SUP Practices

It can be concluded, having reviewed Canada Bay City planning documents and their applications, it can be concluded that Council is moving in the direction of applying sustainable urban planning (SUP) practices (Question 2 above). The State's Greater Sydney planning (including infrastructure projects) and State council amalgamation proposals likely to affect Canada Bay (Question 3) are reviewed next.

8.4 Greater Sydney Planning and Canada Bay

There are planning and developments of Greater Sydney likely to impact on Canada Bay City. In many instances Canada Bay City Council has had the task of providing: planning input on State proposals affecting Burwood. These include making and reviewing public submissions; implementing State planning policies at the local level; joint partnership on service or infrastructure projects; and, monitoring State or private development projects affecting Canada Bay. Other Council planning responsibilities under State legislation that also faces the Council include: engaging the municipality community (e.g. residents, business, services, development interests, and institutions) on State planning policies and development proposals; statutory plan making (e.g. adopting local environment plans (LEPs)); and, reporting back to the State on a range of planning matters (i.e. State proposed roads, state of environment reporting and infrastructure projects).

Greater Sydney planning requirements were set by the NSW State in 2006, with Canada Bay City Council adopting in 2008 the *FuturesPlan20* (NSW 2008). Council's plans in time also had to be consistent with the *Sydney Metropolitan Strategy* (NSW 2011). Council's complimentary town planning documents were the *Canada Bay City Local Environment Plan* (Canada Bay 2013a) and the *Canada Bay City Development Control Plan* (2013b). The State in 2014 adopted its strategy for planning Greater Sydney to 2031 (*A Plan for Growing Sydney* (NSW 2014a)). The Plan forecasts a population growth of 1.6 m (to be added to the Greater Sydney population of 4.84 m at June 2014). In examining this Plan and its impact on Marrickville, three key planning principles as adopted in Chap. 2 are addressed: (1) Developing Sustainable Urban Centres; (2) Developing Sustainable Transport; and, (3) Protecting the Natural Environment. The Plan as it relates to Canada Bay is examined next. The examination will refer to these above key Planning Principles (1–3).

In Developing Sustainable Urban Centres, the Plan notes that Rhodes is to continue to be a strategic growth centre. The Plan singles out Rhodes for this role and notes:

> The State will work with council to protect capacity for long-term employment growth in Rhodes. It will work with council to provide capacity for additional mixed-use development in Rhodes including offices, retail, services and housing. The State will support

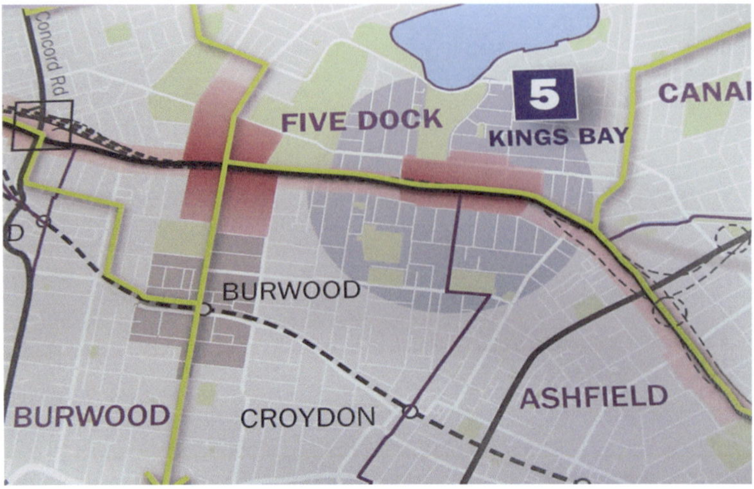

Plate 8.10 Parramatta Rd Urban transformation strategy – Precinct No. 5 Kings Bay (*Source* NSW 2015)

> health-related land uses and infrastructure around Concord Hospital. The State will also work with council to improve walking and cycling connections between Rhodes train station and Concord Hospital. Finally, the State will facilitate construction of a public transport, walking and cycling bridge over Homebush Bay to connect Rhodes to Wentworth Point.

Moving to Developing Sustainable Urban Centres Transport, Canada Bay City Council, along with other Sydney Inner West and Greater Parramatta councils, were asked in late 2014 to comment on the proposed *Parramatta Rd Renewal Project* (NSW 2014b). There are a number of 'precinct renewal' areas under the project, including Precinct 5, the Kings Bay Precinct (Plate 8.10).

The *Draft Parramatta Rd Urban Transformation Strategy* (NSW 2015) (www.newparramattard.com.au) followed in 2015 and identifies eight Precincts, specifically targeted for renewal under Precinct Plans. Precincts were selected due to their access to jobs, infrastructure, public transport, or how well they can support new housing types. The State notes:

> Kings Bay is an industrial precinct dominated by car showrooms, car servicing centers, panel beaters, and small manufacturers and warehousing. As industry moves west, the Precinct's traditional industrial area is declining, opening up land for transformation that can capitalize on the bus Rapid Transit connections to Sydney CBD and Burwood and many large areas of open space. Kings Bay will be a new residential and mixed use urban village on Parramatta Road, with an active main street and strong links to the open space network along Sydney Harbour. Finally, Parramatta Road will be quieter, with two lanes used for public transport and significant tree planting to improve the amenity and environment.

The State's WestConnex roads project, also affecting Canada Bay CityCouncil, is closely tied to the Parramatta Rd renewal project. The State is also to investigate a potential light rail corridor from Parramatta to Sydney CBD via Parramatta Road, including the implications for Canada Bay City. The Parramatta Rd Renewal project, WestConnex project, and the light rail corridor all affect Canada Bay City. They will be addressed further within the Sydney Inner West district context in Chap. 9. Finally, under '3. *Protecting the Natural Environment*' the Plan commits the State to 'protect the health of the subregions waterways and aquatic habitats'. This goal, for example, would apply to the prime conservation challenges in Canada Bay. This would include the protection of bushlands, waterways and foreshores within Canada Bay. The efforts of Canada Bay Council and other councils in the Inner West in undertaking environment protection projects will be outlined in more detail in Chap. 9.

8.4.1 Conclusion on Canada Bay and Greater Sydney Planning

Having reviewed above the State's Greater Sydney planning (including infrastructure projects) it can be concluded the implementation of these plans and developments is likely to have a significant impact on Canada Bay (Question 3). The extent of this impact will be better understood when these State plans are further examined in the context of the whole of the Sydney Inner West District (including Canada Bay) in Chap. 9. Finally, in examining State proposed council amalgamations, options for councils (including *Canada Bay*) working together and in partnership with the State to meet local and district needs will be addressed in Chap. 18 (Planning Reforms and Directions for Planning Sustainable Cities). The subject to amalgamations of councils and how this affects Canada Bay is examined next.

8.5 Amalgamations of Councils and Canada Bay

The subject of amalgamations of Greater Sydney councils (given implications for planning at municipal and Greater Sydney levels) was reviewed in Chap. 2. NSW Premier Mike Baird announced in Dec 2015 the State's preference for Canada Bay City Council to merge with Burwood and Strathfield councils. Looking at these Inner West District councils they compose the western (left) end of the district. Canada Bay City Council stated a preference to merge with Burwood and Strathfield councils. Council also remained opposed to forced amalgamations. The Mayor of Canada Bay urged all residents to have their say on the issue of amalgamations of councils. Council noted that submissions to the new Council Boundary Review were open till the end of February (www.councilboundaryreview.nsw.gov.au). In the interim, the Council noted, the Minister for Local Government had commenced the process of forcibly amalgamating Canada Bay, Burwood and Strathfield councils. Council also noted the Minister said he expected the new council to be 'proclaimed' on 1 July 2016 (note the 12th May 2016 *Local Government (Council Amalgamations) Proclamation 2016* noted in Chap. 2). Results of the review of merger proposals (which includes a new inner west council of Burwood, Canada Bay and Strathfield) would be referred to the Chief Executive of the Office of Local Government for examination and report under the Local Government Act (the Act). In turn, the Chief Executive had appointed a Delegate who would examine the proposals, conduct a public inquiry, call for written submissions, and then report back to the Minister.

8.5.1 Conclusion on Amalgamations

Canada Bay City Council made a clear case, based on research and engaging its community, to merge with Burwood and Strathfield councils. The Minister for Local Government had commenced the process in 2016 (given Strathfield Council desired to stand alone and not amalgamate) to forcibly amalgamate Burwood, Canada Bay and Strathfield councils. At the end of the State review process (Council Boundary Review) the State would determine what amalgamations as proposed would proceed or not proceed. The 12th May 2016 *Local Government (Council Amalgamations) Proclamation 2016* merging Canada Bay and Burwood and Strathfield ccouncils was noted. This question will be further addressed in Chap. 18.

8.6 Overall Conclusions

Having completed an assessment of the development and planning of Canada Bay, there are a number of overall conclusions reached. Firstly, Canada Bay has been experiencing accelerated development and changes to the composition of its population. Given these changes, and examining Canada Bay Council's response, it was also concluded Council is moving in the right direction in increasing its sustainable urban planning (SUP) practices. It was also concluded, however, that there would be significant future impacts on Canada Bay from Greater Sydney growth and development. In reviewing potential Council and State responses to this future scenario, it was concluded that new partnerships between the Council and the State would be needed. A major initiative from the State (2009 onwards) was to investigate local government reform, including a potential amalgamation of councils (e.g. Canada Bay, Burwood and Strathfield). The Council had (late in the State amalgamation process) expressed an interest to amalgamate with these two councils. The 12th May 2016 *Local Government (Council Amalgamations) Proclamation 2016* that merged Canada Bay, Burwood and Strathfield councils was acknowledged. These overall chapter conclusions will be examined against results of assessing planning and development over all six council areas within Sydney Inner West District (Chaps. 3, 4, 5, 6, 7 and 8) and the district as a whole in Chap. 9. Results from the district assessment (and assessment of Parramatta and St George districts in later chapters) will contribute to assessing how to plan for a sustainable Greater Sydney and a sustainable city anywhere (Chap. 18).

8.7 Summary

The chapter addressed the extent Canada Bay Council was applying sustainable urban planning (SUP) practices (see Chap. 1) in planning the municipal area. It was noted that Canada Bay, in Sydney Inner West District, was merged with Burwood and Strathfield councils under the 12th May 2016 *Local Government (Council Amalgamations) Proclamation 2016* (Chap. 2). It was concluded that Canada Bay, one of twelve municipalities within three Greater Sydney Districts (Sydney Inner West, Greater Parramatta and St George) is considered to be in global transition. SUP practices (based on the application of three key SUP principles) were introduced, being: Developing Sustainable Urban Centers; Developing Sustainable Transport; and, Protecting the Natural Environment. A review of Canada Bay background and population changes (i.e. global movement of population) was outlined. In examining the pattern of urban development, four phases of Canada Bay development are examined: Early Canada Bay; Traditional Canada Bay; New Canada Bay; and, Future Canada Bay. With conclusions drawn, key Canada Bay Council planning documents were then critiqued to determine the extent Council applies SUP practices. In addition, given State planning of Greater Sydney affects Canada Bay, a second question was posed: what is the impact of Greater Sydney planning and development on Canada Bay?. In examining this question, examples of State plans (including State proposed council amalgamations) likely to impact on Canada Bay were addressed. By chapter's end, conclusions were drawn on Council planning and impact of Greater Sydney planning on Canada Bay. The chapter noted that these conclusions would be addressed further in: Chap. 9 (examining the extent of SUP practices being applied over the whole of Sydney Inner West District); and, Chap. 18 (examining directions for planning a sustainable Greater Sydney and a sustainable city anywhere).

Notation 8.1 Cultural Diversity Poem – Canada Bay

A poem reflecting the diversity of Canada Bay (Rhodes) follows:

Rhodes' New Order Modernism **(# 13)**

The smell of Berger Paints and chemical plants spread over the houses in 1950s–1970s
The last vestige of these industries belching fumes at Rhodes were snuffed out 1980s
The contamination of soil took years to sterilize to overcome the layers of neglect
The ground was too valued to sit, a master plan blueprint was stamped as new order

In 2015 Rhodes is the modernism come to Homebush Bay, Asia and International touch
The pattern of streets and high rise residential marches down every new paved street
The architecture of clean lines with a dash of green and turquoise, a path to the bay
The streets are empty, all at work to pay the mortgage, but they all love this new home

References

Australian Bureau of Statistics (ABS) (2011) Australian census review. ABS, Canberra

Canada Bay Council (2008) FuturesPlan20. NSW, Canada Bay

Canada Bay Council (2010a) State of environment report. NSW, Canada Bay

Canada Bay Council (2010b) Sustaining our environment program. NSW, Canada Bay

Canada Bay Council (2013a) Canada Bay local environment plan. NSW, Canada Bay

Canada Bay Council (2013b) Canada Bay development control plan. NSW, Canada Bay

Canada Bay Council (2014) Environmental sustainability strategy. NSW, Canada Bay

New South Wales (2008) FuturesPlan20. NSW, Sydney

New South Wales (2011) Sydney metropolitan strategy. NSW, Sydney

New South Wales (2014a) A plan for growing Sydney. NSW, Sydney

New South Wales (2014b) Parramatta Rd renewal project. NSW, Sydney

New South Wales (2015) Draft Parramatta Rd urban transformation strategy. NSW, Sydney

South Sydney Regional Organisation of Councils (2013) Renewable energy master plan. NSW, South Sydney

Chapter 9
Planning Sydney Inner West District, Greater Sydney

Dedication *Many citizens of Sydney Inner West are recognized here by way of dedicating this chapter to them (Plate 9.1).*

Plate 9.1 Past citizens of Sydney inner west *left to right* Maybanke Anderson; Augustus Alt; Henry Parkes; William Arnott; Eleanor Dark; and, Thomas Walker (*Source* Wikipedia and Australian Biographies 2016)

R.C. Rauscher, S. Momtaz, *Cities in Global Transition*,
DOI 10.1007/978-3-319-39865-5_9

Firstly, Maybanke Anderson (Marrickville) was a pioneer in women's suffrage, federation, and education reform in Australia. A historical snippet provides an insight into Anderson's legacy:

> Maybanke Anderson married Edmund Kay Wolstenholme, an Australian timber merchant, built a large house called 'Maybanke' in Marrickville. In 1885 Maybanke opened Maybanke School, a girls' school that she operated in her home preparing girls for the University of Sydney entrance examination. Operating for 10 years, the school was later known as Maybanke College. (Source: Dictionary of Australian Biography 2015)

Augustus Alt (Ashfield) was a surveyor in the new colony of Australia and at the end of his career built an estate home (still standing) in Ashfield (street now bears his name, Alt St.). A historical snippet provides an insight into Alt's legacy:

> Augustus Alt, in 1787, was appointed to the new colony of New South Wales as its first surveyor of lands and as a member of its commission for trial of pirates. He accompanied Governor Arthur Phillip in an excursion to establish settlement at the head of Parramatta River in November 1788. In February 1794 he was granted 100 acres (40 ha), Hermitage Farm, at Petersham (near today's Ashfield). (Source: Wikipedia 2015)

Sir Henry Parkes (Leichhardt) is regarded as the Father of the Australian Federation (1901). Henry lived in two homes in Leichhardt, Hampton Villa (1888–1892) at Balmain and Kenilworth (where he died in 1896) at Annandale. A historical snippet provides an insight into Parkes legacy:

> As the earliest advocate of a Federal Council of the colonies of Australia, a precursor to the Commonwealth of Australia, Sir Henry Parkes was the most prominent of the Australian Founding Fathers. Parkes was described during his lifetime by *The Times* as "the most commanding figure in Australian politics". *Alfred Deakin* described him as "though not rich or versatile, his personality was massive, durable and imposing, resting upon elementary qualities of human nature elevated by a strong mind". (Source: Wikipedia 2015)

William Arnott (Burwood) was a business person with a close association to the Burwood area. A historical snippet provides an insight into Arnott's legacy:

> William Arnott was an Australian biscuit manufacturer, founder of the Arnott's Biscuits Holdings. Shortly after his death, Arnott's sons spread out the business to other parts of the world, including East Asia and South Africa. (Source: Wikipedia 2015)

Eleanor Dark (Strathfield) was a writer who was born in Strathfield. A historical snippet provides an insight into Dark's legacy:

> Eleanor Dark (Strathfield) was an Australian author whose novels included *Prelude to Christopher* (1934) and *Return to Coolami* (1936), both winners of the Australian Literature Society Gold Medal for literature, and her best known work *The Timeless Land* (1941). (Source: Wikipedia 2015)

Thomas Walker (Canada Bay) was an Australian politician and banker with close association with the Canada Bay area. A historical snippet provides an insight into Walker's legacy:

> Thomas Walker was an Australian politician and banker; he and his daughter Dame Edith Walker were important philanthropists. Walker died in 1886 in Concord, New South Wales, and was buried in the cemetery at St John's Ashfield. He left a large fortune, and was survived by his daughter Edith. (Source: Wikipedia 2015)

Abstract Planning of Sydney Inner West District, the first of three case study districts of Greater Sydney, is examined in this chapter. The chapter acknowledges the *Local Government (Council Amalgamations) Proclamation 2016* proposed mergers and expanding the boundaries of nominated councils noted in Chap. 2. Having examined the question of global transition and planning practices of the six municipalities within the District, this chapter addresses: (1). planning the Sydney Inner West District as a whole; and, the extent the State (responsible for district planning) is meeting the planning issues. Key planning principles as adopted (Chap. 2) to apply to Greater Sydney (and this district) are applied to the district. These key principles include: (1). Developing Sustainable Urban Centres; (2). Developing Sustainable Transport; and, (3). Protecting the Natural Environment. Sample plans and development projects for Sydney Inner West are critiqued under each of these planning principles. After reaching a conclusion under each planning principle, an overall conclusion is drawn on the extent the State (planning the District) is meeting the planning issues of the District. Finally, it's noted that these Sydney Inner West District conclusions will be assessed alongside conclusions reached in examining Greater Parramatta District (Chaps. 10, 11, 12, 13) and St George District (Chaps. 14, 15, 16, 17). All these conclusions are examined (Chap. 18) in the context of SUP practices applied to Greater Sydney as a global city (affecting municipal areas within Sydney Inner West and the district as a whole).

9.1 Background to Sydney Inner West

The suburbs of Sydney Inner West are partly located along the southern shore of the Parramatta River and Homebush Bay and partly along the shores on the Cooks River. The Aboriginal clans lived for many thousands of years within the Sydney Inner West district (Chap. 2). Moving from the early Aboriginal land occupation of Greater Sydney, the geography of Sydney Inner West is summarized as follows:

> Sydney Inner West reflects rolling hills intersected by shallow valleys through which waterways including Iron Cove Creek, Hawthorne Canal (formerly Long Cove Creek), Whites Creek and Johnston's Creek flow. These waterways have been drastically altered since the

> late 19th century by the means of concrete lining. For example, in the 1860s Iron Cove Creek was a freely flowing waterway which in places broadened into ponds that made excellent and picturesque swimming holes. It is now barely a trickle lined by residential areas, parkland and fast-food restaurants. The suburbs within the region are characterised by medium to high-density housing and include some of the older constructions in Sydney such as the terrace houses of Glebe, Newtown and Annandale. Parts of the inner west have been subject to gentrification, particularly in Marrickville, Newtown and its surrounding areas. (*Source* Wikipedia 2015)

Looking at the Sydney Inner West district's early development, it is noted:

> The railway was opened from Sydney to Parramatta in 1855. There were four stations – Newtown, Ashfield, Burwood and Homebush. In 1874, Five Dock Station was opened and in 1876 it was renamed Croydon Station, supposedly because the suburb was about the same distance from the old Homebush racecourse as the London suburb of Croydon was from another racecourse. In the same year, Concord Station was opened, renamed Redmyre in 1877 and finally changed to Strathfield in 1885. In 1860, the number of trains daily to and from Sydney was six each way. One hundred years later, the number of trains daily was 255 to Sydney and 259 from Sydney. In 1860, 29,868 passengers made the trip to or from Sydney. In 1960, over two million passengers made the trip over the year. Two level crossings at Burwood Road were replaced by overhead bridges in 1892. The third bridge and set of lines were built in 1926. The opening of the railway was a momentous event for Burwood as it made the suburb easily accessible to the city and brought a number of wealthy merchants and industrialists who built spacious country houses. One of the oldest surviving members of this class of gentleman's residences is Henry Webb's 'Cicado' (1863), 74 Queen Street, Croydon. (*Source* Wikipedia 2015)

Sydney Inner West is noted by the Australian Bureau of Statistics as the Inner Western Sydney statistical region of City of Canada Bay, and municipalities of Ashfield, Burwood and Strathfield. The NSW Government's *Metropolitan Strategy* (NSW 2013a) adds an extra local government area to its Inner West subregion (that is the Municipality of Leichhardt). Also, the local newspaper, the Inner West Courier, adds the local government area of Marrickville to this district. The municipalities thus considered as Sydney Inner West for the purpose of this book include Marrickville, Ashfield, Leichhardt, Burwood, Strathfield and Canada Bay.

The common geography, early history (Aboriginal and European) and common population ties all reinforce Sydney Inner West as a district. A list of the suburbs for each municipality (six in all) of Sydney Inner West is contained in each respective chapter (Chaps. 3, 4, 5, 6, 7, 8). A brief historical background of Sydney Inner West is gleaned from historical photos (Plate 9.2). The opening of a viaduct (top left) for the western train line in the late 1880s shows the strength of public investment at that time. Cross district tram movements were popular in the early 1900s (top middle); Early road transport routes were popular as this Parramatta Rd scene shows (top right). Shipping docking bays and steam trains kept Balmain and Rozelle busy (bottom left). Parramatta Rd forms a critical link in the transport of Sydney (bottom middle). Finally, industry was scattered throughout the district (bottom right).

Plate 9.2 Early Sydney inner West. *Left to right* opening of viaduct for train line; Cross district tram; Early transport; Shipping docking bays and steam trains; Busy Parramatta Rd forms of transport; and, Industry scattered throughout the district (*Sources* Leichhardt and Marrickville Councils Archives Collections 2015)

The State designates the Sydney Inner West district within the Central Subregion (Plate 9.3) under *A Plan for Growing Sydney* (NSW 2014a). In focusing on the Inner West (to the left of 'Sydney' in this plate) note: (a). the 'strategic center' of Burwood and Rhodes (just north of Burwood); (b). WestConnex (black with white stripe) running top-left to center and from center to lower left (south); (c). the four blue bands east-west and north-south representing 'urban renewal corridors' of north, east, south and west parts of Greater Sydney; and, finally, (d). red crossed bottom-right is Sydney Airport Precinct and Port Botany.

With this historical and geographical background of the Sydney Inner West, the selected three key planning principles (Chap. 2) are now examined. These principles include: (1). Developing Sustainable Urban Centres; (2). Developing Sustainable Transport; and, (3). Protecting the Natural Environment.

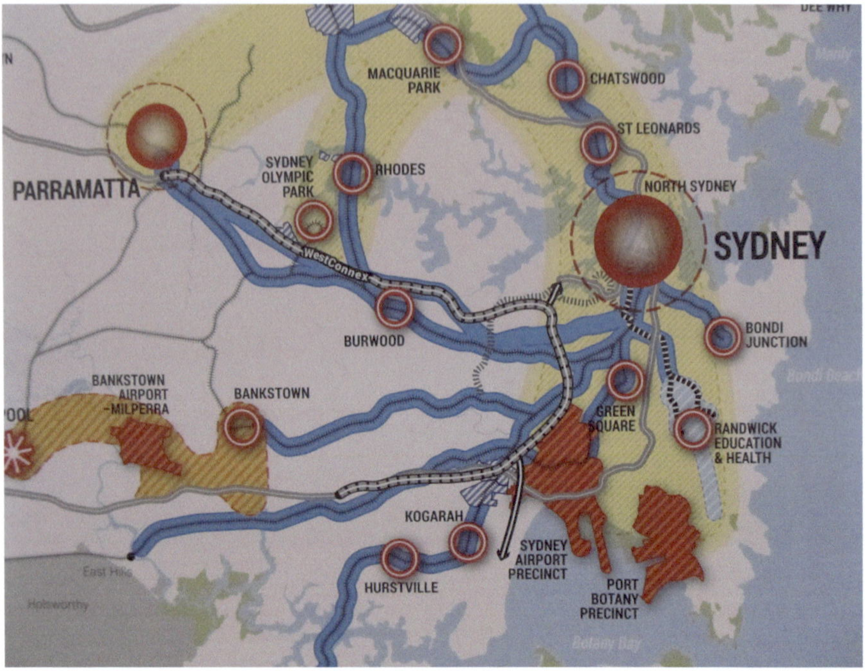

Plate 9.3 Greater Sydney major centres (*Source A Plan for Growing Sydney* NSW 2014a)

9.2 Developing Sustainable Urban Centers

The planning principal Developing Sustainable Urban Centers is the first principal to be examined. Case study examples include: (1) Bays Precinct Renewal; and (2) Burwood District Center; and, (3) Rhodes Growth Center.

9.2.1 Bays Precinct Renewal

The Bays Precinct includes the White Bay Power Station and is designated by the State to provide new homes and jobs and to maintain a working port. The Bays Plans in urban design (and transport planning) will affect Sydney Inner West. The Bays Precinct Taskforce, which was established by an earlier NSW Government, worked for a significant time and produced community plans and suggestions through a cross-agency program. The Bays Precinct is made up of the waterways and foreshores of Johnston's Bay, White Bay, Rozelle Bay and Blackwattle Bay. Just two kilometers west of the Sydney CBD, it is nestled within and around Pyrmont, Ultimo, Annandale, Glebe, Rozelle, Lilyfield and Balmain. Much of the land cannot be accessed by the public and as such it is not reaching its potential, and almost all of the land is owned by the NSW Government. The Bays Precinct in the context of Sydney CBD in illustrated in Plate 9.4.

Plate 9.4 Bays precinct planning in context of Sydney CBD. Top *left clockwise*: Sydney CBD; Rozelle Bay; Glebe Island; and Balmain at *left* (*Source* NSW Government Web 2016)

The State reports:

> Glebe Island, as Sydney Harbour's last deep-water port, feeds our city's demand for sugar, gypsum and cement. Glebe Island has transitioned from the former car import terminal to its current role as the temporary site of the Sydney Exhibition Centre. The silos greet Sydneysiders as they travel across the Anzac Bridge. The historic Glebe Island Bridge, with remediation, could be used once again. White Bay is used for Sydney's second cruise ship terminal, supporting around 1,500 jobs, more than 100 ships and 200,000 passengers each year. Sydney Fish Market is one of Sydney's biggest tourism drawcards and a busy hub of retail and wholesale suppliers. In addition, Rozelle Bay hosts a range of maritime industries including waterway maintenance, Heritage Fleet and the Sydney Superyacht Marina. Finally, land surrounding the White Bay Power Station and the Rozelle Rail Yards is generally unused, with the Power Station buildings empty and derelict. Nearby uses include mixed industries, some working harbour uses and transport connections. (*Source*: Urban Growth Web 2014)

Plate 9.5 illustrates the Bays Precinct Planning Details. An aerial view of Bays (left) shows its proximity to the Sydney CBD (in background). The extent of the Bays Precinct and its foreshores (right) provides challenging land and water planning.

The State outlines this project as follows:

> The Bays Precinct consists of approximately 80 hectares of Government owned land and sits within 2 km of the Sydney CBD. It includes sites such as the heritage-listed White Bay Power Station, Glebe Island, White Bay, Rozelle Bay, Rozelle Rail Yards and Blackwattle Bay, including the Sydney Fish Market. It is the story of Sydney's many parts, combining commerce, industry, leisure, housing and water activities. The Bays Precinct Urban Renewal Program will transform currently underutilized areas into a destination that will contribute significantly to the economic, cultural and social well- being of the city and the state. The Precinct's great heritage values, proximity to the city and waterfront position present exciting new opportunities for entertainment and leisure facilities, as well as residential, retail and commercial development. The Program will regenerate waterfront areas in and around Blackwattle Bay including the Sydney Fish Market, Rozelle Bay, Rozelle Rail Yards and the heritage-listed White Bay Power Station to provide a vibrant new harbor frontage and foreshore access. The Bays Precinct will be revitalized as a world-class, iconic waterfront destination, and will deliver vibrant and dynamic places for the city and the State. (*Source*: Urban Growth Web 2014)

Plate 9.5 Bays precinct planning details. *Left to right*: Aerial view of Bays with Sydney CBD in background; and, Aerial view of Bays precinct (*Source* NSW Government Web 2016)

Parts of The Bays Precinct are used for important maritime and port uses and the State says it is committed to maintaining a working port, noting sections of the port. From Sept 2015 the State started planning priority destinations for the project. In examining destinations within The Bays and transport, the State is proposing a walking scheme called 'the Seven Minute City' as follows:

> Planning for transport and mobility will be based on the idea of the seven-minute city – short distances between housing and workplaces and public transport, bike paths, walkways, water, public spaces, public buildings and commercial facilities. This requires us to think about transport connections, capacity, types of uses, and how we might create mixed use hubs in each Destination, so people don't have to travel far for day to day tasks. The kinds of transport will need to be consistent with the staged rollout of the priority Destinations – vital to support the ambition of job creation and economic development in The Bays Precinct. We will look at opportunities to increase walking and cycling and to make active and public transport so efficient that it is a first-choice option. Changing travel behaviour will have practical benefits such as fewer cars on the road, and social benefits such as healthier lifestyles, cleaner air, and safer, more attractive access between spaces and places. (*Source*: Urban Growth Web 2014)

In late October 2015 the State announced ideas for a Glebe Island-White Bay Tech Precinct (Plate 9.6).

Under the Tech Precinct concept, the White Bay Power Station would become, in the State plans, for headquarters of a global tech company or a campus for university or educational institution. The State announced it wanted to change Glebe Island from a working port to a 'technology innovation hub'. Also, under consideration was a light rail spur from Annandale to White Bay Cruise Ship terminal. The State compared the prospects to 'London's Tech City, New York's Roosevelt island

Plate 9.6 Bays precinct in context of inner west (*Source* NSW Government Web 2016)

Plate 9.7 Bays precinct concepts. *Left to right*: Artist's concept of a Glebe Island Tech Park; and potential for residential precinct (*Source* Daily Telegraph 22 Oct 2015)

and California's Silicon Valley. The State concluded that proposals would be considered into 2016, with construction expected, to begin in 2017. (Source: Daily Telegraph 22 Oct 2015). The concepts (in ideas submitted to the State) included (Plate 9.7) (left to right): a Glebe Island Tech Park; and, b Residential Precinct.

9.2.1.1 Conclusions

Having reviewed the Bays Precinct Project, there are a number of significant effects on council areas within the Sydney Inner West district that this project is likely to have. These can be summarized under a need for: a. more defined allocations of open spaces lands; greater housing affordability provisions; and, more environmental and heritage protection measures. Burwood District Center is examined next.

9.2.2 *Burwood District Center*

The State has worked with Burwood Council to provide capacity for additional mixed- use development in Burwood. This mix includes offices, retail, services and housing. The State is also investigating the potential for a light rail corridor from Parramatta to Sydney CBD via Parramatta Rd, including implications for Burwood. Examples of new development in Burwood reflecting this mix are illustrated in Plate 9.8 (left-right). A concept plan illustrates the levelling of new buildings, incorporating shops below, medium rise and high rise (top-left). A variation in bulk, facades and colour gives medium rise buildings a streetscape character (top-right). Early high rise residential buildings were built adjacent to the western railway (Sydney Central to Parramatta and Western Sydney) (middle-left). The effect of these buildings is overwhelming in scale, within inadequate setbacks. Burwood continues to develop Burwood's district centre commercial offices as illustrated here (middle-right). Medium rise office complexes (bottom-left) are located in streets adjacent to the shopping boulevard. Medium density 6–8 story residential buildings usually contain commercial shops below, convenient and assisting in creating a streetscape (bottom-right).

9.2.2.1 Conclusions on Burwood District Centre

Having reviewed aspects of the Burwood District Center proposal, there are a number of significant issues for councils within the Sydney Inner West district to consider. These can be summarized under needs to consider: a. impacts of the urban design aspects of projects (i.e. height, bulk, and overshadowing); impact of the additional population on services and transport; greater allocations of open spaces lands and pedestrian linkages; greater housing affordability provisions; and, more environmental and heritage protection measures. The Rhodes Urban Center is examined next.

9.2.3 *Rhodes Urban Center*

Urban design in most of the inner west is reaching higher investment levels and bigger projects (i.e. councils developing civic precincts). At the district level the State agencies responsible for developing plans for urban design is UrbanGrowthNSW. Canada Bay's expanding urban growth centre at Rhodes is illustrated in Plate 9.9 (left to right). The high rise residential buildings (top-left)

Plate 9.8 Burwood's expanding district centre (*Source* Ray Rauscher 2015)

Plate 9.9 Canada bay's expanding growth centre (*Source*: Ray Rauscher 2015)

dominate the skyline. Streetscapes (top-right) are generally paved with occasional building art to provide a visual variation. Most precincts within Rhodes blend high rise with medium rise (middle-left). Buildings are often positioned as focal points for roads and green spaces (middle-right). Medium density complexes are often built with shared entries and landscaping between buildings (bottom-left). Finally, prominent high rise buildings are often positioned at entry points into Rhodes (bottom-right).

9.2.3.1 Conclusions on Rhodes Urban Center

Having reviewed aspects of the Rhodes Urban Center proposal, there are a number of significant issues for councils within the Sydney Inner West district to consider. These can be summarized under needs to consider: a. impacts of the urban design aspects of projects (i.e. height, bulk, and overshadowing); impact of the additional population on services and transport; greater allocations of open spaces lands and pedestrian linkages; greater housing affordability provisions; and, more environmental and heritage protection measures.

9.2.3.2 Overall Conclusions on Developing Sustainable Urban Centers

Having reviewed a number of Sydney Inner West strategic centre's projects and author (Rauscher) qualitative research field work (2014–2015), it can be concluded the State needs to further address the impact of development under the principle Developing Sustainable Urban Centers (Chap. 2). The Sydney Inner West Councils (as concluded earlier in Chaps. 3, 4, 5, 6, 7, 8), were addressing: (1) these development trends under each council planning practices; and, (2) the State 'subregional guidelines' (Chap. 2). The strategic center projects (as reviewed above) have seen major advances made in recent years. Many questions, however, on planning at this level remain. In terms of meeting the principle Developing Sustainable Urban Centers within the Sydney Inner West District, it is concluded that Councils and the State need to consider new planning arrangements. Options for these arrangements are canvassed in Chap. 18. The next planning principal to be examined is Developing Sustainable Transport.

9.3 Developing Sustainable Transport

Having examined the first urban planning principal Developing Sustainable Urban Centers within Sydney Inner West, the second principal to be examined is D*eveloping Sustainable Transport*. With an extra 1.6 million people expected to be living in Sydney by 2036, there are many opportunities for transport to meet sustainability principles. All of the local government areas within Sydney Inner West have been

Information Box 9.1 Sydney Inner West Transport Key Documents

City of Sydney WestConnex Appraisal (2015)	http://clovermoore.com.au/westconnex/
GreenWay Active Transport Strategy & Action Plan (2009a)	www.marrickville.nsw.gov.au
NSW Light Rail Program (NSW 2014a)	http://www.transport.nsw.gov.au/lightrail-program
NSW Long Term Transport Master Plan (NSW 2013a)	www.transport.nsw.gov.au
NSW Draft Parramatta Road Transformation Strategy (NSW 2015)	www.transport.nsw.gov.au
Parramatta Rd Urban Renewal Strategy (NSW 2014b)	www.newparramattard.com.au
Sydney City Centre Access Strategy (NSW 2013b)	www.transport.nsw.gov.au
Western Sydney WestConnex Project (NSW 2014c)	www.westconnex.com.au

Source: Webs as noted February 2016

experiencing several decades of growth. As a result all the councils have been working to accommodate this growth and to apply the most effective transport provisions. Thus urban design in most of the council areas is reaching higher investment levels and bigger projects (i.e. councils developing civic precincts). At the district level the two State agencies responsible for developing plans for urban design and transport are UrbanGrowthNSW and the NSW Dept. of Transport and Maritime Services.

A number of transport projects either planned or underway affect Sydney Inner West. These include: (a) District Light Rail Development; (b) Parramatta Rd Corridor Concepts; and, (c) Westconnex Plans. Each of these is examined separately, though they link as a whole. Key transport documents to be addressed below are noted in Information Box 9.1 (a–z) for ease of accessing these documents.

9.3.1 Light Rail

Light Rail Development within Sydney Inner West is predominantly the responsibility of the NSW Dept. of Transport and Maritime Services (http://www.transport.nsw.gov.au/lightrail-program). The original goods line provided a sound infrastructure basis for the light rail (now running from Central Sydney CBD to Dulwich Hill). This goods line has a history going back to the 1800s and continues to operate in parts today (2016).

The operational sections of Sydney Metropolitan Goods Line is shown in Plate 9.10. This includes Dulwich Hill Station on Goods Line to Enfield and Rozelle; and, Marrickville Station on Goods Line going to Port Botany.

Sydney Light Rail Inner West extension as opened in 2014 (Plate 9.11). The extension was from Lilyfield station and ending at Dulwich Hill (GHD 2010).

Plate 9.10 Operational sections of Sydney metropolitan goods line. *Left to right* Dulwich Hill station, with goods line on right; and, Marrickville station with goods line to *left* (*Source* Wikipedia)

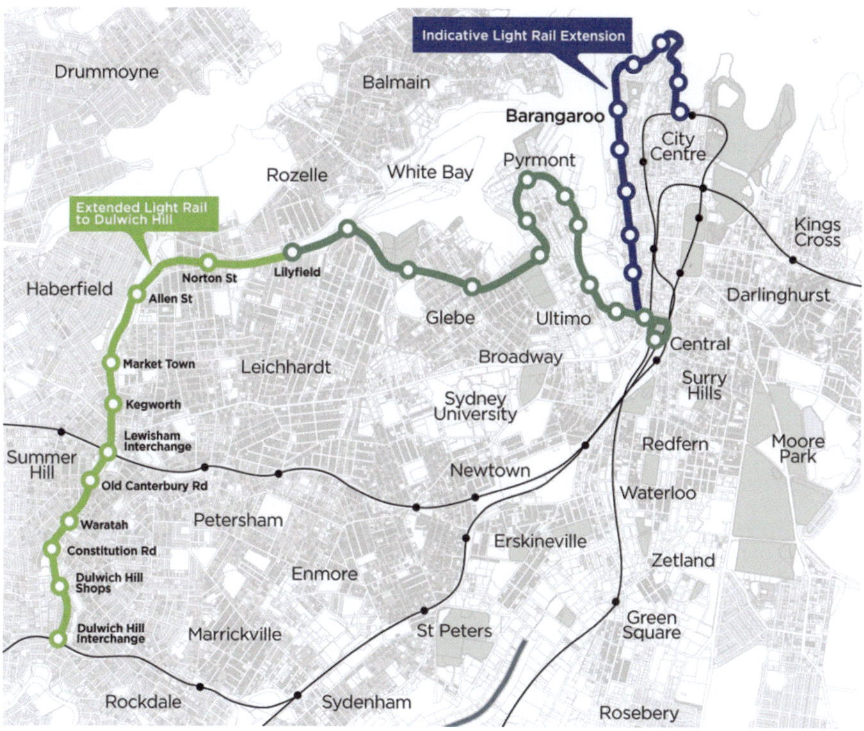

Plate 9.11 Light rail extensions to Dulwich Hill (light *green* on *left*) (*Source* GHD 2010)

A *GreenWay Active Transport Strategy & Action Plan* (Marrickville 2009a) was developed to encourage more people to take up healthier transport options such as walking and cycling in the GreenWay. The MPCS comments on the Plan as follows:

> The Plan sets out a range of short and long-term initiatives that encourage more people to use the GreenWay for walking and cycling and make it attractive and convenient for users of all ages. GreenWay Hubs are proposed as central places for GreenWay activity. During the development of the Strategy, the State Government announced its decision to defer the GreenWay shared pathway as part of the Inner West Light Rail Extension. The initiatives in this strategy can operate independently of the GreenWay shared pathway. They seek to enable a broad active transport culture, foster greater community engagement with the environment and increase patronage of the planned light rail extension. (Source: Marrickville 2009a, b)

The connection between the Inner West light rail (commented above under the GreenWay corridor) and the whole of the Sydney light rail planning is important for the Sydney Inner West District. A visual review of the Inner West light rail is provided in Plate 9.12. A pair of 'Urbos 2 s' operating here on the Dulwich Hill line (top-left). A remaining unused section of the heritage listed 1800s railway goods

Plate 9.12 Inner west light rail detail *left to right* Pair of 'Urbos 2 s'; historic 1800s railway goods line; Conversion of goods line on Dulwich Hill line; Light rail station platform; Light rail customer station lift; and, Approaching tram on Dulwich Hill line (*Source* Raymond Rauscher 2014 and NSW Transport Web 2016)

line at Summer Hill within GreenWay corridor here (top-middle). The conversion of Goods Line to the Dulwich Hill light rail line shown here (top-right). New light rail station platform opened in 2014 along Dulwich Hill line (bottom-left). The State provided a number of station lifts along the new Dulwich Hill line (bottom-middle). An approaching tram on the Dulwich Hill line is seen here (bottom-right).

By early 2015 there was considerable debate on a SIW public and local businesses call for an extension of a light rail from Strathfield to Parramatta. A coalition of big business and landowners called WestLine Partnership (WLP) released a report (Deloitte Access Economics 2015) in late February 2015 for a $2.9 billion light rail route. The project offers some changes to components of WestConnex (later in this chapter) that would affect properties and open spaces in SIW. The WLP argues for a light rail starting at Westmead to Parramatta, then to Olympic Park, Rhodes (Canada Bay Council) and Strathfield (Strathfield Council). The report shows that development in a 1.5 km corridor around the light rail line had the potential to create 35,000 jobs and homes for 70,000 people. The members of the WLP include ANZ Stadium, Australian Turf Club, Accor, Dexus, Sydney Olympic Park, Sydney Business Chamber and Auburn and Canada Bay councils. At the same time the State is investigating a route to Strathfield from Parramatta (called the 'Western Sydney Light Rail Line'). Here the State would use funds from the proposed privatization of the State's electricity assets in private sector development

9.3.1.1 Conclusions on Light Rail

Having reviewed the light rail planning proposals, there are a number of significant effects on council areas within the Sydney Inner West District that these proposals are likely to have. These can be summarized under a need for: a. increased opportunity for expanding the public transport system; stimulation of greater housing opportunities along the corridors of the light rail system; and, opportunities for expansion of connecting open space connectors.

9.3.2 Parramatta Rd and WestConnex

Looking at long term transport planning, the NSW government produced the *NSW Long Term Transport Master Plan* (NSW 2013a) as a strategy to address Sydney's transport challenges and to deliver an integrated and modern transport system. The Master Plan is supported by a series of integrated transport delivery plans and regional strategies which address specific needs for buses, heavy rail and light rail. The Government states:

> Over the next 25 years, the population of Greater Metropolitan Sydney is forecast to increase from 5.6 million to 7.4 million. By 2031, the number of trips made around the city each day will increase by 31 per cent, from 16 to 21 million trips. Over the next 20 years, nearly 150,000 workers will be located in the city center. These changes will place greater

Plate 9.13 Parramatta Rd Urban renewal project. *Left* to *right*: Granville to Camperdown (*Source* NSW 2014b)

> demands on our road, rail and bus networks. The existing transport system does not have sufficient capacity to accommodate the sheer volume of demand that converges on the CBD and inner Sydney on a daily basis. (Source: NSW 2013a, b)

9.3.2.1 Parramatta Rd

UrbanGrowth NSW released the Draft *Parramatta Road Urban Renewal Project* (NSW 2014b) (www.newparramattard.com.au) in late 2014, which aims to facilitate the revitalization of the Parramatta Road Corridor running across ten council areas from Parramatta to Sydney CBD, including Homebush. (Plate 9.13).

The State says:

> The Draft Strategy proposes renewal in precincts along Parramatta Road that they have identified as suitable for large increases in housing density. It identifies the Homebush Precinct for the highest proportion of high rise residential development in the corridor, with targets of 10,350 to 16,200 new dwellings or 33 percent of the overall proposed population growth for the Parramatta Rd Corridor project. UrbanGrowth NSW released the Draft Parramatta Road Urban Renewal Strategy in late 2014 (with the draft Strategy on public exhibition till mid-February 2015). The Strategy aims to facilitate the revitalization of the Parramatta Road corridor running across ten council areas from Parramatta to Sydney CBD, including Homebush. The Draft Strategy proposes renewal in precincts along Parramatta Road that they have identified as suitable for large increases in housing density. It identifies the Homebush Precinct for the highest proportion of high rise residential development in the corridor, with targets of 10,350 to 16,200 new dwellings or 33 percent of the overall proposed population growth. The State note that this project will bring new life to local communities and global Sydney. There will be investment in infrastructure, more housing and jobs, and better links within and between the diverse neighborhoods along the corridor. The Draft Parramatta Road Urban Renewal Strategy could result in significant high density residential development in the Homebush precinct and the other precincts identified for increased densities in the Inner West and West regions. This project will bring new life to local communities and global Sydney. There will be investment in infrastructure, more housing and jobs, and better links within and between the diverse neighbourhoods along the corridor. (*Source* NSW 2014a)

The *Draft Parramatta Road Urban Renewal Project* (NSW 2014b) has identified eight locations or Precincts, close to public transport and other community infrastructure, that are suitable for long term growth and change. Within Sydney Inner West these Precincts are located at Homebush, Burwood, Kings Bay (part of Five Dock), Taverners Hill, Leichhardt and Camperdown. The State notes that 'renewal in some areas may happen quite quickly, but renewal in other areas will happen over 20 years or more'. The *Draft Parramatta Road Urban Renewal Project* (NSW 2014b) includes an Integrated Land Use and Transport Concept. A key element of this concept is focusing growth and change where there is good access to public transport. The Draft Concept also suggests areas where transport linkages and active transport could be introduced or enhanced to support urban renewal. People living in revitalized places are encouraged to walk or cycle safely and conveniently to bus stops, rail stations, ferry wharfs and key destinations. In the long term the corridor has the potential to support the creation of 50,000 new dwellings and 50,000 jobs, and is forecast to contribute an additional $28 billion of investment, Urban Growth NSW notes. In examining where the new dwellings would go, The State notes: ' the Precincts have been chosen because they: (a) are close to public transport; (b) are near employment zones; (c) have capacity for housing growth; and, (d) have access to city infrastructure. The final version of this Parramatta Rd renewal was released as the *Parramatta Rd Urban Transformation Strategy* (NSW 2015) (www.newparramattard.com.au) and was adopted in late 2015. The Strategy sets a long term vision for the transformation of the 20-km Parramatta Road Corridor.

9.3.2.2 Conclusions on Parramatta Rd Urban Transformation

Having reviewed the Parramatta Rd urban transformation proposals, there are a number of significant effects on council areas within the Sydney Inner West district that these proposals are likely to have. These can be summarized under a need for: a. more details on the urban design aspects of projects (i.e. the height, bulk, overshadowing guidelines for new residential buildings within the redevelopment areas); impact of the additional population on services and current Parramatta Rd transport; integration of open spaces lands and pedestrian linkages; greater housing affordability provisions; and, details on environmental and heritage protection measures. The WestConnex project is examined next.

9.3.2.3 WestConnex Planning

The Parramatta Rd urban renewal project is integral to the planning of the NSW Government long term transport master planning of the WestConnex Motorway project.

The WestConnex motorway project, as illustrated, commences with Stage 1, Parramatta to Haberfield (left-top green) to middle-left). This project is a motorway M4 widening to Strathfield and an M4 East Motorway (dotted green) to Haberfield.

The next stage (dark blue dotted) will continue to Camperdown, St Peters and Sydney Airport. The final stage (light blue dotted) continues as the M5 Motorway from Sydney Airport to Bexley and Beverly Hills (St George district). The total project is estimated to cost of $11–$11.5 billion, with $200 million to "seed funding" for Urban Revitalization along the corridor. Incorporating the Parramatta Rd Corridor project. The Westconnex project www.westconnex.com.au is a 33 km motorway linking Sydney's west and south, being Australia's largest road transport project. While the WestConnex project will have the potential to significantly reduce regional traffic congestion on local roads, a number of Sydney Inner West councils object to several aspects of the proposal. Other councils (i.e. Canada Bay Council) are supportive, but want to minimize the impacts during the construction. To further the level of cooperation and strong partnerships, Local Government is included and represented at all levels of the project. Collaboration includes: a. Memorandums of Understanding (MOUs); Parramatta Road Mayors Group to provide recommendations and advice to State Government Ministers; Parramatta Road State/Local Partnering Groups to provide advice on the Draft Integrated Land Use and Transport Concept and other plans; and, an Integrated Project Team to prepare the plans that are submitted to government for approval.

The urban design of the identified Precincts (Plate 9.14) is central to projects such as WestConnex. This Precinct planning is critical as noted above (and addressed in each of the chapters on Sydney Inner West municipalities) (Chaps. 3, 4, 5, 6, 7, 8). The precincts are Granville (1); Auburn (2); Homebush (3); Burwood (4); Kings Bay (Five Dock) (5); Taverners Hill (6); Leichhardt (7); and Camperdown (8). Plans for each of the Precincts (1–8) were detailed with a timescale of 30 years for final renewal of Parramatta Rd.

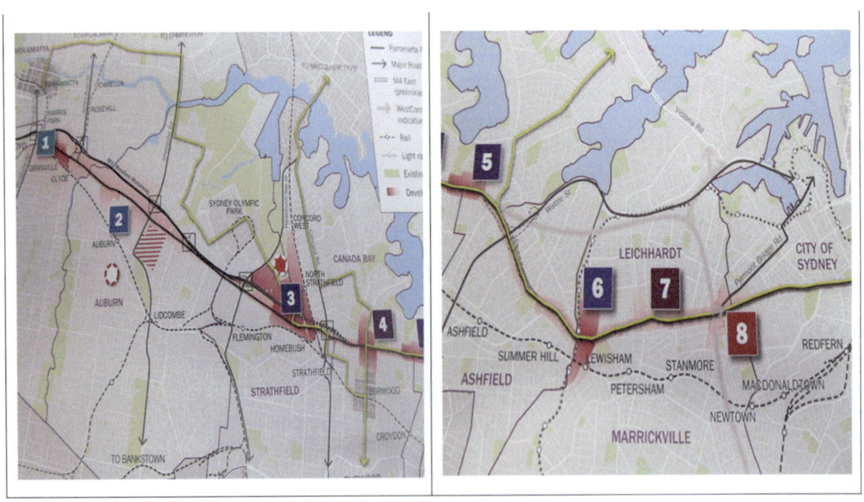

Plate 9.14 Parramatta Rd urban renewal project. *Left to right:* Precincts 1–4; and, precincts 5–8 (*Source* NSW 2014b)

9.3.2.4 Conclusions on WestConnex Proposals

Having reviewed the WestConnex proposals, there are a number of significant effects on council areas within the Sydney Inner West district that these proposals are likely to have. These can be summarized under a need for: a. more details on the cost-benefits of these proposals (especially given the size of project expenditures anticipated); more details on road vs. public transport alternatives; more information on and timing aspects of proposals; extent that the proposals will be a catalyst for further redevelopment projects; long term effect of potentially inducing additional car orientated demands; impact of the additional population on services and housing that the road proposals will have; greater integration of open spaces lands and pedestrian linkages; and, more environmental (i.e. not alienating public open spaces such as in the Sydney Park impact) and heritage protection measures.

9.3.2.5 Conclusions on Developing Sustainable Transport

Having reviewed a number of Sydney Inner West strategic centre's projects and author (Rauscher) qualitative research field work (2014–2015), it can be concluded the State needs to further address transport issues under Developing Sustainable Transport. The Sydney Inner West councils, it was concluded earlier in Chaps. 3, 4, 5, 6, 7, 8, were addressing: (1) transport trends within each council's planning practices; and, (2) the State 'subregional guidelines' (Chap. 2). The transport projects (as reviewed above) have seen major advances made in recent years, however many questions on transport planning at this level remain. In terms of meeting the State principle Developing Sustainable Transport within the Sydney Inner West District, it is concluded that new planning arrangements among Councils and the State need consideration. Options for these new planning arrangements are canvassed in Chap. 18. The next planning principal to be examined is Protecting the Natural Environment.

9.4 Protecting the Natural Environment

Having examined the first two urban planning principals (Developing Sustainable Urban Centers and Developing Sustainable Transport) within Sydney Inner West District, the third planning principal to be examined is Protecting the Natural Environment. Conservation protections within the Sydney Inner West District (municipalities of Marrickville, Ashfield, Leichhardt, Burwood, Strathfield and city of Canada Bay) has to date been a cooperative effort among the six councils and the State (refer to Chaps. 3, 4, 5, 6, 7, 8). At times, on a common planning basis, councils outside Sydney Inner West have become involved (i.e. where waterways or green spaces affect these other councils). Two conservation protection features of the district are selected to be examined, being: (1) the Cooks River; and, (2) Inner West Greenway.

9.4.1 Cooks River

An examination of planning and restoration of the prime waterway in the district, the Cooks River, is a district natural and recreation resource. Cooks River has a major catchment across the SIW and a number of other councils (16 councils in all). There are a number of links to projects that focus on the Cooks River as a natural and community resource as detailed in Information Box 9.2 (alphabetical).

The Cooks River Alliance (CRA) (http://cooksriver.org.au/resources/plans) is a partnership between eight councils bordering the Cooks River. The CRA focuses on the long-term improvement of the Cooks River catchment. It commenced in October 2011 and follows the successful Cooks River Sustainability Initiative (CRSI). The CRSI was a grant-funded project that focused on long-term improvement of the Cooks River catchment, with the grant finishing in 2011. The Cooks River Foreshore Working Group, a team of representatives from local councils and NSW State Government agencies making improvements to the health and amenity of the Cooks River system, completed its work also in 2011. The Marrickville Cooks River Committee was established to work with the community in improving the environment for the Cooks River and promote the value of the river's value as a natural system. The monitoring of progress in improving the Cooks River has been undertaken by the River Science project. This project is an ecological monitoring program conducted jointly between Marrickville, Kogarah, Rockdale and Canterbury Councils. The program aims to increase the scientific knowledge available to the community about the Cooks River. This program is jointly funded by Marrickville, Canterbury and Rockdale Councils with support from the Wolli Creek Preservation Society and the Department of Environment and Climate Change (DECC). In addition, the Cooks River Health Report Cards are used to monitor (since 2012) water quality, vegetation and macro-invertebrates providing an assessment of the Cooks River and Georges River (these rivers feed into Botany Bay). This project is facilitated by the Georges River Combined Councils' Committee with its nine member councils, the Cooks River Alliance's eight member councils, Streamwatch group, and the community. Finally, the *Cooks River Integrated Interpretation Strategy* has an ultimate aim of introducing research and education interpretation into the forward planning, management and conservation of the river.

An examination of the Cooks River (Plate 9.15) as part of the Cooks River Catchment area shows progress and challenges confronting the local councils and the State. There have been pedestrian platforms built along the river (top left). In

Information Box 9.2 Cooks River Waterway and Greenway Documents

Cooks River Alliance	http://cooksriver.org.au/
Cooks River Ecological Monitoring Program	http://www.marrickville.nsw.gov.au/en/environment/in-your-community/cooks-river/riverscience/
Cooks River Foreshore Working Group	http://www.marrickville.nsw.gov.au/en/environment/in-your-community/cooks-river/cooks-river-foreshore-working-group/
Cooks River Health Report Cards	http://www.georgesriver.org.au/River-Health-Report-Cards.html
Cooks River Integrated Interpretation Strategy	http://www.marrickville.nsw.gov.au/en/environment/in-your-community/cooks-river/cooks-river-integrated-interpretation-strategy/
Cooks River Sustainability Initiative	http://cooksriver.org.au/
Cooks River to Iron Cove GreenWay Master Plan and Coordination Strategy	http://cooksriver.org.au/
GreenWay	http://www.marrickville.nsw.gov.au/en/community/transport-and-infrastructure/greenway/
Marrickville Cooks River Committee	http://www.marrickville.nsw.gov.au/en/environment/in-your-community/cooks-river/marrickville-cooks-river-committee/
Our River - Cooks River Sustainability Initiative (CRSI)	http://www.marrickville.nsw.gov.au/en/environment/in-your-community/cooks-river/cooks-river-sustainability-initiative/
River Science	www.cooksriver.org.au

Source Webs as noted February 2016

addition, considerable signage has created defined recreation areas (top middle). An innovative project has been the provision of a river frontage cafe (top right). Also, along the river there are numerous examples of parkland up-gradings (bottom left). The State Government has contributed to river banks restoration (bottom center). Finally, improved grasslands are visible along the foreshores of the river (bottom right).

Plate 9.15 Cooks river restoration. *Left to right* Pedestrian platform; Signage; Cafe; Parkland upgrading; River banks restoration; and, grasslands with water authority viaduct in background (*Source* Raymond Rauscher 2015)

9.4.2 Inner West GreenWay

Having examined planning for the main water resource (Cooks River) in Sydney Inner West, green space planning examines the Inner West 'GreenWay' (Cooks River to Iron Cove). This greenway is the main urban space green corridor in the Inner West. The GreenWay follows the route of the Rozelle to Dulwich Hill light rail passenger service that opened in 2014 and connects two of Sydney's most important waterways, the Cooks River and Iron Cove Bay (part of Sydney Harbour). A *Cooks River to Iron Cove GreenWay Master Plan and Coordination Strategy* (Marrickville 2009b) (www.greenway.org.au) (herein referred to as the MPCS) was a joint effort of Marrickville, Ashfield and Leichhardt councils and adopted to guide the GreenWay project (Plate 9.16). The concept of the GreenWay is illustrated (left) showing (right to left): native vegetation; light rail track; walkway; and distant water way. The GreenWay (right) location incorporates four municipalities of Marrickville, Ashfield, Leichhardt and Canterbury.

The greenway corridor is 5 km long and extends from the Cooks River, Earlwood, in the south to Iron Cove in the north. The corridor passes through four local government areas (Marrickville, Ashfield, Leichhardt and Canterbury) following the

Plate 9.16 GreenWay corridor. *Left to right* Cooks river to Iron Cove greenWay; and, GreenWay location in relation to local government areas (*Source* Marrickville 2009a, b)

line of the disused Dulwich Hill light rail line. The district waterways and green spaces planning to date has seen major advances made throughout the district (as reviewed above). To move ahead further, the authorities (local, State and Federal) will need to develop the governance structures and programs to meet waterways and green spaces sustainability criteria. Questions on future planning and maintenance to be addressed include: (a) Type of governance arrangement overseeing the whole of the catchment of the Cooks River, Iron Cove and associated waterways will be required; (b) Role of community groups to look after sections of the Cooks River and other green spaces and foreshores of waterways; (c) Defined and regular State and Federal governments funding of green spaces and waterways will be needed on a rolling and long term commitment.

9.4.2.1 Conclusion

Having reviewed the Cooks River project and the Inner West Greenway proposals, there are a number of significant effects on council areas within the Sydney Inner West district that these proposals are likely to have. These can be summarized under a need for: a. wider State agreement to fund aspects of these projects, especially the Greenway as a concept; expanding the design aspects of further project stages; impact of the additional population on these projects as planned for the district under State plans; further integration of open spaces lands and pedestrian linkages; and, more environmental and heritage protection measures.

9.4.2.2 Conclusions on Protecting the Natural Environment

Having reviewed two major Sydney Inner West environment protection projects (i.e. Cooks River and Inner West Greenway) and author (Rauscher) qualitative research field work (2014–2015), conclusions can be drawn. The environment protection projects (as reviewed above) have seen major advances made in recent years, however many questions on environmental protection at this level remain. It is concluded the State, in partnership with councils, needs to further address natural environment protection. Options for these new planning arrangements are canvassed in Chap. 18 (Directions for Planning Sustainable Cities). The subject to amalgamations of councils and how this affects Sydney Inner West examined next.

9.5 Amalgamations of Councils and Inner West District

The subject of amalgamations of Greater Sydney councils (given implications for planning at municipal and Greater Sydney levels) was reviewed in Chap. 2. NSW Premier Mike Baird announced in Dec 2015, in examining the Sydney Inner West District, the State's preference: a. for Marrickville, Ashfield and Leichhardt councils to merge; and, in addition for Burwood, Canada Bay City, and Strathfield councils to merge. The position on the councils' merger question was reviewed in the respective chapter on each council. Each council adopted its preference on the merger question, some councils to stand alone (Marrickville, Ashfield, Leichhardt and Strathfield) and others to merge (Burwood and Canada Bay). All councils remained opposed to forced amalgamations. The Mayors of all councils urged residents to have their say on the issue of amalgamations of councils. Councils noted that submissions to the new Council Boundary Review were open till the end of February (www.councilboundaryreview.nsw.gov.au). In the interim, the Councils noted, the Minister for Local Government had commenced the process of forcibly amalgamating the two groups of councils in accordance to State preference as noted earlier. Councils noted the Minister said he expected the new merged councils to be 'proclaimed' on 1 July 2016 (note the 12th May 2016 *Local Government (Council Amalgamations) Proclamation 2016* noted in Chap. 2). Results of the review of merger proposals would be referred to the Chief Executive of the Office of Local Government for examination and report under the Local Government Act (the Act). In turn, the Chief Executive had appointed a Delegate for each of the proposed merged council groups. The Delegates (2) would examine the proposals, conduct public inquiry, call for written submissions, and then report back to the Minister on the two merger proposals.

9.5.1 Conclusion on Amalgamations

The State, in examining the Sydney Inner West District, expressed a preference: a. for Marrickville, Ashfield and Leichhardt councils to merge; and, in addition for Burwood, Canada Bay City, and Strathfield councils to merge. Four of the six councils (Marrickville, Ashfield Leichhardt and Strathfield) indicated to the State that they prefer to stand alone following research and public engagement. Only Burwood and Canada Bay City councils agreed to merge. At the end of the State review process (Council Boundary Review) the State would determine what amalgamations as proposed would proceed or not proceed. The 12th May 2016 *Local Government (Council Amalgamations) Proclamation 2016* proposed mergers and boundary changes to nominated Sydney Inner West councils was noted. This question will be further addressed in Chap. 18.

9.6 Overall Conclusions

Having completed an assessment of the development and planning of Sydney Inner West District (critiquing plans and developments against the adopted three planning principles), there are a number of overall conclusions reached. Firstly, examining conclusions reached in assessing development and planning in each of the six municipalities (Marrickville, Ashfield, Leichhardt, Burwood, Strathfield, and Canada Bay), it was concluded that each of these areas have been experiencing accelerated development and changes in population composition. Given these changes, and examining these councils' responses, it was also concluded these councils were all moving in the right direction in increasing their sustainable urban planning (SUP) practices. It was also concluded, however, that there would be significant future impacts on all these municipalities from Greater Sydney growth and development. In reviewing potential councils' and State responses to this future scenario, it was concluded that new partnerships between these councils and the State are needed. A major initiative from the State (2009 onwards) was to investigate local government reform. This included a potential amalgamation of councils (e.g. Marrickville, Ashfield and Leichhardt; and, Burwood, Strathfield and Canada Bay councils). Many of the councils had during this State amalgamation process, however, expressed an intent to stand alone and not amalgamate (Marrickville, Ashfield, Leichhardt, and Strathfield). Two other councils (Burwood and Canada Bay) agreed to amalgamate. The State started a forced amalgamation process in early 2016, results to date not determined. The 12th May 2016 *Local Government (Council Amalgamations) Proclamation 2016* proposed mergers and boundary changes of nominated Sydney Inner West councils was noted. From the municipal level to the district, the conclusions reached from the assessment of the planning and development of Sydney Inner West District as a whole will be examined next.

In examining the first planning principle, Developing Sustainable Urban Centers, three centers were examined: The Bays Precinct (Leichhardt); Burwood District Centre (Burwood); and, Rhodes Growth Center (Canada Bay). With The Bays Precinct planning, there were a number of significant effects this project is likely to have on some council areas within the Sydney Inner West. These can be summarized under a need for: more defined allocations of open spaces lands; greater housing affordability provisions; and, more environmental and heritage protection measures. Moving to the Burwood District Centre, there are a number of significant issues for the State and councils within the Sydney Inner West District to consider. These can be summarized under: impacts of the urban design aspects of projects (i.e. height, bulk, and overshadowing); impact of the additional population on services and transport; greater allocations of open spaces lands and pedestrian linkages; greater housing affordability provisions; and, more environmental and heritage protection measures. Moving to the Rhodes Growth Area (in Canada Bay), there are a number of significant issues for the State and councils within the Sydney Inner West district to consider. These can be summarized under needs to consider: impacts of the urban design aspects of projects (i.e. height, bulk, and overshadowing); impact of the additional population on services and transport; greater allocations of open spaces lands and pedestrian linkages; greater housing affordability provisions; and, more environmental and heritage protection measures. Having reviewed these urban centers and examined author's (Rauscher) qualitative research field work (2014–2015) (Chaps. 3, 4, 5, 6, 7, 8), it can be concluded the State, in partnership with councils, needs to further address the impact of development under the State planning principal Developing Sustainable Urban Centers (Chap. 2). The urban centres (as reviewed above) have seen major advances made in recent years. Many questions, however, of the State and Councils conferring on planning at this level remain. The planning principle Developing Sustainable Transport in Sydney Inner West is assessed next.

In examining the second planning principle, Developing Sustainable Transport, three transport projects within Sydney Inner West were assessed: light rail; Parramatta Rd renewal; and WestConnex. Having reviewed the light rail planning proposals, there are a number of significant benefits to the district that these proposals are likely to have. These can be summarized under a need for: increased opportunity for expanding the public transport system; stimulation of greater housing opportunities along the corridors of the light rail system; and, opportunities for expansion of connecting open space connectors. Moving to Parramatta Rd renewal, and having reviewed the Parramatta Rd urban transformation proposals, there are a number of significant effects within the Sydney Inner West District that these proposals are likely to have. These can be summarized under a need for: a. more details on the urban design aspects of projects (i.e. the height, bulk, overshadowing guidelines for new residential buildings within the redevelopment areas); impact of the additional population on services and current Parramatta Rd transport; integration of open spaces lands and pedestrian linkages; greater housing affordability provisions; and, details on environmental and heritage protection measures. Finally, having reviewed the WestConnex proposals, there are a number of significant effects within the Sydney Inner West district that these proposals are likely to have. These can be summarized under a need for: more details on the cost-benefits of these proposals (especially given the size of project expenditures anticipated); more details on road vs. public transport alternatives; more information on and timing aspects of proposals; extent that the proposals will be a catalyst for further redevelopment projects; long term effect of potentially inducing additional car orientated demands; impact of the additional population on services and housing that the road proposals will have; greater integration of open spaces lands and pedestrian linkages; and, more environmental (i.e. not alienating public open spaces such as in the Sydney Park impact) and heritage protection measures. The transport projects (as reviewed above) have seen major advances made in recent years, however many questions on transport planning at this level remain. Having reviewed these transport projects, and author (Rauscher) qualitative research field work (2014–2015), it can be concluded the State, in partnership with councils, needs to further address transport issues under the planning principal principle Developing Sustainable Transport. The results of assessing the planning principle Protecting the Natural Environment *was* examined next.

Having reviewed two major Sydney Inner West environment protection projects (i.e. Cooks River and Inner West Greenway) and author (Rauscher) qualitative research field work (2014–2015), conclusions were drawn. The environment protection projects (as reviewed above) have seen major advances made in recent years, however many questions on environmental protection at this level remain (as noted in the review). It is concluded the State, in partnership with councils, needs to further address natural environment protection.

Finally, having addressed the question of planning the Sydney Inner West District (critiquing plans and developments against the adopted three planning principles) an overall conclusion is drawn that new planning arrangements among Councils and the State (responsible for planning the District) are needed. This conclusions will be further addressed within Chap. 18 (Directions for Planning Sustainable Cities). Having examined the Sydney Inner West District, the Greater Parramatta District is examined next (Chaps. 10, 11, 12, 13).

9.7 Summary

Planning of Sydney Inner West District, the first of three case study districts of Greater Sydney, was examined in this chapter. Having examined the question of global transition and planning practices of the six municipalities within the District (Chaps. 3, 4, 5, 6, 7, 8), the chapter addressed: (1) planning the Sydney Inner West District as a whole; and, (2) the extent the State (responsible for district planning) was meeting the planning issues. Key planning principles as adopted (Chap. 2) to apply to Greater Sydney (and this district) were reviewed. These key principles include: (1) Developing Sustainable Urban Centres; (2) Developing Sustainable Transport; and, (3) Protecting the Natural Environment. Sample plans and developments within Sydney Inner West were critiqued under each of these principles. After reaching a conclusion under each principle, an overall conclusion was drawn that new planning arrangements among Councils and the State (planning the District) were needed in planning the District. Finally, it's noted that these Sydney Inner West District conclusions will be assessed alongside conclusions reached in examining Greater Parramatta District (Chaps. 10, 11, 12, 13) and St George District (Chaps. 14, 15, 16, 17). All these conclusions are examined (Chap. 18) in the context of SUP practices applied to Greater Sydney as a global city (affecting municipal areas within Sydney Inner West and the district as a whole). Having examined the Sydney Inner West District, the Greater Parramatta District is examined next (Chaps. 10, 11, 12, 13).

References

City of Sydney (2015) WestConnex appraisal. NSW, Sydney

Gutteridge Hastings and Davies (GHD) (2010) Dulwich hill light rail assessment. NSW, Sydney

Marrickville Council (2009a) A GreenWay active transport strategy & action plan. NSW, Marrickville

Marrickville Council (2009b) Cooks river to Iron Cove GreenWay master plan and coordination strategy. NSW, Marrickville

New South Wales (2013a) NSW long term transport master plan. NSW, Sydney

New South Wales (2013b) Sydney city centre access strategy. NSW, Sydney

New South Wales (2014a) A plan for growing Sydney. NSW, Sydney

New South Wales (2014b) Parramatta Road urban renewal project. Sydney, NSW

New South Wales (2014c) Western Sydney WestConnex project. NSW, Sydney

New South Wales (2014d) Light rail program. NSW, Sydney

New South Wales (2015) Draft Parramatta Road urban transformation strategy. NSW, Sydney

Part III
Planning Greater Parramatta District

Part I introduced the planning of global cities and sustainable urban planning (SUP) (Chap. 1). A methodology for measuring the extent SUP practices are being applied at municipal and district levels was then presented in Chap. 2. The basis of this methodology is the selection and application of three key SUP principles (herein called planning principles). These planning principles are: Developing Sustainable Urban Centers; Developing Sustainable Transport; and, Protecting the Natural Environment. With these planning principles in mind, three questions are posed for each municipality (in separate chapters: (1) what is the pattern of urban transition the municipality is experiencing; (2) what is the extent SUP practices are being applied by that municipal council; and, (3) what is the impact of Greater Sydney planning on the development of the municipality, including proposed council amalgamations. Part II involved the application of this methodology to the Sydney Inner West District (Chaps. 3,4, 5, 6, 7, 8 and 9). This Part III examines Greater Parramatta district cities (Parramatta, Auburn and Holroyd) (Chaps. 10, 11 and 12), followed by an examination of the whole of the district (Chap. 13). Results from this district assessment (and results from Sydney Inner West and St George assessments) will contribute to addressing directions for planning sustainable districts over the whole of Greater Sydney and in any city worldwide (Chap. 18).

Chapter 10
Planning in Greater Sydney – Parramatta City Case Study

Dedication *This chapter is dedicated to Dowell Philip O'Reilly (18 July 1865 – 5 November 1923) (Plate 10.1).*

Plate 10.1 Dedication: Parramatta – Dowell Philip O'Reilly (*Source* Wikipedia 2015)

O'Reilly was an Australian poet, short story writer and politician. A snippet on O'Reilly's life reveals the legacy he left:

> Dowell O'Reilly was educated at Sydney Grammar School, and when his father died he assisted his mother in keeping a preparatory school for boys at Parramatta. In 1884 O'Reilly published a small volume, *Australian Poems*, and in 1888 a larger volume of verse, *A Peddler's Pack*. In 1894 O'Reilly was elected a member of the Legislative Assembly for Parramatta and sat for four years. He moved the first motion in favor of women's suffrage carried in the New South Wales parliament. (Source: Wikipedia 2015)

Abstract The chapter addresses the extent Parramatta Council (herein referred to as Parramatta) is applying sustainable urban planning (SUP) practices (see Chap. 1) in planning the municipal area. The chapter acknowledges the *Local Government (Council Amalgamations) Proclamation 2016* proposed expanding the boundaries of Parramatta council noted in Chap. 2. Parramatta (in Greater Parramatta District) is one of twelve municipalities within three Greater Sydney Districts (Sydney Inner West, Greater Parramatta and St George), all considered to be in global transition

R.C. Rauscher, S. Momtaz, *Cities in Global Transition*,
DOI 10.1007/978-3-319-39865-5_10

(see Chap. 2). Examining SUP practices is based on the application of three key SUP principles (as adopted in Chap. 2), being: Developing Sustainable Urban Centers; Developing Sustainable Transport; and, Protecting the Natural Environment.

Firstly, a background to Parramatta development and population changes (e.g. global movement of population) is outlined. To examine the pattern of urban development, four phases of Parramatta development are examined: Early Parramatta; Traditional Parramatta; New Parramatta; and, Future Parramatta. Given conclusions reached here, key Parramatta Council planning documents are then critiqued to determine the extent Council applies SUP practices. In addition, given State planning of Greater Sydney affects Parramatta, a second question is posed: what is the impact of Greater Sydney planning and development on Parramatta? In examining this question, examples of State plans (including council amalgamations under the Proclamation noted above) likely to affect Parramatta are addressed. By chapter's end, conclusions are drawn on Council planning and impact of Greater Sydney planning on Parramatta. The chapter notes that these conclusions will be addressed further in: Chap. 13 (examining the extent of SUP practices being applied over the whole of Greater Parramatta District); and, Chap. 18 (examining directions for planning a sustainable Greater Sydney or a sustainable city anywhere).

10.1 Background and Population Change

Parramatta City (herein called Parramatta) in the context of Greater Parramatta (as introduced in Chap. 2) is adjacent to Sydney Inner West and west of the City of Sydney.

Under the *Local Government (Council Amalgamations) Proclamation 2016* Parramatta was proposed to expand its boundaries as noted in Chap. 2. The area now known as Parramatta was home to the Wangal people, inhabiting a large portion of the Greater Parramatta district. A background of Aboriginal occupation of Greater Parramatta is contained in Chap. 13 (examining the 'whole of the district'). The City of Parramatta is a demographically diverse community. Prior to European settlement, the Dharug Aboriginal people lived around and beyond the Greater Parramatta District (cities of Parramatta, Auburn and Holroyd). European settlement began in the 1790s and Parramatta Municipality was formed in 1861, eventually becoming a City in 1938. Parramatta City was amalgamated in 1948 with councils of Ermington, Rydalmere, Dundas, and Granville under Parramatta City. The population of Parramatta is 166,858 (Census 2011). Parramatta prides itself as one of the most multicultural communities in Australia. In respect to linguistically diverse communities (Census 2011) Parramatta City has a significantly higher than average proportion (52.5 %) of family members speaking two or more languages at home (national average is 20.4 %). Finally, the proportion of Parramatta City residents who stated a religious affiliation with Islam was in excess of four times the national average respectively, and Hinduism was in excess of six times.

Other important demographics of Parramatta include:

> Parramatta has an over-representation of young adults when compared to the country as a whole. Nearly half (45.1 %) of the suburb's population is aged 20–34 years, more than double the national average of 20.6 %. About one quarter (27.5 %) of residents were born in Australia, down from 42.4 % in the previous census five years earlier. The next most common countries of birth were India 21.5 %, China 14.7 %, South Korea 1.9 %, Philippines 1.8 % and New Zealand 1.7 %. (Source: Wikipedia 2015)

Parramatta City is 25 km (16 mi) from the Sydney CBD and embraces the suburbs of (note some suburbs shared with another LGA): Baulkham Hills (shared with The Hills Shire); Camellia; Carlingford (shared with Hornsby Shire and The Hills Shire); Clyde; Constitution Hill; Dundas; Dundas Valley; Eastwood (shared with Hornsby Shire and City of Ryde); Epping (shared with Hornsby Shire and City of Ryde); Ermington; Granville (shared with City of Holroyd); Guildford (shared with City of Fairfield and City of Holroyd); Harris Park (shared with City of Holroyd); Kingsdene; Merrylands (shared with City of Holroyd); North Parramatta (shared with The Hills Shire); Northmead (shared with The Hills Shire); Oatlands (shared with The Hills Shire), Old Toongabbie; Parramatta (shared with City of Holroyd); Pendle Hill (shared with City of Holroyd); Rosehill; Rydalmere; Sefton (shared with City of Bankstown); South Granville; Telopea; Toongabbie (shared with City of Blacktown and City of Holroyd); Winston Hills; Wentworthville (shared with City of Holroyd); and, Westmead (shared with City of Holroyd).

Looking closer at the development and infrastructure of Parramatta City:

(a) Parramatta, founded in the same year as Sydney by the British in 1788, is the oldest inland European settlement in Australia, the economic capital of Greater Western Sydney and the sixth largest central business district in Australia. Since 2000, Parramatta has seen the consolidation of its role as a government center with the relocation of agencies from the Sydney CBD. Simultaneously, major upgrades have occurred around the railway station with the expansion of Westfield Parramatta, the creation of a new transport interchange, and the ongoing development of the Parramatta Square local government precinct.

(b) Parramatta is a major business and commercial center, and the second largest CBD in the State of New South Wales. Parramatta has many high density commercial and residential developments. Church Street is home to many shops and restaurants. The northern end of Church Street, close to Lennox Bridge, features *al fresco* dining with a diverse range of cuisines.

(c) Parramatta Square is a proposed civic precinct located in the heart of the city, adjacent to Parramatta Town Hall. The proposal includes a redevelopment of the Parramatta Civic Centre, a culture and arts centre and a new plaza. The designs of the first two projects, a 65 story residential skyscraper and an office building were announced on 20 July 2012. Bicentennial Square, faces the 1883 Town Hall and St John's Cathedral.

(d) Parramatta railway station is a major transport interchange on the Sydney rail network. Parramatta is also serviced by a major bus interchange located on the southern side of the railway station. The Parramatta ferry wharf is at the Charles Street Weir, on the eastern boundary of the Central Business District. The wharf is the westernmost destination of the Sydney Ferries River Cat ferry service which runs on Parramatta River.

(e) Parramatta Road has always been an important thoroughfare for Sydney from its earliest days. From Parramatta the major western road for the state is the Great Western Highway The M4 Western Motorway, running parallel to the Great Western Highway has taken much of the traffic away from these roads, with entrance and exit ramps close to Parramatta. (Source: Wikipedia 2015)

Moving to population changes, the Australian Bureau of Statistics (ABS) comments on Parramatta City ancestry and languages spoken at home:

> At the 2011 Census, the Parramatta local government area was linguistically diverse, with a significantly higher than average proportion (52.5 %) where two or more languages are spoken (national average was 20.4 %); and a significantly lower proportion (43.4 %) where English only was spoken at home (national average was 76.8 %). The proportion of residents who stated a religious affiliation with Islam and Hinduism was in excess of four times and six times the national average respectively. About one quarter (27.5 %) of residents were born in Australia, down from 42.4 % in the previous census five years earlier. The next most common countries of birth were India 21.5 %, China 14.7 %, South Korea 1.9 %, Philippines 1.8 % and New Zealand 1.7 %. However, only 7.6 % identify their ancestry as Australian; the other common self-identified ancestries were Indian 19.2 %, Chinese 18.9 %, English 9.5 %, and Lebanese 2.9 %.

Finally, author's (Rauscher) qualitative research (2014/2015), there is a poem written to reflect the cultural diversity of Parramatta (end of chapter under Notation 10.1): *Parramatta's – A Global Hub* (#14). With this background introduction to Parramatta, the extent of Parramatta's development phases and populations change are examined next.

10.2 Phases of Development

An examination of the development phases of any locality provides a basis of understanding current trends and likely future directions of growth of that locality. Following a look at the background of Parramatta four phases of Parramatta development will be examined. These phases include: Early Parramatta; Traditional Parramatta; New Parramatta; and, Future Parramatta. To assist in addressing this question, author (Rauscher) qualitative research field work was completed in Parramatta Municipality (2014–2015). Early Parramatta (Plate 10.2) settlement was by Aboriginal clans of the Dharug (top-left). The early Kings School is shown here (top-right). Another significant school in the district was Newington School (middle-left). Parramatta housed a number of institutions, including the George St Asylum (middle-right). There was significant industry in the district, including Robert Ritchies Iron Works (bottom-left). Finally, the Morris' stables served the horse transport needs at the time (bottom-right).

Plate 10.2 Early Parramatta. *Left to right*: Aboriginal Clans; Kings School; Newington School; George St Asylum; Robert Ritchies Iron Works; and, Morris' stables (*Source* Terry Kass, *A Past Revealed*, 1996)

The transformation of Parramatta City downtown has been occurring over several decades (1980s onwards) (Plate 10.3). There have been a number of commercial developments such as this one opposite the rail-bus interchange (top-left). A popular water feature in Parramatta Square has recently been upgraded (top-right). Renewal of commercial buildings is matched with art works (middle-left). Expanded

Plate 10.3 Traditional Parramatta. *Left to right*: Commercial development opposite the Parramatta rail and interchange; Water feature in Parramatta Square; Renewal of commercial buildings and art work; Expanded footpath and paving; new office building; and, CBD car park (*Source* Raymond Rauscher 2015)

Plate 10.4 New Parramatta. *Left to right*: six story residential; High rise residential; Indoor shopping center (Westfield's); Railway station entry; New commercial building; and, Commercial connecting shops in railway precinct (*Source* Raymond Rauscher 2015)

footpath and paving attracts more pedestrians (middle-right). A new office building contains commercial shops (often eateries) (bottom-left). Finally, there are a number of car parks scattered around the CBD (bottom-right).

The new Parramatta City Directions (Plate 10.4) reflects a continuing change of the urban landscape. A section of the inner Parramatta City area is zoned for six story residential complexes such as this one (top-left). There has been considerable high rise residential (top-center) around the periphery of the CBD. The commercial draw of Parramatta downtown has increased significantly over the last two decades as illustrated here at the Westfield's building (top-right). The Parramatta railway-bus interchange is a major one serving Western Sydney, as shown here at the station entry (bottom-left). The growing commercial positioning of the CBD is illustrated here in this new commercial building (bottom-center). Finally, Parramatta has many connecting laneways and passageways connecting shops as shown in this railway precinct (bottom-right).

Finally, Future Parramatta (Plate 10.5) reflects the continuing building boom in residential and commercial development at increasing densities throughout the

Plate 10.5 Future Parramatta. *Left to right*: Current new and proposed buildings (*Source* Ray Rauscher and Parramatta Council Web 2016)

municipality. Commercial buildings (left) spread over the whole of the CBD. A new high rise residential on the Parramatta River downtown is under construction (middle). The southern side (over the rail line) of the Parramatta CBD will also see further commercial (right) and residential buildings. Factors of design, such as height, density, housing affordability, overshadowing and open space provisions, are likely to remain issues of development importance and contention within communities at the local government level into the future.

10.2.1 Conclusion

Given the review of development phases of Parramatta (above) and changes in population demographics, it can be concluded that Parramatta has been experiencing accelerated global transition (2000s) (Question 1). The planning of Parramatta in meeting development demands now and in the future is examined next. Particular attention is paid to Question 2 above, whether Parramatta Council planning is moving in the direction of sustainable urban planning (SUP) practices (Chaps. 1 and 2). The planning of Parramatta (proposed to have its boundaries expanded as noted above under the Proclamation) in meeting development demands and applying sustainable urban planning (SUP) practices is examined next.

10.3 Parramatta and Sustainable Urban Planning

Planning of Parramatta City has been taking place since it became a local government body (1861 as noted earlier). A review of planning in Parramatta will pay particular attention to applications of sustainable urban planning (SUP) (as outlined in Chap. 1) practices (including use of indicators of sustainability) (Chap. 2).

Information Box 10.1 Parramatta Municipality Planning

Design Parramatta (2012)	www.parracity.nsw.gov.au
Parramatta 2038 – Community Strategic Plan (2013)	www.parracity.nsw.gov.au
Parramatta Local Environment Plan (2011a)	www.parracity.nsw.gov.au
Parramatta Development Control Plan (2011b)	www.parracity.nsw.gov.au

Source Webs as noted February 2016

Parramatta Council key planning documents (a-z with date and web reference) are listed in Information Box 10.1. A summary of these documents and a brief critique assist with this review of Parramatta planning (documents can be sourced under the web link as noted in the right column).

10.3.1 Parramatta Local Environment Plan

The Parramatta Local Environment Plan (Parramatta City 2011a) *was adopted as the overriding plan for the City. Complimenting this plan is the Parramatta Development Control Plan* (Parramatta City 2011b).

10.3.2 Parramatta 2038 – Community Strategic Plan

The physical strategic planning focus of the Parramatta City Council is contained within *Parramatta 2038* (Parramatta 2013). The Plan also links to the long-term future of the Sydney Metropolitan Region. Components of that plan including environment; transport; and urban design aspects. Council, in its environmental objective, states that 'Parramatta will be an eco-efficient city that effectively manages and uses the City's growth to improve and protect the environment.' Council notes:

> Council recognizes that a rapidly growing population and increasing development in Parramatta requires a shifting focus on the environment. While protection of natural assets remains a priority, strategies to address the efficiency of our built environment and management of environmental risks are also vital to ensuring a prosperous and vibrant city. (*Source* Parramatta City Web 2015)

Council notes that the key strategies are:

1. Natural Environment:: To improve, protect and value our natural heritage and systems, including the extensive network of parks and bushland reserves; continue to protect biodiversity while improving connections between these areas and people; and focus on: a. waterways rehabilitation; b. biodiversity and bushland; management; c. local air quality; and, d. land and soil management

2. Built environment: To invest in and create a low-impact, eco-efficient urban environment that supports Parramatta as an area of significant jobs and residential growth; develop creative responses to improve the efficiency of the City, diversify the source of our resources and help manage increasing energy, water and waste costs; and focus on: a. efficiency and adaptability of precincts and buildings; and, b. growth in green jobs and services
3. Risks and resilience: To minimize and manage environmental risks; increase resilience; improve recovery times; and focus on: a. preparation for extreme weather events and/or other extreme events that disrupt food, water, energy or other resource supply; b. identification of risk and putting plans in place to better deal with events when they happen; and c. flooding risk. (*Source* Parramatta City 2013)

In addition to the planning documents noted above there are other planning projects the Council and other parties are engaged in. These include (not inclusive of all projects), for example: transport planning; and, design of Parramatta as examined below.

10.3.3 Transport Planning

With these strategies in place, on the subject of transport, Council notes:

> As the major transport hub in Greater Sydney, Parramatta's transport interchange already connects more than 50,000 people to Parramatta every day. Of these workers, 82 % live outside Parramatta with only 20 % using public transport to get to work. The NSW 2021 plan has specifically identified the aim to increase this share to 50 % by 2016. Parramatta City Council is proposing major road and transport infrastructure projects that will alleviate pressure on Sydney's road and rail networks, reduce traffic congestion and create more investment, employment and residential opportunities in Greater Sydney. One of these projects the Council is proposing is a Western Sydney Light Rail Network. Such a network could compliment the current Sydney Inner West light rail (and any extension). (*Source* Parramatta City Web 2015)

10.3.4 Design Parramatta

Having examined transport planning of Parramatta City another major planning project is urban design. The City of Parramatta outlines its approach to designing the City via a project team the Council has established, as follows:

> The Design Parramatta project aims to improve city amenity and character by creating quality visions for key streets, parks and public spaces and combining these visions into a city wide Public Domain Framework Plan. It is envisaged that as individual projects are implemented over the next decade, Parramatta will be progressively strengthened by a mosaic of exciting public space improvements that will combine to transform the city into a truly memorable place. The public space renewal will improve amenity and help attract residents and employees to drive the city's sustainable growth. (*Source* Parramatta City Web 2015)

Plate 10.6 Details of Parramatta city urban designs. *Left to right*: CBD shopping area; church St Mall; Clay Cliff Parklands; and, Wentworth Car park (*Source* Parramatta Council Web 2016)

There are a number of City design projects (Plate 10.6) being considered by Parramatta Council. Firstly, the CBD shopping streets would be upgraded (top left). Secondly, the Church St Mall (top right) would have more recreation space. The Clay Cliff Parklands would have both residential and open space components (bottom left). Finally, the Wentworth Car Park (bottom right) would be built with utilization of the rooftop.

Design Parramatta is a project for improving Parramatta City Centre's public domain by enhancing design quality in Parramatta. It was jointly coordinated by Parramatta Council in association with NSW Government (Architect's Office).

Commenting on the public spaces in the City, the Council notes:

> The public spaces of a city; its streets, parks, lanes and squares are collectively known as the public domain. These vital areas provide space for social interaction, cultural activity

and provide access to buildings and utilities. Many global cities including Paris, London, New York, Melbourne and Sydney CBD are engaging designers in public domain projects to improve their city's image, amenity, attractiveness and competitiveness. Currently much of Parramatta's public domain lacks amenity and a memorable image. The Design Parramatta project addresses these issues by creating quality visions for key streets, parks and public spaces and then combining these visions into a city wide Public Domain Framework Plan that will be implemented over the next 16 years. The Parramatta City Centre Public Domain Framework Plan 2012 will coordinate and combine individual projects into a unified city. Design Parramatta focuses on creating designs and working with designers from a range of disciplines to create a rich palette of projects and approaches and maximize creativity and innovation in Parramatta City Centre. (*Source* Parramatta City Web 2015)

10.3.5 Conclusion on Council Planning and SUP Practices

Having reviewed Parramatta City planning documents and their applications, it can be concluded that Council is moving in the direction of applying sustainable urban planning (SUP) practices (Question 2 above). The State's Greater Sydney planning (including infrastructure projects) and State council amalgamation proposals likely to affect Parramatta (Question 3) are reviewed next.

10.4 Greater Sydney Planning and Parramatta

There are planning and developments of Greater Sydney likely to impact on Parramatta City. In many instances Parramatta City Council has had the task of providing: planning input on State proposals affecting Parramatta. These include making and reviewing public submissions; implementing State planning policies at the local level; joint partnership on service or infrastructure projects; and, monitoring State or private development projects affecting Parramatta City. Other Council planning responsibilities under State legislation that also faces the Council include: engaging the municipality community (e.g. residents, business, services, development interests, and institutions) on State planning policies and development proposals; statutory plan making (e.g. adopting local environment plans (LEPs); and, reporting back to the State on a range of planning matters (i.e. State proposed roads, state of environment reporting and infrastructure projects).

Greater Sydney planning requirements were set by the NSW State in 2006, with Parramatta Council adopting the *Parramatta 2038 – Community Strategic Plan* (Parramatta City 2013). Council's plans had to be consistent with the *Sydney Metropolitan Strategy* (NSW 2011).

The *West Central Subregion Draft Subregional Strategy* (WCSDSS) (NSW 2012) is the overarching strategic document that outlines a vision for the West Central Subregion, which comprises Parramatta, Bankstown, Fairfield, Holroyd and

Auburn Local Government Areas (LGA). The State's *A Plan for Growing Sydney* (NSW 2014) is the strategic document that outlines a vision for Sydney to 2031 as reviewed in Chap. 2. The strategy outlines how the subregion will accommodate growth over 20 years to 2031.

The Plan forecasts a population growth of 1.6 m (to be added to the Greater Sydney population of 4.84 m at June 2014). In examining this Plan and its impact on Marrickville, three key planning principles as adopted in Chap. 2 are addressed: (1). Developing Sustainable Urban Centres; (2). Developing Sustainable Transport; and, (3). Protecting the Natural Environment. The principle Developing Sustainable Urban Centres is examined first. The Plan notes that Parramatta is to continue to be the second CBD of Greater Sydney. The State notes:

> The Plan will create lively new neighborhoods with access to local jobs and first-class local amenities by renewing the area between Greater Parramatta and the Olympic Peninsula. The Plan will continue to grow Parramatta CBD as a major center for jobs and world-class shopping and entertainment. Greater Parramatta will grow by integrating the precincts of Parramatta CBD, Westmead Health Precinct, Parramatta North, Rydalmere Education Precinct and Camellia, and encouraging growth across a range of employment types. Finally, the Plan supports an agglomeration of high value jobs in Marsden Park, supported by transport corridors to provide easy access.

One major project in Parramatta City is the *Parramatta North Urban Transformation Program* (Urban Growth NSW web) (www.planning.nsw.gov.au/parranorth). This Parramatta North land designated transformation is about 31 ha of State owned land. The State seeks "to transform the land it into a vibrant area that respects and preserves some of our most important heritage, whilst providing housing and employment opportunities for Parramatta and Greater Sydney." (*Source* Urban Growth NSW web Oct 2015)

The State explains the proposal:

> The original framework master plan comprised an area of 146 ha of NSW Government land across four precincts: Parramatta Park Precinct, the Cumberland Precinct, Old Kings School Precinct and the Sports and Leisure Precinct. However, the focus of the current proposal is on two precincts covering an area of 31 ha. The 'Land Subject to Rezoning Plan' shows the area that is subject to proposed changes to the planning controls. Parramatta Gaol was excluded from the project site as it was subject to a NSW Aboriginal Land Claim. This land claim was successful and ownership will be transferred to the Deerubbin Aboriginal Land Council. (NSW Dept. of Planning and Environment Council 2014)

The Heritage significance of the site is illustrated in Plate 10.7.

In summarizing the project the State concludes:

> The project proposes to deliver: creation of a new, vibrant hub in Parramatta North; restoration of heritage buildings and recognition of the social history associated with the site; approximately 3,900 new dwellings over the life of the project; around 2,600 new jobs over the life of the project, including during construction; protection of areas with high environmental value, such as the riverbank corridors; improved connections between the Parramatta CBD and Westmead; and, new or improved social, cultural and recreational infrastructure for the Parramatta community and Greater Sydney. (NSW Dept. of Planning and Environment Council 2014)

Plate 10.7 North Parramatta Heritage (Source: Raymond Rauscher 2015)

The planning principle Development Sustainable Transport is examined next. The Plan notes the importance of the transport connections between Parramatta and Sydney as highlighted in the strategy, including the WestConnex roads project from Sydney to Parramatta. Parramatta is also closely tied to the Parramatta Rd renewal project. The State also sees the North West Rail Link and WestConnex acting as catalysts to grow local centers and improve accessibility to employment opportunities. Increasing rail links around all renewal areas of Parramatta, as noted above, is an intention of the Plan. The rail link, Westconnex and the Parramatta Rd Renewal projects, as they affect Greater Parramatta District, will be addressed further in Chap. 13. The State notes:

> The Parramatta Road Corridor is 20 km long, strategically connecting the two largest concentrations of jobs in Sydney – Sydney CBD and Greater Parramatta. The construction of WestConnex will allow for significant improvements to local amenity by reducing through traffic on surface roads and allowing the enhanced north-south local connectivity. The Government will investigate the feasibility of light rail along Parramatta Road for the length of the corridor. (Source: NSW 2014)

In late 2015 the State released *The Draft Parramatta Rd Urban Transformation Strategy* (NSW 2015) (www.newparramattard.com.au). The Strategy identifies eight Precincts, specifically targeted for renewal under Precinct Plans. The State explains:

> Precincts were chosen due to their access to jobs, infrastructure, public transport, or how well they can support new housing types. The eight Precincts are places with unique character – diverse places that attract visitors and new residents and that have potential to be revitalized. Each Precinct is subject to a Precinct Implementation Plan, detailed on the following pages. These Plans include: a vision for the Precinct, why it has been identified as a growth Precinct, and what it will be like in the future – including population, housing and jobs projections three plans that illustrate how the vision will be delivered: A. Structure Plan that outlines the way land uses will be distributed and how the Precinct will change over time. B. Built Form Plan will show: the potential height of buildings; and C. Access and Movement Plan that details how people will move in and around the Precinct; and, how the vision can be delivered. (*Source NSW* Dept. of Planning and Environment 2015)

There are a number of 'precinct renewal' areas under the project, including Granville (#1) (Plate 10.8).

The State spells out more detail for the Granville Precinct:

> The Granville Precinct, close to Sydney's second CBD and supported by two rail stations, will enjoy better connections between its north and south, currently hindered by Parramatta Road and the rail line. New higher-density residential communities will be centered on open space and taller buildings will be located on a tree-lined Parramatta Road. These new developments will be supported by activity around the active main street of Good Street, a new urban plaza just off the main street, and more options for walking or accessing open space. Granville is a bustling multicultural town center with a mix of social, retail and dining areas to the south of the Station. The urban transformation will focus on the area to the north of the station and will make the most of the easy access to two rail stations and the Parramatta CBD. Close to Sydney's second CBD at Parramatta, Granville will be a vibrant mix of new housing, shops and commercial spaces, linked by a much improved network of streets and attractive new parks and public spaces. (*Source NSW* Dept. of Planning and Environment web 2016)

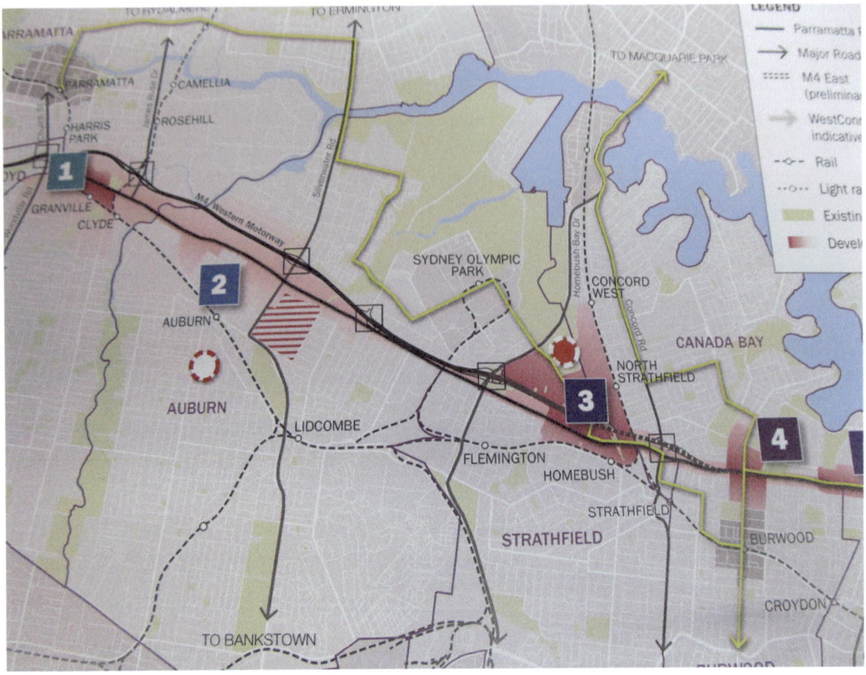

Plate 10.8 Parramatta Rd urban transformation strategy – Precinct 1 Granville (to *left*) (*Source* NSW 2015)

The planning principle Protecting the Natural Environment is examined next. The Plan commits the State to 'protect the health of the subregions waterways and aquatic habitats'. This goal, for example, would apply to the prime conservation challenges in Parramatta. This would include the protection of bushlands, waterways and foreshores within Parramatta. The efforts of Parramatta City in undertaking environment protection or the Parramatta River and its foreshores, for example, will be outlined in more detail in Chap. 13.

10.4.1 Conclusion on Greater Sydney Planning and Parramatta

Having reviewed above the State's Greater Sydney planning (including infrastructure projects) it can be concluded the implementation of these plans and developments is likely to have a significant impact on Parramatta City (Question 3). The extent of this impact will be better understood when these State plans are further examined in the context of the whole of the Greater Parramatta District (including Parramatta City) in Chap. 9. Finally, in examining State proposed council amalgamations, options for councils (including Parramatta) working

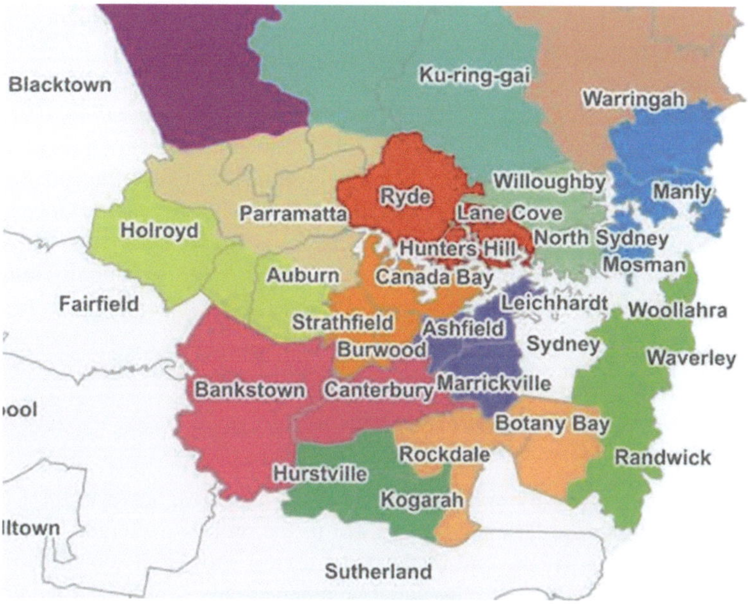

Plate 10.9 Amalgamations and Parramatta council proposed boundaries (*Source*: NSW Department of Local Government Web 2016)

together and in partnership with the State to meet local and district needs will be addressed in Chap. 18 (Planning Reforms and Directions for Planning Sustainable Cities). The subject to amalgamations of councils and how this affects Parramatta examined next.

10.5 Amalgamations of Councils and Parramatta

The subject of amalgamations of Greater Sydney councils (given implications for planning at municipal and Greater Sydney levels) was reviewed in Chap. 2. NSW Premier Mike Baird announced in Dec 2015 the State's preference for Parramatta City Council to have its boundaries changed (Plate 10.9) to absorb parts of Auburn (Olympic Park and Wentworth Point); Hornsby; and Baulkham Hills. The Council would lose the Granville area ward to a proposed merger of Auburn and Holroyd councils.

Parramatta Council did not object to the boundary changes overall, but did not want to lose the Granville area ward as proposed. Council also remained opposed to forced amalgamations. The Mayor of Parramatta urged all residents to have their say on the issue of amalgamations of councils. Council noted that submissions to the new Council Boundary Review were open till the end of February (www.councilboundaryreview.nsw.gov.au). In the interim, the Council noted, the Minister for

Local Government had commenced the process of forcibly amalgamating Auburn and Holroyd councils. Council also noted the Minister said he expected the new council (with adjusted boundaries) to be 'proclaimed' on 1 July 2016 (note the 12th May 2016 *Local Government (Council Amalgamations) Proclamation 2016* noted in Chap. 2). The results of the review of proposed boundary adjustments and merger (Auburn and Holroyd councils) would be referred to the Chief Executive of the Office of Local Government for examination and report under the Local Government Act (the Act). In turn, the Chief Executive had appointed a Delegate who would examine the proposals, conduct a public inquiry, call for written submissions, and then report back to the Minister.

10.5.1 Conclusion on Amalgamations

Parramatta Council appears to accept the boundary changes, however the Council did not want to lose the Granville area ward to the proposed merged councils of Auburn and Holroyd. At the end of the State review process (Council Boundary Review) the State would determine what amalgamations as proposed would proceed or not proceed. The 12th May 2016 *Local Government (Council Amalgamations) Proclamation 2016* proposed a merger of Rockdale and Botany councils was noted. This question will be further addressed in Chap. 18.

10.6 Overall Conclusions

Having completed an assessment of the development and planning of Parramatta, there are a number of overall conclusions reached. In examining the patterns of urban development of Parramatta it was concluded that Parramatta has been experiencing accelerated development and changes to the composition of its population. Given these changes, and examining Parramatta Council's response, it was also concluded Council is moving in the right direction in increasing its sustainable urban planning (SUP) practices. It was also concluded, however, that there would be significant future impacts on Parramatta from Greater Sydney growth and development. In reviewing potential Council and State responses to this future scenario, it was concluded that new partnerships between the Council and the State would be needed. A major initiative from the State (2009 onwards) was to investigate local government reform, including a potential amalgamation of councils (e.g. Parramatta boundary changes; and merging of Auburn and Holroyd councils). The 12th May 2016 *Local Government (Council Amalgamations) Proclamation 2016* proposed expanded boundaries of Parramatta Council was acknowledged. These overall chapter conclusions will be examined against results of assessing planning and development of the remainder of the three cities of Greater Parramatta (Chaps. 10, 11, 12) and the district as a whole (Chap. 13). Results from the district assessment (and assessment of Sydney Inner West and St George districts in later chapters) will contribute to assessing how to plan for a sustainable Greater Sydney and a sustainable city anywhere (Chap. 18).

10.7 Summary

The chapter addressed the extent Rockdale Council was applying sustainable urban planning (SUP) practices (see Chap. 1) in planning the municipal area. It was noted that Parramatta, in the Greater Parramatta District, was proposed to have its boundaries expanded under the 12th May 2016 *Local Government (Council Amalgamations) Proclamation 2016* (Chap. 2). It was concluded that Parramatta, one of twelve municipalities within three Greater Sydney Districts (Sydney Inner West, Greater Parramatta and St George) is considered to be in global transition. SUP practices (based on the application of three key SUP principles) were introduced, being: Developing Sustainable Urban Centers; Developing Sustainable Transport; and, Protecting the Natural Environment. A review of Parramatta background and population changes (i.e. global movement of population) was outlined. In examining the pattern of urban development, four phases of Parramatta development are examined: Early Parramatta; Traditional Parramatta; New Parramatta; and, Future Parramatta. With conclusions drawn, key Parramatta Council planning documents were then critiqued to determine the extent Council applies SUP practices. In addition, given State planning of Greater Sydney affects Parramatta, a second question was posed: what is the impact of Greater Sydney planning and development on Parramatta?. In examining this question, examples of State plans (including State proposed council amalgamations) likely to impact on Parramatta were addressed. By chapter's end, conclusions were drawn on Council planning and impact of Greater Sydney planning on Parramatta. The chapter noted that these conclusions would be addressed further in: Chap. 13 (examining the extent of SUP practices being applied over the whole of Greater Parramatta District); and, Chap. 18 (examining directions for planning a sustainable Greater Sydney and a sustainable city anywhere).

Notation 10.1 Cultural Diversity Poem – Parramatta Poem (Rauscher 2015 unpublished) reflecting the diversity of Parramatta:

***Parramatta's – A Global Hub** (# 14)*

The creation of Parramatta from 1700s pioneer center to a new second city to Sydney
A center of strip shops, in 1940s of smelling diesel on the edge of Sydney's western reach
The cosmopolitan Parramatta today sports most languages of the world, numbering 300+
A stretch of Parramatta River under Lenox Bridge sees strollers of Church St, cosmopolitan hub

It was the 70's plans of PM Whitlam, Uren, and Jack Mundey who laid foundations
Wise planners and architects redrew the streetscapes, cars were rerouted
Now a shimmering Novotel and reflective steel and glass buildings symbolize Parramatta
An oasis of the west as Sydney's geographic center settles in this new global hub

References

Australian Bureau of Statistics (ABS) (2011) Australian census review. ABS, Canberra
Kass T (1996) A past revealed. Parramatta City Council, NSW
New South Wales (2011) Sydney metropolitan strategy. NSW, Sydney
New South Wales (2012) West central subregion draft subregional strategy. Sydney
New South Wales (2014) A plan for growing Sydney. NSW, Sydney
New South Wales (2015) draft Parramatta Rd urban transformation strategy. NSW, Sydney
Parramatta City Council (2011a) Parramatta local environment plan. Parramatta, NSW
Parramatta City Council (2011b) Parramatta development control plan. NSW, Parramatta
Parramatta City Council (2012) Design Parramatta. NSW, Parramatta
Parramatta City Council (2013) Parramatta 2038 – community strategic plan. NSW, Parramatta

Chapter 11
Planning Auburn, Greater Sydney

Dedication *This chapter is dedicated to John Thomas Lang (1876–1975) (known as 'Jack Lang') was an Australian politician serving as Premier of NSW for two terms (Plate 11.1).*

Plate 11.1 Dedication: Auburn – Jack Lang (*Source* Wikipedia 2015)

A snippet on Jack Lang provides an insight into his legacy:

> During his first term as Premier, Lang carried out many social programs, including state pensions for widowed mothers with dependent children under fourteen, a universal and mandatory system of workers' compensation for death, illness and injury incurred on the job, funded by premiums levied on employers, the abolition of student fees in state-run high schools and improvements to various welfare schemes such as child endowment (which Lang's government had introduced). Lang spent his long retirement editing his newspaper *The Century*, and wrote several books about his political life, including *The Great Bust*, *I Remember* and *The Turbulent Years*. (*Source* Wikipedia 2015)

R.C. Rauscher, S. Momtaz, *Cities in Global Transition*,
DOI 10.1007/978-3-319-39865-5_11

Abstract The chapter addresses the extent Auburn Council (herein referred to as Auburn) is applying sustainable urban planning (SUP) practices (see Chap. 1) in planning the municipal area. The chapter acknowledges the *Local Government (Council Amalgamations) Proclamation 2016* merging Auburn with Holroyd council noted in Chap. 2). Auburn (in Greater Parramatta District) is one of twelve municipalities within three Greater Sydney Districts (Sydney Inner West, Greater Parramatta and St George), all considered to be in global transition (see Chap. 2). Examining SUP practices is based on the application of three key SUP principles (as adopted in Chap. 2), being: Developing Sustainable Urban Centers; Developing Sustainable Transport; and, Protecting the Natural Environment.

Firstly, a background to Parramatta development and population changes (e.g. global movement of population) is outlined. To examine the pattern of urban development, four phases of Auburn development are examined: Early Auburn; Traditional Auburn; New Auburn; and, Future Auburn. Given conclusions reached here, key Auburn Council planning documents are then critiqued to determine the extent Council applies SUP practices. In addition, given State planning of Greater Sydney affects Auburn, a second question is posed: what is the impact of Greater Sydney planning and development on Auburn? In examining this question, examples of State plans (including council amalgamations under the Proclamation noted above) likely to affect Auburn are addressed. By chapter's end, conclusions are drawn on Council planning and impact of Greater Sydney planning on Auburn. The chapter notes that these conclusions will be addressed further in: Chap. 13 (examining the extent of SUP practices being applied over the whole of Greater Parramatta District); and, Chap. 18 (examining directions for planning a sustainable Greater Sydney or a sustainable city anywhere).

11.1 Background and Population Change

Auburn City (herein called Auburn) in the context of Greater Parramatta (as introduced in Chap. 2) is south of Parramatta and adjacent to Sydney Inner West to the west of the City of Sydney. Under the *Local Government (Council Amalgamations) Proclamation 2016* Auburn was merged with Holroyd council as noted in Chap. 2. The suburbs of Auburn include: Auburn, Berala, Lidcombe, Newington, Regents Park, Rookwood, Silverwater, Sydney Olympic Park, and Wentworth Point. With this background introduction to Auburn, the extent of Auburn's development phases and changing demographics are examined next. The area now known as Auburn was home to the Wangal people, inhabiting a large portion of the Greater Parramatta District (Chap. 2). The City of Auburn is a local government area in the Greater Western Sydney region of New South Wales, Australia. The area under administration is located about 15 km (9.3 mi) west of the Sydney central business district. The suburb was named after Oliver Goldsmith's poem *The Deserted Village*, which describes 'Auburn' in England as the *loveliest village of the plain.* Auburn is culturally very diverse. Suburbs of Auburn City include: Auburn, Berala, Lidcombe,

Newington, Regents Park, Rookwood, Silverwater, Sydney Olympic Park, and Wentworth Point. A look at the development of Auburn is a starting point:

> Auburn has a mixture of residential, commercial and industrial areas. A commercial area is located close to Auburn railway station. There are many multicultural restaurants and cafes. South of the railway station, the commercial area stretches for approximately 1 km and features many Middle Eastern & Asian shops, restaurants and supermarkets. This has made Auburn a focal point for various immigrant groups in Sydney, especially for significant events such as football matches or political developments in the Middle East. North of the railway station, commercial and industrial developments are also located along the length of Parramatta Road and surrounding streets. The Sri Mandir in Auburn is Australia's oldest Hindu temple, having opened in 1977. The Auburn Gallipoli Mosque took thirteen years to construct and was largely funded by the Turkish community in the area. The name of the mosque reflects the legacy of Gallipoli in Turkey and the shared bond between Australian society and the Australian Turkish Muslim Community who constructed the mosque. (*Source* Wikipedia 2015)

The Auburn area was a farming area, known as "Liberty Plains". Auburn was proclaimed as a Borough in 1892 and achieved city status in 2009. Auburn City's population at 2011 Census was 73,738. Auburn City also had higher that the national average population growths over the 2003–2011 period. At the 2011 Census, Auburn City proportion of residents who stated their ancestry as Chinese, or as Lebanese, was in excess of six times the national average. The proportion of residents who stated an affiliation with Islam was in excess of 11 times the national average. Meanwhile, the Census noted, the area was also (as with Parramatta) linguistically diverse. Arabic, Cantonese, Mandarin, Turkish, and Korean languages spoken in households ranged from five times to 22 times the national averages. The proportion of residents who stated a religious affiliation with Islam excess of four times the national average and Hinduism was in excess of six times. Finally, the traditionally Anglo-Celtic European population has slowly been replaced by a high percentage of immigrants from Turkish, Lebanese and Vietnamese backgrounds. Auburn also has fast growing smaller populations of Somali, Bosnian, Iraqi, Iranian, Afghan, Pakistani, and Sudanese communities. With this background introduction to Auburn Municipality, the extent of Auburn's development phases and changing population are examined next.

Moving to population changes the Australian Bureau of Statistics comments on Auburn City ancestry and languages spoken at home:

> Auburn prides itself as one of the most multicultural communities in Australia. The traditionally Anglo-Celtic European population has slowly been replaced by a high percentage of immigrants from Turkish, Lebanese and Vietnamese backgrounds. Auburn also has fast growing Somali, Bosnian, Iraqi, Iranian, Afghan, Pakistani, and Sudanese communities. At the 2011 Census, the proportion of residents in the Auburn local government area who stated their ancestry as Chinese, or as Lebanese, was in excess of six times the national average. The proportion of residents who stated an affiliation with Islam was in excess of eleven times the national average. Meanwhile, as at the Census date, the area was linguistically diverse, with Arabic, Cantonese, Mandarin, Turkish, and Korean languages spoken in households, and ranged from five times to 22 times the national averages. According to the 2011 census of Population, there were 33,122 residents in Auburn. 31.9 % of residents were born in Australia. The most common other countries of birth were China 13.3 %, Turkey

6.2 %, India 5.0 % and Lebanon 4.0 %. 13.5 % of residents spoke only English at home. Other languages spoken at home included Arabic 15.7 %, Turkish 11.5 %, Mandarin 10.3 % and Cantonese 8.1 %. The most common responses for religious affiliation were Islam 42.0 %, Catholic 12.6 % and No Religion 12.3 %. (Source ABS 2011)

Finally, in author's (Rauscher) qualitative research (2014/2015), there is a poem written to reflect the cultural diversity of Auburn (end of chapter under Notation 11.1): *Tripoli Rises in Auburn*. Phases of Auburn development are examined next.

11.2 Phases of Development

An examination of the development phases of any locality provides a basis of understanding current trends and likely future directions of growth of that locality. Following a look at the background of Auburn four phases of Burwood development will be examined. These phases include: Early Auburn; Traditional Auburn; New Auburn; and, Future Auburn. To assist in addressing this question, author (Rauscher) qualitative research field work was completed in Auburn City (2014–2015). Early Auburn City is reviewed in Plate 11.2. Auburn Town Hall is shown as it was in 1892 (top left). The community of Auburn used the Auburn Hall (top right) from 1902 onwards. The area has always been subjected to heavy commercial transport movements, as illustrated here (middle-left) in 1920; Operating out of Mortuary Station (middle right) Rookwood 1869, is a departing train. The Auburn Rail Stn. is seen here in 1920 (bottom left). Finally, the decorations for the Auburn Jubilee 1928 is shown here (bottom-right).

A survey around Auburn City centre (in Auburn suburb) reveals significant growth (Plate 11.3). The Gallipoli Mosque (top left) is a focal attraction over a large portion of western Sydney. The City Council chambers are housed in a modern building with the library in one of the wings (top-left). The traditional values in Auburn City are reflected in this war memorial within a memorial park in the Auburn CBD (middle left). A portion of a City Center road has been reclaimed for a children's soft fall area (middle-right). A new commercial below and residential above is illustrated here (bottom-left). The Chinese commercial community occupy a strip opposite Auburn rail interchange (bottom-right).

A new suburb of Auburn City is Sydney Olympic Park**,** a large sporting, cultural and leisure complex. The area was redeveloped for the 2000 Olympics. The facilities built continue to be used for sporting and cultural events, including the Sydney Royal Easter Show and, the Sydney Festival. and a number of world-class sporting fixtures. The suburb also contains commercial development and extensive parklands. The area was originally part of the suburb of Homebush Bay, but was designated a suburb in its own right in 2009.

Examining the New Auburn (Plate 11.4) gives an indication of the extent that the CBD is growing (Chapter explores the planning of the CBD later). There are a number of welfare Islamic service centers in Auburn (top-left). Several office complexes have been developed in downtown Auburn (top-center). A large new educational

Plate 11.2 Early Auburn. *Left to right* Town Hall 1892; Auburn Hall 1902; Heavy commercial transport c.1920; Mortuary Station, Rookwood 1869; Auburn Rail Stn. 1920; Auburn Jubilee 1928 (*Source: Liberty Plains,* Auburn Council 1982)

Plate 11.3 Traditional Auburn. *Left to right*: Gallipoli Mosque; Auburn Council Administration Building; War Memorial in Memorial Park; Soft Fall area in reclaimed footpath; Commercial below and residential above; and, colorful Chinese sector (*Source*: Raymond Rauscher 2015)

Plate 11.4 New Auburn. *Left to right*: Community Islamic Centre; New office complex; New educational center; High rise around a plaza; Staggered high rise; and, Pedestrian connector to CBD (*Source:* Raymond Rauscher 2015)

Plate 11.5 Future Auburn. *Left to right*: Development and proposed developments (*Source*: Auburn Council Web 2016)

center occupies a CBD site (top-right). The Auburn plaza has commercial outlets on most of its circumference (bottom-left). The high rise residential complexes, often with shops below, continue to be built at Wentworth Point (bottom-center). Finally, a new pedestrian connector links the Council administration area to the shops (bottom-right).

Finally, Future Auburn (Plate 11.5) reflects the continuing building boom in residential development at increasing densities throughout the municipality. Building in growth areas such as Wentworth Point (left) is likely to see whole streets built as a precinct. The opportunities for future recreation innovation within building design is shown here (centre) within an archery area at Sydney Olympic Park. In additional to residential future structures the demand for tourist accommodation (right) is likely to continue as illustrated here at Sydney Olympic Park. These building design factors (such as height, density, housing affordability, overshadowing and open space provisions) are likely to remain issues of development importance and contention within communities at the local government and State levels into the future.

A glimpse at Auburn of old and new is captured in the poem *Tripoli Rises in Auburn* (Appendix 1 #15).

11.2.1 Conclusion

Given the review of development phases of Auburn (above) and changes in population demographics, it can be concluded that Auburn has been experiencing accelerated global transition (2000s) (Question 1). The planning of Auburn (merged with Holroyd council as noted above under the Proclamation) in meeting development demands and applying sustainable urban planning (SUP) practices is examined next.

11.3 Auburn and Sustainable Urban Planning

Planning of Auburn City has been taking place since it became a local government body in 1861. A review of planning in Auburn will pay particular attention to applications of sustainable urban planning (SUP) (as outlined in Chap. 1) practices (including use of indicators of sustainability) (Chap. 2). Auburn Council key planning documents (a-z with date and web reference) are listed in Information Box 11.1. A summary of these documents and a brief critique assist with this review of Parramatta planning (documents can be sourced under the web link as noted in the right column).

Information Box 11.1 Auburn Municipality Planning

Auburn Local Environment Plan (2010a)	www.auburn.nsw.gov.au
Auburn Development Control Plan (2010b)	www.auburn.nsw.gov.au
Auburn Town Centre Strategy Plan 2031 (2009)	www.auburn.nsw.gov.au
Auburn City Community Strategic Plan 2013–2023 (2013)	www.auburn.nsw.gov.au

Source Webs as noted February 2016

11.3.1 Auburn Local Environment Plan

The Auburn Local Environment Plan (Auburn 2010a) *was adopted in 2010. This plan is supplemented by the Auburn Development Control Plan* (Auburn 2010b).

11.3.2 Auburn Town Center Strategy Plan 2031

Auburn planning is built around the *Auburn Town Center Strategy Plan 2031* (Auburn Council 2009). Council notes within the Strategy:

> This Strategy provides the planning context for future development in the town centre. It establishes a vision and coordinated directions addressing a range of planning, community, transport, economic and environment issues to guide short, medium and long term strategic planning policies. The Auburn Town Centre Strategy is guided by urban design principles. Urban design is a multidisciplinary process which involves the design, development and application of physical, regulatory, financial, organizational, cultural, environmental and promotional strategies to achieve the common design objective of the making of place. This design includes several "place-making" policies, such as Public Domain Plan, Civic Improvement Program, updated Infrastructure Manual, Outdoor Dining Policies and a Facade Upgrade Program.

Council continues:

> An active, community oriented and economically thriving town centre competing as a sub-regional centre, with well-designed buildings, attractive streetscapes, parks and public spaces that reflect the urban vitality of its culturally diverse population. Six Strategic Themes will guide the short and long term action plan that will enable Council and the community to achieve our Vision for the Auburn town centre, including: urban structure; built form and character, revitalization; circulation, transport and parking, public domain and culture; and, sustainable economic development and employment. (*Source* Auburn Council 2009)

Examples of these key actions are contained in Plate 11.6 below.

Council further notes, in examining the Auburn Town Centre:

> The Auburn town center is located at the heart of what is an exceptionally rich area – in culture, language and ethnic diversity. It is the largest centre in the Auburn Council area and is located at a major railway station and transport node. The centre presently has a distinctive Middle Eastern and Chinese character. This distinctive character provides the centre

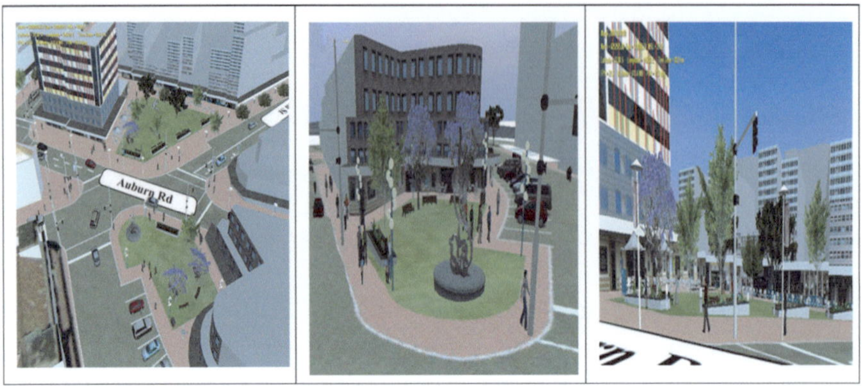

Plate 11.6 Auburn town center upgrade. *Left to right:* A 5-ways intersection; Civic plaza; and, Auburn Village Park (*Source* Auburn City Council Web 2016)

> with a strong identity and distinguishes it from other centers in Auburn and within the sub-region. It provides an opportunity for interpretative design in architecture, streetscape and associated urban elements, public domain and public art, lighting and signage. This Strategy provides the planning context for future development in the town centre. It establishes a vision and coordinated directions addressing a range of planning, community, transport, economic and environmental issues to guide short, medium and long term strategic planning policies. The strategic planning approach is especially relevant to the Auburn town centre which is poised to accommodate high density residential and mixed use developments. (*Source* Auburn Council web 2016)

11.3.3 Auburn Town Centre Heritage Review

The *Auburn Town Centre Heritage Review* (Auburn Council 2007) found that there is a representative group of buildings and places that reflect the urban form, human endeavor and pattern of settlement of the area which is intrinsic to the cultural significance of Auburn, including: early twntieth Century retail development on South Parade and Rawson Street; Railway Park; unique street and subdivision pattern; and, remnant late nineteenth Century/early twentieth Century buildings along Auburn Road and South Parade.

11.3.4 Sustainable Development

Finally, examining its policies on sustainable development the Council notes:

> Auburn Council is committed to promoting sustainable development. A number of sustainable development benchmarks are currently in use across Australia for various building types. Most of these schemes are voluntary. However, there is potential to incorporate sustainable development requirements to meet relevant benchmarks within built form planning controls via the Draft Auburn DCP 2009. New residential development within the town

centre must meet the requirements under the State Environmental Planning Policy (Building Sustainability Index: BASIX) 2004. The application of BASIX is also mandatory for residential alterations and additions. There are two main sustainable development benchmarks for assessment of commercial office building across Australia. There is the opportunity to require a particular rating level under one of these benchmarks within built form planning controls for the town centre. These are the Australian Building Greenhouse Rating Scheme and the Green Building Greenhouse Rating Scheme. Further investigative work will be done at the development of the new DCP controls to determine the appropriateness of these benchmarks, or any other form of sustainable development control. Council must be mindful of achieving a balance between sustainable built form while promoting development investment and future commercial/retail growth. There is no benchmark sustainable tool yet for retail development. (*Source* Auburn Council 2009)

11.3.5 Conclusion on Council Planning and SUP Practices

Having reviewed Auburn City Council planning documents and their applications, it can be concluded that Council is moving in the direction of applying sustainable urban planning (SUP) practices (Question 2 above). The State's Greater Sydney planning (including infrastructure projects) and State council amalgamation proposals likely to affect Auburn (Question 3) are reviewed next.

11.4 Greater Sydney Planning and Auburn

There are planning and developments of Greater Sydney likely to impact on Auburn City. In many instances Auburn City Council has had the task of providing: planning input on State proposals affecting Auburn. These include making and reviewing public submissions; implementing State planning policies at the local level; joint partnership on service or infrastructure projects; and, monitoring State or private development projects affecting Auburn City. Other Council planning responsibilities under State legislation that also faces the Council include: engaging the municipality community (e.g. residents, business, services, development interests, and institutions) on State planning policies and development proposals; statutory plan making (e.g. adopting local environment plans (LEPs)); and, reporting back to the State on a range of planning matters (i.e. State proposed roads, state of environment reporting and infrastructure projects).

Greater Sydney planning requirements were set by the NSW State in 2006, with Auburn City Council adopting the Auburn City Community Strategic Plan 2013–2023 (Auburn Council 2013). Council's plans had to be consistent with the *Sydney Metropolitan Strategy* (NSW 2011). Overview of strategic planning of Auburn is determined by the State Government. In planning for Auburn, the NSW Government notes:

> Whilst the West Central Subregion Draft Subregional Strategy (WCSDSS) identifies a growth of 12,000 jobs for the LGA for 2031, the Department of Planning has noted that the 12,000 jobs are solely for the Olympic Park-Rhodes Strategic Centre. The WCSDSS forecasts an additional 17,000 new dwellings. While 6000 dwellings are allocated to the Olympic Park – Rhodes Strategic Centre, the remaining 11,000 dwellings are targeted for the rest of the Auburn LGA. (*Source* NSW 2011)

The State in 2014 adopted its strategy for planning Greater Sydney to 2031 (*A Plan for Growing Sydney* (NSW 2014a)). The Plan forecasts a population growth of 1.6 m (to be added to the Greater Sydney population of 4.84 m at June 2014). In examining this Plan and its impact on Auburn, three key planning principles as adopted in Chap. 2 are addressed: (1) Developing Sustainable Urban Centres; (2) Developing Sustainable Transport; and, (3) Protecting the Natural Environment. The planning principle *Developing* Sustainable Urban Centres is examined first. The Plan notes that Olympic Park (within Auburn City) will continue as a growth area (within Auburn City). The Plan indicates:

> The State will work with council to provide capacity for additional mixed-use development in Sydney Olympic Park including offices, retail, services and new homes. The State will facilitate delivery of Carter Street Priority Precinct, Lidcombe Priority Precinct and Wentworth Point Priority Precinct. The State will also work with council to improve walking and cycling connections from Sydney Olympic Park train station towards Newington, Wentworth Point, Concord West train station and Lidcombe train station. Finally, the State will investigate a potential light rail corridor from Parramatta CBD to Sydney Olympic Park. (*Source Plan for Growing Sydney* 2014)

The State goes further under the Sydney CBD to Parramatta – Strategic Transport Plan (Transport NSW 2015a)

> A Plan for Growing Sydney identifies the Greater Parramatta to Olympic Peninsula priority growth area, which includes Greater Parramatta along with precincts at Silverwater, Carter Street, Sydney Olympic Park, Wentworth Point and Rhodes. The area will be targeted for mixed use development and new supporting transport infrastructure. In addition to these initiatives, the Strategic Transport Plan is also focused on the role of Parramatta as Sydney's second CBD, and the possibility of up to 100,000 jobs (from around 78,000 jobs now) in Greater Parramatta over the next 20 years, as outlined in A Plan for Growing Sydney. Other major transport investments are being considered. These proposals will all be subject to standard government funding considerations including business cases and assurance reviews: A. Four light rail routes linking to Parramatta CBD are under investigation – one could link Parramatta CBD with Strathfield and Burwood via Sydney Olympic Park. B. Public transport scenarios for the section of Parramatta Road between Burwood and the Sydney CBD are also being investigated. C. A range of Parramatta River ferry initiatives, including 80 extra weekly services, a new wharf at Rhodes, new river ferries and a wharf upgrade program are being progressed.

Another project in the Auburn City area is The Wentworth Point Urban Activation Precinct (UAP) (Plate 11.7).

This precinct (18.6 ha) is located at the northern end of Wentworth Point, Homebush. The mixed use precinct is surrounded by Sydney Olympic Park to the west, Parramatta River to the north, Homebush Bay to the east and privately owned land to the south. The State proposes:

Plate 11.7 Wentworth point urban activation precinct. (*Source*: NSW Urban Growth Web 2016)

> Two high density residential neighborhoods, home to approximately 2300 apartments, a 3.9 ha Peninsula Park and three additional pocket parks, new primary school, pedestrian and cycle promenade around the foreshore of both Parramatta River and Homebush Bay and maritime facilities that may include dry boat storage, rowing/kayaking facilities. There is demand for people to live in this area due to the easily accessible ferry transport links and employment, public space and recreation facilities at nearby Sydney Olympic Park. UrbanGrowth NSW is acting in the role of development manager on behalf of Roads and Maritime Services. Our focus is on obtaining the required planning approvals, undertaking enabling infrastructure major works and coordinating the sale to the private sector to facilitate housing supply. This is a Parramatta River foreshore project adjoining Sydney Olympic Park, featuring a vibrant residential and mixed use development, a high quality Peninsula Park and a dedicated Maritime precinct. The rezoning for the Wentworth Point UAP was announced by the Minister for Planning in June 2014. (*Source* NSW Urban Growth 2016)

Planning principle Developing Sustainable Transport is examined next. Auburn City Council, along with other Greater Parramatta and Sydney Inner West councils, were asked in late 2014 to comment on the *Parramatta Rd Renewal Project* (NSW 2014b) and the proposed WestConnex Project. In late 2015 the State released *The Draft Parramatta Rd Urban Transformation Strategy* (NSW 2015a, b) (www.newparramattard.com.au). Both the Westconnex and the Parramatta Rd Renewal projects, as they affect Auburn and the Greater Parramatta District, will be addressed further in Chap. 13. There are a number of 'precinct renewal' areas under the project, including the Auburn Precinct (Plate 11.8).

The Strategy identifies eight Precincts, specifically targeted for renewal under Precinct Plans. The State explains:

> Auburn is a major employment area on Parramatta Road and a true mixed use area. As well as industrial, commercial and retail uses, Auburn is home to the Auburn Botanic Gardens on the edge of the Duck River and many residential areas housing a diverse population, 60 per cent of whom were born overseas. Taking advantage of its location close to major employment areas such as Parramatta and Sydney Olympic Park, Auburn can be a location

Plate 11.8 *Draft Parramatta* Rd Urban transformation strategy – Precinct 2 Auburn (*Source*: NSW 2015a, b)

> for significant employment growth, supported by moderate scale residential development and an improved streetscape. Auburn is an ideal place for new jobs across industrial, commercial and retail sectors, taking advantage of its access to the transport network as well as connections to Sydney Olympic Park and Parramatta. A new creative and residential hub will evolve around the junction of Parramatta Road, St Hilliers Road and the M4. The majority of new housing will be between Parramatta Road and Auburn Station around Auburn Park. New street trees and landscaping will help to green Parramatta Road, as will links between Haslams Creek and Duck River to Parramatta River. (*Source* WestConnex 2015)

Finally, planning principle Protecting the Natural Environment is examined next. The Plan commits the State to 'protect the health of the subregions waterways and aquatic habitats'. This goal, for example, would apply to the prime conservation challenges in Auburn. This would include the protection of bushlands, waterways and foreshores within Auburn. The efforts of Auburn Council in protecting waterways and foreshores of the Parramatta River and Homebush Bay in cooperation with other councils in Greater Parramatta will be outlined in more detail in Chap. 13.

11.4.1 Conclusion on Auburn and Greater Sydney Planning

Having reviewed above the State's Greater Sydney planning (including infrastructure projects) it can be concluded the implementation of these plans and developments is likely to have a significant impact on Auburn (Question 3). The extent of this impact will be better understood when these State plans are further examined in the context of the whole of the Greater Parramatta District (including Auburn) in Chap. 13. Finally, in examining State proposed council amalgamations, options for councils (including Auburn) working together and in partnership with the State to meet local and district needs will be addressed in Chap. 18 (Planning Reforms and Directions for Planning Sustainable Cities). The subject to amalgamations of councils and how this affects Auburn examined next.

11.5 Amalgamations of Councils and Auburn

The subject of amalgamations of Greater Sydney councils (given implications for planning at municipal and Greater Sydney levels) was reviewed in Chap. 2. NSW Premier Mike Baird announced in Dec 2015 the State's preference for: a. Auburn to merge with Holroyd; and b. Parramatta City Council to have its boundaries changed (Chap. 10) to absorb parts of Auburn (Olympic Park and Wentworth Point); Hornsby; and Baulkham Hills. The early proposed merged councils of Parramatta, Auburn and Holroyd. (Plate 11.9).

Auburn Council preferred merging with Burwood and Canada Bay and not losing Olympic Park and Wentworth Point to Parramatta Council. Council also remained opposed to forced amalgamations. The Mayor of Auburn City urged all residents to have their say on the issue of amalgamations of councils. Council noted that submissions to the new Council Boundary Review were open till the end of February (www.councilboundaryreview.nsw.gov.au). In the interim, the Council noted, the Minister for Local Government had commenced the process of forcibly

Plate 11.9 Early proposed merger of Parramatta, Auburn and Holroyd Councils (*Source*: NSW Dept. of Local Government Web 2015)

amalgamating Auburn and Holroyd councils. Council also noted the Minister said he expected the new council (with adjusted boundaries) to be 'proclaimed' on 1 July 2016 (note the 12th May 2016 *Local Government (Council Amalgamations) Proclamation 2016* noted in Chap. 2). The results of the review of proposed boundary adjustments and merger (Auburn and Holroyd councils) would be referred to the Chief Executive of the Office of Local Government for examination and report under the Local Government Act (the Act). In turn, the Chief Executive had appointed a Delegate who would examine the proposals, conduct a public inquiry review, call for written submissions, and then report back to the Minister.

11.5.1 Conclusion on Amalgamations and Auburn City Council

Auburn City Council stated its case to the State of the Council have a preference to merge with Burwood and Canada Bay councils and not to lose Olympic Park and Wentworth Point. At the end of the State review process (Council Boundary Review) the State would determine what amalgamations as proposed would proceed or not proceed. The 12th May 2016 *Local Government (Council Amalgamations) Proclamation 2016* merging Auburn and Holroyd councils was noted. This question will be further addressed in Chap. 18.

11.6 Overall Conclusions

Having completed an assessment of the development and planning of Auburn, there are a number of overall conclusions reached. In examining the patterns of urban development of Auburn it was concluded that Auburn has been experiencing accelerated development and changes to the composition of its population. Given these changes, and examining Auburn Council's response, it was also concluded Council is moving in the right direction in increasing its sustainable urban planning (SUP) practices. It was also concluded, however, that there would be significant future impacts on Auburn from Greater Sydney growth and development. In reviewing potential Council and State responses to this future scenario, it was concluded that new partnerships between the Council and the State would be needed. A major initiative from the State (2009 onwards) was to investigate local government reform, including a potential amalgamation of councils (e.g. Parramatta boundary changes; and, merging of Auburn and Holroyd councils). The 12th May 2016 *Local Government (Council Amalgamations) Proclamation 2016* that merged Auburn and Holroyd councils was acknowledged. These overall chapter conclusions will be examined against results of assessing planning and development of the remainder of the three cities of Greater Parramatta (Chaps. 10, 11 and 12) and the district as a whole (Chap. 13). Results from the district assessment (and assessment of Sydney Inner West and St George districts) will contribute to addressing directions for planning sustainable districts over the whole of Greater Sydney (Chap. 18).

11.7 Summary

The chapter addressed the extent Auburn Council was applying sustainable urban planning (SUP) practices (see Chap. 1) in planning the municipal area. It was noted that Auburn, in Greater Parramatta District, was merged with Holroyd council under the 12th May 2016 *Local Government (Council Amalgamations) Proclamation 2016* (Chap. 2). Planning of Auburn Municipality, second of three local government areas in the Greater Parramatta District of Greater Sydney, was examined in this chapter. The chapter focused on patterns of Auburn development, and how Auburn Council in its planning was addressing this development. Firstly, a background on Auburn development over time was presented of the municipality. Three key questions were addressed in examining this background: (1). what is the pattern of urban transition the municipality is experiencing; (2). what is the extent sustainable urban planning (SUP) practices are being applied by the council; and, (3). what is the impact of Greater Sydney planning on the development of the municipality, including proposed council amalgamations. In addressing Question 1, development patterns, four phases of Auburn development were examined: Early Auburn; Traditional Auburn; New Auburn; and, Future Auburn. In addition, recent population demographic changes within Auburn were examined to gauge the extent overseas born national groups were settling in Auburn (a global transition factor). In addressing Question 2, sustainable urban planning (SUP), Auburn Council's key planning documents were critiqued and conclusions drawn on the extent Council applies SUP practices. In examining Question 3, Greater Sydney Region planning, key State plans affecting Auburn were addressed (including amalgamation of councils). Conclusions were drawn on each of these questions, and overall conclusions on: the pattern of urban transition in Auburn; the extent Council is applying SUP practices; and, the impact of Greater Sydney planning on Auburn. The chapter notes that these conclusions will be addressed further in Chap. 13, examining the pattern of development of the whole of Greater Parramatta District (including Auburn City). The next chapter will examine the final city (City of Holroyd) within the Greater Parramatta district.

Notation 11.1 Cultural Diversity Poem - Auburn
Poem (Rauscher 2015 unpublished) reflecting the diversity of Auburn:

Tripoli Rises in Auburn (#15)

The Gallipoli Mosque signals your arrival at Auburn City, red sun setting in West Sydney
The enchantment of the Middle East, China and Europe in Sydney streets
Just 9 km from Sydney's heart you walk in Turkish Delight Wonderland
Hallah butcheries, Iraqi sweets, Ethiopian coffees, Lebanon fellah breads

A casual setting in a new open arena with edges of cafe crowd of mixed nationals
Old ambulance station sitting proud in plaza middle, still serving the community
Elders feel at home here with multi-cultural support services and religious edifices
Orderly, confident, personal, enterprising with happy smiles, Tripoli rises in Auburn

References

Auburn Council (1982) Liberty plains. NSW, Auburn
Auburn Council (2007) Auburn town centre heritage review. NSW, Auburn
Auburn Council (2009) Auburn town centre strategy 2031. NSW, Auburn
Auburn Council (2010a) Auburn local environment plan. NSW, Auburn
Auburn Council (2010b) Auburn development control plan. NSW, Auburn
Auburn Council (2013) Auburn city community strategic plan 2013–2023. NSW, Auburn
Australian Bureau of Statistics (ABS) (2011) Australian census review. ABS, Canberra
New South Wales (2011) Sydney metropolitan strategy. NSW, Sydney
New South Wales (2014a) A plan for growing Sydney. NSW, Sydney
New South Wales (2014b) Parramatta Rd renewal project. NSW, Sydney
New South Wales (2015a) Sydney CBD to Parramatta – strategic transport plan. Transport NSW, Sydney
New South Wales (2015b) Draft Parramatta Rd urban transformation strategy. NSW, Sydney

Chapter 12
Planning in Greater Sydney – Holroyd City Case Study

Dedication *This chapter is dedicated to Betty Cuthbert, a popular and successful athlete of Australia (Plate 12.1).*

Plate 12.1 Dedication: Holroyd – Betty Cuthbert (*Source*: Australian Biography 2016)

A brief profile of Betty Cuthbert (1938–) provides an insight into her legacy:

> Elizabeth ("Betty") Cuthbert AM, MBE (born Merrylands, was an Australian athlete, and a fourfold Olympic champion. Cuthbert had a distinctive running style, with a high knee lift and mouth wide open. (Source Wikipedia 2015)

R.C. Rauscher, S. Momtaz, *Cities in Global Transition*,
DOI 10.1007/978-3-319-39865-5_12

Abstract The chapter addresses the extent Holroyd Council (herein referred to as Holroyd) is applying sustainable urban planning (SUP) practices (see Chap. 1) in planning the municipal area. The chapter acknowledges the *Local Government (Council Amalgamations) Proclamation 2016* merging Holroyd with Auburn council noted in Chap. 2). Holroyd (in Greater Parramatta District) is one of twelve municipalities within three Greater Sydney Districts (Sydney Inner West, Greater Parramatta and St George), all considered to be in global transition (see Chap. 2). Examining SUP practices is based on the application of three key SUP principles (as adopted in Chap. 2), being: Developing Sustainable Urban Centers; Developing Sustainable Transport; and, Protecting the Natural Environment.

Firstly, a background to Holroyd development and population changes (e.g. global movement of population) is outlined. To examine the pattern of urban development, four phases of Holroyd development are examined: Early Holroyd; Traditional Holroyd; New Holroyd; and, Future Holroyd. Given conclusions reached here, key Holroyd Council planning documents are then critiqued to determine the extent Council applies SUP practices. In addition, given State planning of Greater Sydney affects Holroyd, a second question is posed: what is the impact of Greater Sydney planning and development on Holroyd? In examining this question, examples of State plans (including council amalgamations under the Proclamation noted above) likely to affect Holroyd are addressed.by chapter's end, conclusions are drawn on Council planning and impact of Greater Sydney planning on Holroyd. The chapter notes that these conclusions will be addressed further in: Chap. 13 (examining the extent of SUP practices being applied over the whole of Greater Parramatta District); and, Chap. 18 (examining directions for planning a sustainable Greater Sydney or a sustainable city anywhere).

12.1 Background and Population Change

Holroyd City (herein called Holroyd) in the context of Greater Parramatta District (as introduced in Chap. 2) is adjacent and west of Parramatta and Auburn, and to the west of the City of Sydney. Under the *Local Government (Council Amalgamations) Proclamation 2016* Holroyd was merged with Auburn council as noted in Chap. 2). Suburbs of Holroyd include: Girraween, Granville, Greystanes, Guildford, Guildford West, Holroyd, Harris Park, Mays Hill, Merrylands, Merrylands West, Pemulwuy, Pendle Hill, South Wentworthville, Toongabbie, Wentworthville, Westmead, Woodpark, and, Yennora. The area now known as Holroyd was home to the Wangal people, inhabiting a major portion of the Greater Parramatta district. Merrylands, the main centre of Holroyd City, was named after the former English home of Arthur Todd Holroyd (1806–1887), who acquired land in this area in 1855 and who gave his name to the adjoining suburb of Holroyd. Originally known as the Municipality of Prospect and Sherwood, Holroyd was incorporated in 1872. Holroyd was renamed a Municipality in 1927 and was declared the City of Holroyd in 1990. The administration centre and commercial heart of Holroyd City is Merrylands, located 25 km west of Sydney CBD. Merrylands, covers 2 local government areas, with the railway being the boundary between Parramatta City Council (eastern side of railway) and Holroyd City Council (western side of the railway). During the last two Census periods (2003–2011) Holroyd had higher that the national average population growths over this period. The population of Holroyd City is at 99,163 (census 2011). Key features of Merrylands include:

a. Merrylands has a commercial area around Merrylands railway station. Stockland Merrylands Mall is a shopping center with supermarkets, discount department stores and specialty shops.
b. The Merrylands railway station is serviced by the Airport, Inner West & South Line and the Cumberland Line of the Sydney Trains network. Merrylands is also on the bus route from Liverpool to Parramatta.
c. Several main cycle routes pass through Merrylands and surrounding suburbs and linking up to Liverpool, Penrith and Sydney Harbour. M4 has an exit in Crown Street, Parramatta, Merrylands and Holroyd access.
d. As a sample of residents born in Australia, Merrylands recorded 53.4 % as born overseas.

Of these people the most common countries of birth were Lebanon 8.4 %, China 3.8 %, India.7 %, Afghanistan 3.2 % and Philippines 1.9 %. (*Source* Wikipedia 2015)

The 2011 Census showed the strength of multiculturalism in Holroyd. The proportion of residents in the Holroyd local government area, for example, who stated their ancestry as Lebanese was in excess of 16 times the national average. Likewise, the proportion of Indian residents was in excess of five times the national average. In the same census, the proportion of residents who stated a religious affiliation with Islam was in excess of eight times the national average and Hinduism was in excess seven times. Meanwhile, as at the Census date, the area was linguistically diverse. Arabic, Tamil, and Hindi languages were spoken in a high proportion of households, ranging from five times to 17 times (among the different languages) the national averages.

The Australian Bureau of Statistics (ABS 2011) comments on Holroyd City ancestry and languages spoken at home:

> At the 2011 Census, the proportion of residents in the Holroyd local government area who stated their ancestry as Lebanese, was in excess of sixteen times the national average; and the proportion of Indian residents was in excess of five times the national average. The proportion of residents who stated a religious affiliation with Islam or Hinduism was in excess of eight times and seven times the national average respectively; and the proportion of residents with no religion more than half the national average. Meanwhile, as at the Census date, the area was linguistically diverse, with Arabic, Tamil, or Hindi languages spoken in a high proportion of households, and ranged from five times to seventeen times the national averages. Within the suburb of Holroyd, at the 2011 census, almost 70 % of people were born outside of Australia, with the top countries of birth being India 8.4 %, China 7.2 %, Afghanistan 5.5 %, Lebanon 3.9 % and Philippines 3.7 %. About three quarters of people spoke a language other than English at home. The most common other languages spoken were Arabic 10.2 %, Mandarin 8.3 %, Dari 4.0 %, Hindi 3.8 % and Cantonese 3.3 %. (*Source* ABS 2011)

Finally, in author's (Rauscher) qualitative research (2014/2015), there is a poem written to reflect the cultural diversity of Holroyd (end of chapter under Notation 12.1): *Multicultural Meets Suburban Edge at Holroyd.* With this background to Holroyd Municipality, the extent of Holroyd's development phases is examined next.

12.2 Phases of Development

An examination of the development phases of any locality provides a basis of understanding current trends and likely future directions of growth of that locality. Following a look at the background of Holroyd four phases of Holroyd development will be examined. These phases include: Early Holroyd; Traditional Holroyd; New Holroyd, and, Future Holroyd. To assist in addressing this question, author (Rauscher) qualitative research field work was completed in Holroyd City (2014–2015). Early Holroyd (Plate 12.2) was centred on St Bartholomew Anglican Church 1842 (top-left). The town of Prospect built its Town Hall in 1900 (top-right). Dunmore illustrates to extent of solid home construction (middle-left) at Pendle Hill 1927. A works construction crew completes drainage works (middle-right); Depression relief workers are shown in this suburban street (bottom-left). Finally, an early motorbus at Merrylands is shown here (bottom-right).

Holroyd renewal is occurring (2015) throughout most of the City's suburbs. Plate 12.2 illustrates examples of that renewal in Merrylands. Holroyd maintains its suburban roots with cottage housing, as illustrated here (top-left). A mural in the CBD depicts early rural Holroyd (top-right). The City library is housed in a modern building (middle-left). Attached to the library is the City Administration Center (middle-right). The cafe scene is illustrated here in a stretch of Merrylands (bottom-left). Finally, the Holroyd City residents are served by a central Hotel (bottom-right) in Merrylands.

The New Holroyd (Plate 12.3) reflects increasing higher densities and mall shopping, complimented by town shopping as illustrated in Merrylands. Shown here is a three story residential building with commercial below (top-left). A new medium rise 6-story with commercial below (top-center) under construction one street from the rail station. The major mall retailer in Holroyd is Stockland's (top-right). An 8-story residential anchors a busy intersection (bottom-left). In the 1970s-2000 many 2 and 3-story walk-ups were built (bottom-center). Finally, the Merrylands rail interchange has been modernized to meet future growth of Holroyd (bottom-right).

Finally, Future Holroyd (Plate 12.4) reflects the continued building boom in residential development at increasing densities throughout the municipality. Building within many of the Holroyd suburbs is likely to see this trend, such as Merrylands. Older heritage structures can be restored to present educational tourism such as here (left) at the old brickworks. New six-eight story residential buildings (centre) are likely to be spread over a wider area of Holroyd. Finally, the demand for smaller clusters of medium density (right) built within precincts close to open spaces and shops is likely to continue well into the future of Holroyd. These building design factors (such as height, density, housing affordability, overshadowing and open space provisions) are likely to remain issues of development importance and contention within communities at the local government and State levels into the future.

Plate 12.2 Early Holroyd. *Left to right* St Bartholomew Anglican Church 1842; Prospect's Town Hall 1900; Dunmore House, Pendle Hill 1927; Works Construction Crew; Depression relief workers; Early motorbus at Merrylands 1934 (*Source Holroyd – A Social History of Western Sydney*, Grace Karslens 1991)

Plate 12.3 Traditional Holroyd *Left to right* Merrylands Rail Interchange; Mural of Early Holroyd; City Council Library; Holroyd Administration Center; Merrylands cafe row; and, Hotel (*Source* Raymond Rauscher 2015)

Plate 12.4 New Holroyd. *Left to right*: The Forum; Infill residential; Factory conversion; Bus depot; Orphan School Creek complex; Residential around square (*Source* Raymond Rauscher 2014)

12.2.1 Conclusion on Development Phases and Population Change

Given the review of development phases of Holroyd (above) and changes in population demographics, it can be concluded that Holroyd has been experiencing accelerated global transition (2000s) (Question 1). The planning of Holroyd (merged with Auburn council as noted above under the Proclamation) in meeting development demands and applying sustainable urban planning (SUP) practices is examined next.

12.3 Holroyd and Sustainable Urban Planning

Planning of Holroyd City has been taking place since it became a local government body in 1861. A review of planning in Holroyd will pay particular attention to applications of sustainable urban planning (SUP) (as outlined in Chap. 1) practices

Information Box 12.1 Holroyd Municipality Planning

Holroyd Local Environment Plan (2013a)	www.holroyd.gov.nsw.au
Holroyd Development Control Plan (2013b)	www.holroyd.gov.nsw.au
Wentworthville Centre Revitalization Planning Project (2013c)	www.holroyd.gov.nsw.au

Source Webs as noted February 2016

(including use of indicators of sustainability) (Chap. 2). Holroyd Council key planning documents (a–z with date and web reference) are listed in Information Box 12.1. A summary of these documents and a brief critique assist with this review of Holroyd planning (documents can be sourced under the web link as noted in the right column).

12.3.1 Holroyd Local Environment Plan

The *Holroyd Local Environment Plan* (Holroyd 2013a) is the main planning document for Holroyd Council. This Plan is supplemented by design guidelines under the *Holroyd Development Control Plan* (Holroyd 2013b).

12.3.1.1 Wentworthville Centre Revitalization Planning Project

Holroyd Council is undertaking a number of strategic studies. Firstly, Council engaged in 2014 consultants to examine and report back on revitalizing Wentworthville, called the *Wentworthville Centre Revitalization Planning Project* (Holroyd 2013c). Architecture, economic and transport studies were expected to be adopted by Council in mid-late 2015. Council has an ongoing program to review its 160 heritage places within the City area (referenced in the *Holroyd Local Environment Plan* 2013). In addition, Council's Living Holroyd program was created to focus the City on the future. Council anticipates instituting integrated planning and balanced sustainable strategies. Council in addition notes:

a. The Division of Local Government's Integrated Planning and Reporting structure has been developed to replace existing Council planning requirements and will enable Council to strategically respond to current and future community needs.
b. Council is required to develop a long-range planning system which refines its focus through community engagement and seeks to align the provision of Council Services with resources at financial, asset and human resource level.
c. The Living Holroyd Community Strategic Plan has been developed to map the future of Holroyd City, representing the voice of over 2000 participants in the Living Holroyd Program to date.
d. Establishing a central vision for Holroyd and a series of supporting long-term strategies, the Community Strategic Plan balances a commitment to social cohesion, the local economy, our natural and built environments and all members of the Holroyd Community.

e. The Community Strategic Plan has been written from the perspective of the community. It outlines what we as a collective see as important to the future of our city and highlights what we can expect to see and experience as Holroyd grows through the next 20 years. The plan itself is built upon a set of 5 Directions, which will form the basis of all Council operations through a supporting Delivery Program, strategic resource plan and additional operational and project plans currently under development.
f. Living Holroyd has established a set of plans to outline the long-term vision for the future of the City, created in partnership with the community. (Source Holroyd City Council Web 2015)

Finally, the Holroyd City Council plans include:

a. The 2013–2017 Delivery Program (incorporating the 2014/2015 Operational Plan which establishes a rolling program of actions and priorities that outline how Council will deliver on the outcomes proposed in the Community Strategic Plan,
b. The Resource Strategy – includes the Long Term Financial Plan, the Asset Management Strategy and the Workforce Strategy
c. Council's Asset Management Plans include Roads and Bridges, Storm Water, Parks and Recreation and Buildings. A list of indicative works for each category over the next 10 years has also been provided. (*Source* Holroyd City Council Web 2015)

12.3.2 Conclusion on Council Planning and SUP Practices

It can be concluded, having reviewed Holroyd City Council planning documents and their applications, it can be concluded that Council is moving in the direction of applying sustainable urban planning (SUP) practices (Question 2 above). The State's Greater Sydney planning (including infrastructure projects) and State council amalgamation proposals likely to affect Holroyd (Question 3) are reviewed next.

12.4 Greater Sydney Planning and Holroyd

There are planning and developments of Greater Sydney likely to impact on Holroyd. In many instances the Holroyd City Council has had the task of providing: planning input on State proposals affecting Holroyd. These include making and reviewing public submissions; implementing State planning policies at the local level; joint partnership on service or infrastructure projects; and, monitoring State or private development projects affecting Holroyd City. Other Council planning responsibilities under State legislation that also faces the Council include: engaging the municipality community (e.g. residents, business, services, development interests, and institutions) on State planning policies and development proposals; statutory plan making (e.g. adopting local environment plans (LEPs); and, reporting back to the State on a range of planning matters (i.e. State proposed roads, state of environment reporting and infrastructure projects).

Greater Sydney planning requirements were set by the NSW State in 2006, with Holroyd Council adopting in 2013 the *Holroyd Local Environment Plan* (Holroyd

2013a). Council's plans had to be consistent with the *Sydney Metropolitan Strategy* (NSW 2011). The State in 2014 adopted its strategy for planning Greater Sydney to 2031 (*A Plan for Growing Sydney* (NSW 2014a). The Plan forecasts a population growth of 1.6 m (to be added to the Greater Sydney population of 4.84 m at June 2014). In examining this Plan and its impact on Holroyd, three key planning principles as adopted in Chap. 2 are addressed: (1). Developing Sustainable Urban Centres; (2). Developing Sustainable Transport; and, (3). Protecting the Natural Environment. The planning principle *Developing* Sustainable Urban Centres is examined first. The Plan notes a number of projects, including Parramatta CBD (within Parramatta City) and Olympic Park (within Auburn City) strategic growth centres, will affect Holroyd. It is likely therefore that Holroyd will continue to experience residential, commercial and services growth. The planning principle Development Sustainable Transport is examined next. Holroyd Council, along with other Greater Parramatta District councils, were asked in late 2014 to comment on the proposed *Parramatta Rd Renewal Project* (NSW 2014b). The State's WestConnex roads project, also affecting Holroyd Council, is closely tied to the Parramatta Rd renewal project. The key transport projects will be further addressed in reviewing the whole Greater Parramatta district (including Holroyd) in Chap. 13.

Finally, planning principle Protecting the Natural Environment is examined next. The Plan commits the State to 'protect the health of the subregions waterways and aquatic habitats'. This goal, for example, would apply to the prime conservation challenges in Holroyd. This would include the protection of bushlands, waterways and foreshores within Holroyd. The efforts of Holroyd Council in protecting these natural areas in cooperation with other councils in Greater Parramatta will be outlined in more detail in Chap. 13.

12.4.1 Conclusion on Holroyd and Greater Sydney Planning

Having reviewed above the State's Greater Sydney planning (including infrastructure projects) it can be concluded the implementation of these plans and developments is likely to have a significant impact on Holroyd (Question 3). The extent of this impact will be better understood when these State plans are further examined in the context of the whole of the Greater Parramatta District (including Holroyd) in Chap. 13. Finally, in examining State proposed council amalgamations, options for councils (including Holroyd) working together and in partnership with the State to meet local and district needs will be addressed in Chap. 18 (Planning Reforms and Directions for Planning Sustainable Cities). The subject to amalgamations of councils and how this affects Holroyd examined next.

12.5 Amalgamation of Councils and Holroyd

The subject of amalgamations of Greater Sydney councils (given implications for planning at municipal and Greater Sydney levels) was reviewed in Chap. 2. An early proposed merger of Holroyd, with Parramatta and Auburn is illustrated in Plate 12.5. NSW Premier Mike Baird announced in Dec 2015 the State's preference for: (a) Holroyd to merge with Auburn; and (b) Parramatta City Council to have its boundaries changed (Chap. 10) to absorb parts of Auburn (Olympic Park and Wentworth Point); Hornsby; and Baulkham Hills. The proposed merged councils of Auburn and Holroyd would include the Granville area ward currently part of the City of Parramatta.

Plate 12.5 Future Holroyd (*Source*: Holroyd Council Web 2016)

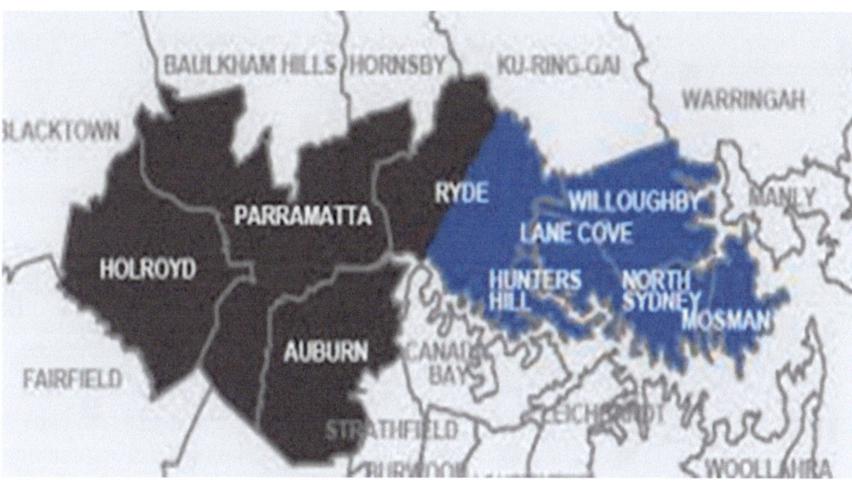

Plate 12.6 Early Proposed Merger of Parramatta, Auburn and Holroyd Councils (*Source* NSW Dept. of Local Government Web 2015)

Holroyd Council preferred to stand-alone as a council and not to merge. Council also remained strongly opposed to forced amalgamations. The Mayor of Holroyd City urged all residents to have their say on the issue of amalgamations of councils. Council noted that submissions to the new Council Boundary Review were open till the end of February (www.councilboundaryreview.nsw.gov.au). In the interim, the Council noted, the Minister for Local Government had commenced the process of forcibly amalgamating Holroyd and Auburn councils. Council also noted the Minister said he expected the new council (with adjusted boundaries) to be 'proclaimed' on 1 July 2016 (note the 12th May 2016 *Local Government (Council Amalgamations) Proclamation 2016* noted in Chap. 2). The results of the review of proposed boundary adjustments and merger (Holroyd and Auburn councils) would be referred to the Chief Executive of the Office of Local Government for examination and report under the Local Government Act (the Act). In turn, the Chief Executive had appointed a Delegate who would examine the proposals, conduct a public inquiry review, call for written submissions, and then report back to the Minister.

12.5.1 Conclusion on Amalgamations and Holroyd City Council

Holroyd City Council stated its case to the State for the Council to stand-alone and not be merged. The State recommended Holroyd Council merge with Auburn City Council. Holroyd Council stated its case from the start of the State initiated amalgamation process that Council wanted to stand alone and not amalgamate. At the end of the State review process (Council Boundary Review) the State would determine what amalgamations as proposed would proceed or not proceed. The 12th May 2016 *Local Government (Council Amalgamations) Proclamation 2016* merging Holroyd and Auburn councils was noted.

This question will be further addressed in Chap. 18.

12.6 Overall Conclusions

Having completed an assessment of the development and planning of Holroyd, there are a number of overall conclusions reached. In examining the patterns of urban development of Holroyd it was concluded that Holroyd has been experiencing accelerated development and changes to the composition of its population. Given these changes, and examining Holroyd Council's response, it was also concluded Council is moving in the right direction in increasing its sustainable urban planning (SUP) practices. It was also concluded, however, that there would be significant future impacts on Holroyd from Greater Sydney growth and development. In reviewing potential Council and State responses to this future scenario, it was concluded that new partnerships between the Council and the State would be needed. Finally, a major initiative from the State (2009 onwards) was to investigate local government reform, including a potential amalgamation of councils (e.g. Parramatta boundary changes; and merging of Auburn and Holroyd councils). Holroyd Council stated its case from the start of the State initiated amalgamation process that Council wanted to stand alone and not amalgamate. The 12th May 2016 *Local Government (Council Amalgamations) Proclamation 2016* that merged Holroyd and Auburn councils was acknowledged. These overall chapter conclusions will be examined against results of assessing planning and development of the remainder of the three cities of Greater Parramatta (Chaps. 10, 11 and 12) and the district as a whole (Chap. 13). Results from the district assessment (and assessment of Sydney Inner West and St George districts in later chapters) will contribute to applying SUP practices within districts (including municipalities and cities) over the whole of Greater Sydney (Chap. 18).

12.7 Summary

The chapter addressed the extent Holroyd Council was applying sustainable urban planning (SUP) practices (see Chap. 1) in planning the municipal area. It was noted that Holroyd, in Greater Parramatta District, was merged with Auburn council under the 12th May 2016 *Local Government (Council Amalgamations) Proclamation 2016* (Chap. 2). It was concluded that Holroyd, one of twelve municipalities within three Greater Sydney Districts (Sydney Inner West, Greater Parramatta and St George) is considered to be in global transition. SUP practices (based on the application of three key SUP principles) were introduced, being: Developing Sustainable Urban Centers; Developing Sustainable Transport; and, Protecting the Natural Environment. A review of Holroyd background and population changes (i.e. global movement of population) was outlined. In examining the pattern of urban development, four phases of Holroyd development are examined: Early Holroyd; Traditional Holroyd; New Holroyd; and, Future Holroyd. With conclusions drawn, key Holroyd Council planning documents were then critiqued to determine the extent Council applies SUP practices. In addition, given State planning of Greater Sydney affects Holroyd, a second question was posed: what is the impact of Greater Sydney planning and development on Holroyd?. In examining this question, examples of State plans (including State proposed council amalgamations) likely to impact on Holroyd were addressed. By chapter's end, conclusions were drawn on Council planning and impact of Greater Sydney planning on Holroyd. The chapter noted that these conclusions would be addressed further in: Chap. 9 (examining the extent of SUP practices being applied over the whole of Greater Parramatta District); and, Chap. 18 (examining directions for applying SUP practices in Greater Sydney or in any city worldwide).

Notation 12.1 Cultural Diversity Poem – Holroyd

A poem (Rauscher 2015 unpublished) reflecting the diversity of Holroyd follows.

Multicultural Meets Suburban Edge at Holroyd (**# 16**)

Holroyd City is a metropolis where multicultural meets traditional suburbia in a city mix
The traditions of Australia are this city's core as the waves of new settlers arrive
The city of Mediterranean symbols, lifestyles and Arabian lore enchanting the young
The quietness and order of Merrylands is as suburban reaches beyond the metropolis

The energy to blend culture past with new cultures abounding in the streets
Crisp decorative Town Hall and all Nations Library welcome
A memorial park anchors the railway providing a blanket of green
A vision of further challenges of balancing the traditional with the modern

References

Australian Bureau of Statistics (ABS) (2011) Australian census review. ABS, Canberra
Holroyd City Council (2013a) Holroyd local environment plan. NSW, Holroyd
Holroyd City Council (2013b) Holroyd development control plan. NSW, Holroyd
Holroyd City Council (2013c) Wentworthville centre revitalization planning project. NSW, Holroyd
Karslens G (1991) Holroyd – a social history of Western Sydney. NSW, Sydney
New South Wales (2011) Sydney metropolitan strategy. NSW, Sydney
New South Wales (2014a) A plan for growing Sydney. NSW, Sydney
New South Wales (2014b) Parramatta Rd renewal project. NSW, Sydney

Chapter 13
Planning Greater Parramatta District – Greater Sydney

Dedication *This part of the book is dedicated to three people who have left legacies in the Greater Parramatta district. A historical snippet on each person gives an insight into that person's legacy (Plate 13.1).*

Plate 13.1 Past Citizens of Greater Parramatta. John Lewin; David Lennox; and, Allan Cunningham. (Source Wikipedia 2015)

John William Lewin was an English-born artist active in Australia from 1800. The first professional artist of the colony of New South Wales, he illustrated the earliest volumes of Australian natural history. Lewin was the son of a professional scientific artist, William Lewin, who was the author of an eight-volume work *The Birds of Great Britain* (1789–1794).

David Lennox was a Scottish-Australian bridge-builder and master stonemason born in Ayr, Scotland. Prior to this time, the young colony of New South Wales had no skilled stonemasons. Lennox, in time, became Superintendent of Bridges, NSW.

R.C. Rauscher, S. Momtaz, *Cities in Global Transition*,
DOI 10.1007/978-3-319-39865-5_13

Allan Cunningham was an English botanist and explorer, primarily known for his travels in Australia to collect plants. He liaised closely with Australian botanist Sir Joseph Banks. (*Source* Wikipedia 2015)

Abstract Planning of the Greater Parramatta District, the second of three case study districts of Greater Sydney, is examined in this chapter. The chapter acknowledges the *Local Government (Council Amalgamations) Proclamation 2016* proposed mergers and expanding the boundaries of nominated councils. Having examined the question of global transition and planning practices of the three municipalities within the District (Chaps. 10, 11 and 12), the chapter addresses: (1). planning the Greater Parramatta District as a whole; and, (2). the extent the State (responsible for district planning) is meeting the planning issues of the District. Key planning principles as adopted (Chap. 2) to apply to Greater Sydney (and this district) are reviewed. These key principles include: (1). Developing Sustainable Urban Centres; (2). Developing Sustainable Transport; and, (3). Protecting the Natural Environment. Sample plans and developments within Greater Parramatta are critiqued under each of these principles. After reaching a conclusion under each principle, an overall conclusion will be drawn on the extent the State (planning the District) is meeting the planning issues of the District. Finally, it's noted that these Greater Parramatta District conclusions will be assessed alongside conclusions reached in examining Sydney Inner West District (Chaps. 3, 4, 5, 6, 7, 8 and 9) and St George District (Chaps. 14, 15, 16 and 17). All these conclusions are examined (Chap. 18) in the context of SUP practices applied to Greater Sydney as a global city (affecting municipal areas within Greater Parramatta and the district as a whole).

13.1 Background to Greater Parramatta

The Greater Parramatta district (Cities of Parramatta, Auburn and Holroyd), as with Sydney Inner West (Chap. 9) the geography of this land was also favorable to early European settlers. The settlement and valued land of the district can be summarized as follows:

> Parramatta was founded in 1788, the same year as Sydney. As such, Parramatta is the second oldest city in Australia, being only 10 months younger than Sydney. The British Colonists, which had arrived in January 1788 on the First Fleet at Sydney Cove, had only enough food to support themselves for a short time and the soil around Sydney Cove proved too poor to grow the amount of food that 1000 convicts, soldiers and administrators needed to survive. During 1788, Governor Arthur Phillip had reconnoitred several places before choosing Parramatta as the most likely place for a successful large farm. Parramatta was the furthest navigable point inland on the Parramatta River (i.e. furthest from the thin, sandy coastal soil) and also the point at which the river became freshwater and therefore useful for farming (*Source* Wikipedia 2015).

The common geography, early history (Aboriginal and European) and common population ties reinforce Greater Parramatta as a metropolitan district. The suburbs (from a real estate basis) contained in each of the municipalities with the district (Parramatta, Auburn and Holroyd) are noted within the respective Chaps. 10, 11 and 12). A brief historical background of Greater Parramatta is gleaned from historical photos of the district (Plate 13.2). The rotunda (top-left) in Memorial park Parramatta reflects early community civil pride. The centres and Auburn, Holroyd and Parramatta were connected with horse transport (top-middle). Improvement to road (top-right) and rail infrastructure allowed commercial expansion of the centres in the district. Two institutions, among many located in Parramatta, were the goal (bottom left) and the Female Factory (bottom right). Most early development (bottom right) took place along the Parramatta River, connecting river craft to the City of Sydney and Sydney Harbour.

Greater Parramatta (Cities of Parramatta, Auburn and Holroyd) is today the geographical center of the Sydney Metropolitan Region. The population of Greater Parramatta (2011 Census) is 339,759 (Parramatta at 166,858, Auburn at 73,738 and

Plate 13.2 Early Greater Parramatta District. *Left to right* Rotunda; Auburn Early Transport; Parramatta early roads; Goal at Parramatta; Female Factory, Parramatta; and, Early Development along Parramatta River (*Sources* Parramatta City Archives and Ray Rauscher 2015)

Holroyd at 99,163). Population growth in Parramatta has been significantly higher than the national average over the last two Census periods (2003–2011). Both Auburn and Holroyd also had higher that the national average population growths over this period. Greater Parramatta prides itself as one of the most multicultural communities in Australia. Greater Parramatta, under the A *Plan for Growing Sydney* (NSW 2014a, b) (Chaps. 2 and 10), plays a strategic role in Sydney's growth to 2031 (Plate 13.3).

The State designates this district within the West Central Subregion under this Plan. In examining this Plate note: (a). The CBD of Parramatta (center red circle); (b). the strategic centers (clockwise from left) of Blacktown, Norwest, Castle Hill; Sydney Olympic Park, and Bankstown; (c). WestConnex at middle-left (Olympic Park to Parramatta); (d). the blue band right to left representing 'urban renewal corridor'; (d). The broken grey lines representing alternative 'light rail investigations' from Parramatta north to Castle Hill to north, east to Epping and Carlingford, east to Olympic Park and Strathfield, and south to Bankstown; blue crossed lines top-left (Norwest and Castle Hill) resenting a 'priority precinct'; and, finally, bottom center red striped is 'Bankstown Airport Precinct' and orange striped 'Transport Gateway'. The Plan notes, in addressing the 'West Central Subregion':

> The West Central subregion will be a significant focus for infrastructure investment and intensive growth over the next 20 years. Greater Parramatta will continue to be Sydney's second CBD and focus for jobs growth and services delivery Sydney's west. A growing and prosperous Grater Parramatta will be supported by a network of centres providing job and services closer to home for many of the subregions residents. This will improve liveability and contribute to strong, resilient communities throughout the subregion. Within 25 years Western Sydney will be home to more than half of all Sydneysiders. A significant population growth will occur in the North West and South West Growth Centres and around Parramatta. The government's actions to create jobs across Western Sydney will leverage off the significant number of committed and proposed infrastructure projects across Western Sydney (*Source* NSW 2014)

The State predicts that just under a million (900,000) of the project 1.6 m population growth of Greater Sydney will occur in Western Sydney. The key urban planning principles as outlined in Chap. 2 are now addressed for Greater Parramatta, including: (1). Developing Sustainable Urban Centres; (2). Developing Sustainable Transport; and, (3). Protecting the Natural Environment.

Plate 13.3 Greater Parramatta Subregion (*Source* NSW 2014a, b)

13.2 Developing Sustainable Urban Centers

The planning principal *Developing Sustainable Urban Centers* is the first principal to be examined. Case study examples include: (1). Parramatta District Center; and (2). Olympic Park Growth Center.

13.2.1 Parramatta District Center

The State has worked with Parramatta City Council to provide Parramatta with an updated district centre (Plate 13.4). This would provide Parramatta with the capacity for additional mixed commercial, service and residential uses to serve a Greater Parramatta. The State agreed in Dec 2015 for a light rail corridor from Parramatta to Strathfield and a link to Olympic Park (partly utilising Parramatta Rd as part of that road's renewal, as addressed later).

Examples of new development in Parramatta City reflecting centre renewal are illustrated in Plate 13.5 (left-right). A concept plan illustrates the levelling of new buildings, incorporating shops below, medium rise and high rise (top-left). A variation in bulk, facades and colour gives medium rise buildings a streetscape character (top-right). Early high rise residential buildings were built adjacent to the western railway (Sydney Central to Parramatta and Western Sydney) (middle-left). The effect of these buildings is overwhelming in scale, within inadequate setbacks. Parramatta continues to develop its district centre commercial offices as illustrated here (middle-right). Medium rise office complexes (bottom-left) are located in streets adjacent to the shopping boulevard. Medium density 6–8 story residential buildings usually contain commercial shops below, convenient and assisting in creating a streetscape (bottom-right).

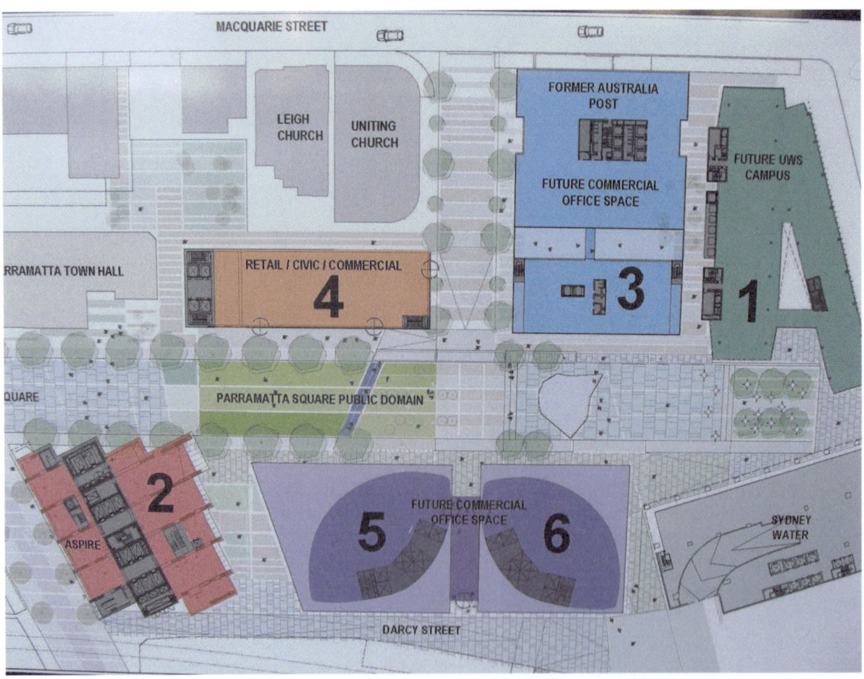

Plate 13.4 Parramatta's Planned District Centre (*Source* Parramatta City Council Web 2016)

Plate 13.5 Parramatta's Expanding District Centre (*Sources* Ray Rauscher and Parramatta City Council Web 2016)

13.2.1.1 Conclusions on Parramatta District Centre

Having reviewed the Parramatta District Center proposal, there are a number of significant effects on council areas within Greater Parramatta district that this project is likely to have. These can be summarized under a need for: (a). more details on the urban design aspects of projects (i.e. height, bulk, overshadowing); impact of the additional population on services and transport; greater allocations of open spaces lands and pedestrian linkages; greater housing affordability provisions; and, more environmental and heritage protection measures.

13.2.2 Olympic Park Growth Center

At the district level the State agencies responsible for developing plans for urban design is Urban Growth NSW. Olympic Park Suburb is concentrated around the site of the year 2000 Australia hosting of the Olympics. Urban design of Olympic Park Center is reaching higher investment levels and bigger projects. The State has worked with Auburn City Council to develop Olympic Park as a major urban centre. This provides Auburn City and the Greater Parramatta District with the capacity for additional mixed commercial, service and residential uses. Examples of new development of Olympic Park are contained in Plate 13.6 (left to right). A major three building residential with support services will become a central feature of Olympic Park (top-left). The proximity (foreground) of the Olympic Park water features and wetlands provide valued environmental areas. The State has developed an environmental educational studies centre (top-right). The formal plantings (middle-left) of the Olympic Park has provided many treed walkways. The wetlands section of Olympic Park contain one of the richest cross sectional of wildlife in the Sydney metropolitan region. An example of some of the bird species is shown here on a poster at the ‘bird hutch’ for bird watchers. The formal structures at Olympic Park, such as this observation tower provide focal points. Finally, elevated walkways (bottom-right) have been created through the wetlands with educational signs along the route.

13.2.2.1 Conclusions on Sydney Olympic Park Growth Centre

Having reviewed aspects of the Olympic Park Growth Centre, there are a number of significant issues for councils within the Greater Parramatta. These can be summarized under needs to consider: (a). impacts of the urban design aspects of projects (i.e. height, bulk, and overshadowing); impact of the additional population on services and transport; greater allocations of open spaces lands and pedestrian linkages; greater housing affordability provisions; and, more environmental and heritage protection measures.

Plate 13.6 Olympic Park Center (*Source*: Ray Rauscher 2015)

13.2.2.2 Conclusions on Developing Sustainable Urban Centers

Having reviewed two Greater Parramatta urban centres and author qualitative research field work (Rauscher 2014–2015), conclusions can be drawn. The urban center projects (as reviewed above) have seen major advances made in recent years, however many questions on planning at this level remain. It can be concluded the State, in partnership with councils, needs to adopt new planning arrangements to meet the planning principle *Developing Sustainable Urban* Centers. Options for these new planning arrangements are canvassed in Chap. 18.

13.3 Developing Sustainable Transport

Having examined the first urban planning principle *Developing Strategic Urban Centers* within Greater Parramatta, the second principle to be examined is D*eveloping Sustainable Transport*. With an extra 1.6 million people expected to be living in Sydney by 2036, there are many opportunities for transport to meet sustainability principles. The three municipalities within Greater Parramatta have been experiencing several decades of growth. As a result these councils have been working to accommodate this growth and to apply the most effective transport provisions. At the district level the two State agencies responsible for developing plans for urban design and transport are UrbanGrowthNSW and the NSW Dept. of Transport and Maritime Services.

A number of transport projects either planned or underway affect Greater Parramatta. These will be examined under: Long Term Transport Plans; (b). Light Rail Development; (c). Parramatta Rd Corridor; and, WestConnex.

13.3.1 Long Term Transport Plans

The key documents focusing on these plans are noted in Information Box 13.1.

Information Box 13.1 Greater Parramatta Transport Key

NSW Long Term Transport Master Plan (NSW 2013)	www.transport.nsw.gov.au
NSW Light Rail Program	http://www.transport.nsw.gov.au/lightrail-program
Parramatta Rd Urban Transformation Strategy	www.newparramattard.com.au
Western Sydney WestConnex Project	www.westconnex.com.au
Comments on WestConnex Project	http://clovermoore.com.au/westconnex

Source Webs as noted February 2016

Plate 13.7 Transport Plan – Parramatta to Olympic Park (*Source* Dept. of Transport and Maritime Services Web 2016)

Stepping back from individual transport projects to long term transport planning, the NSW government produced the *NSW Long Term Transport Master Plan* (NSW 2013) (www.westconnex.com.au) as a strategy to address Sydney's transport challenges and to deliver an integrated and modern transport system. The Master Plan is supported by a series of integrated transport delivery plans and regional strategies which address specific needs for buses, heavy rail and light rail. A portion of the plan is incorporated in Plate 13.7. The State agreed in Dec 2015 for a light rail corridor from Parramatta to Strathfield and a link to Olympic Park. This corridor would partly utilise Parramatta Rd (below the hatched motorway from Parramatta at middle-left to Olympic Park lower-right corner). The Parramatta Rd's renewal project will provide the opportunity to incorporate this light rail project. The map shows other key road and public transport connection in a north and south directions. Note the proximity of Olympic Park to the centres of Rhodes and Wentworth Point (middle-right) as discussed earlier.

Before examining the Master Plan, a background on transport planning for the whole of the Greater Sydney Region is offered by NSW peak public transport advocacy group, EcoTransit (www.ecotransit.org.au). The group has put forward alternative transport proposals that the group states are based on sustainability principles. EcoTransit argues broadly that the less resources used by the transport sector, the

Plate 13.8 Sydney Light Rail Orbital Proposal (*Source* EcoTransit Web 2015)

more efficient our economy is and the less damage is done to the environment. The group sees public transport, walking and cycling fitting these criteria, not urban freeways which entrenches car use.

13.3.2 Developing Light Rail

The development of light rail within the Greater Parramatta District has been placed into a priority category by the State, after considerable years of lobbying by the municipal councils and citizen groups of the district. For example, Eco Transit Inc. has developed a *Sydney Light Rail Orbital* proposal (Plate 13.8). The proposal provides a connected light rail system, incorporating (clockwise from left): a line to Parramatta Road extending to Olympic Park and City of Parramatta; a north-west line (Victoria Rd); White Bay Green Link; South East link (under construction commencing in George St, 2015); East-West connecting St George district; and, the Inner West.

Moving from this background to Greater Sydney public transport and light rail planning put forward by EcoTransit to Greater Parramatta, there are a number of key transport development projects affecting this district. These include: (a). District Light Rail Development; (b). Parramatta Rd Corridor Concepts; and, (c). Westconnex Plans. Each of these is examined separately, accepting they link as a whole. Given the above light rail comments and those in Chap. 9 (Sydney Inner West), rail proposals for Greater Parramatta are now examined. Light rail development, within Greater Parramatta is predominantly the responsibility of the Dept. of Transport and

Maritime Services (http://www.transport.nsw.gov.au/lightrail-program). The emphasis on that light rail system was the recent extension of light rail from Lillyfield to Dulwich Hill, and potential for further extensions. The initiative for a Western Sydney Light Rail Network has been taken by Parramatta City, in cooperation with those councils that would benefit from such a network. The Council notes:

> Parramatta City Council proposes a Western Sydney Light Rail Network to manage the planned growth of Western Sydney and connect people to jobs. A light rail offers a viable, quick to build solution for Western Sydney. It is more reliable than bus transport and cheaper and quicker to implement than heavy rail and can facilitate faster, more regular services. (*Source* Parramatta City Web 2015)

The light rail lines would include (Plate 13.9) a number of key stations. The Macquarie Park Line would connect Westmead health precinct, Parramatta CBD, University of Western Sydney at Rydalmere, Eastwood town centre and Macquarie Park. Secondly, the Parramatta to Castle Hill rail would connect: Castle Hill, the gateway to the northwest; Baulkham Hills town center to Parramatta CBD; Rydalmere; and an interchange to the North West Rail link.

Examining the proposed light rail project closer (Plate 13.10) gives a look at how the service would operate on the ground. On the right is a sketch of a landscaped section of the light rail. The tram would also service the Parramatta CBD as illustrated on the right.

Having reviewed the light rail planning proposals, there are a number of significant effects on council areas within the Greater Parramatta district that these proposals are likely to have. These can be summarized under a need for: (a). increased opportunity for expanding the public transport system; stimulation of greater hous-

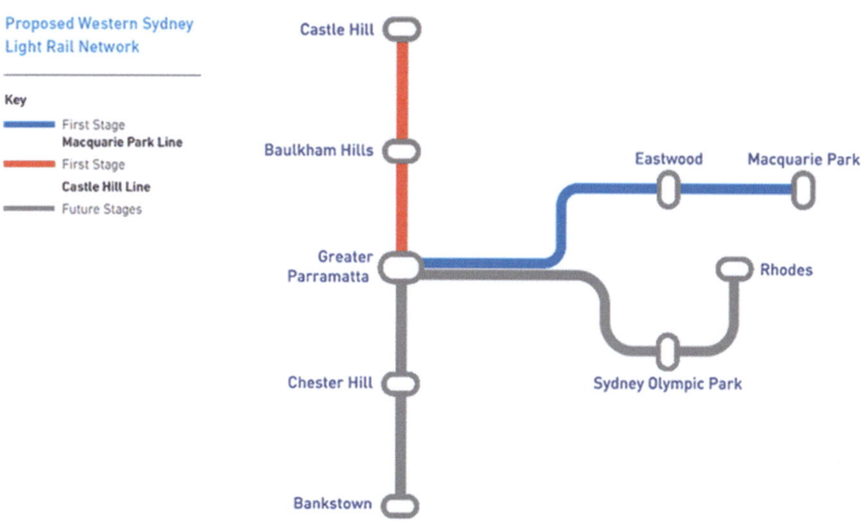

Plate 13.9 Proposed Western Sydney Light Rail Key Stations (*Source* Parramatta Council Web 2016)

Plate 13.10 Schematics of Proposed Western Sydney Light Rail. *Left to right*: Landscaped section; Serving the Parramatta CBD (*Source* Parramatta Council Web 2016)

ing opportunities along the corridors of the light rail system; and, opportunities for expansion of connecting open space connectors.

13.3.2.1 Conclusions on Light Rail Development

Having reviewed the light rail planning proposals, there are a number of significant benefits to the district that these proposals are likely to have. These can be summarized under a need for: increased opportunity for expanding the public transport system; stimulation of greater housing opportunities along the corridors of the light rail system; and, opportunities for expansion of connecting open space connectors.

13.3.3 Parramatta Rd Corridor

UrbanGrowth NSW released the *Draft Parramatta Road Urban Renewal Project (NSW* 2014a, b*)* (www.newparramattard.com.au) in late 2014. The strategy aims to facilitate the revitalization of the Parramatta Road Corridor running across ten council areas from Parramatta to Sydney CBD, including Auburn and Parramatta municipalities. The strategy identifies eight locations or 'precincts', close to public transport and other community infrastructure, that are suitable for long term growth and change. Within Greater Parramatta the selected precinct is Granville (Plate 13.11).

Several details of the Parramatta Rd Granville Precinct are provided in Plate 13.12. There is potential for Parramatta Rd to be below ground level, depending on final designs (left). The proposed Parramatta to Olympic Park and Strathfield light rail (middle) could parallel Parramatta Rd and tie in with the road upgrade. Finally, the proposed Metro Train line from Parramatta to Baulkham Hills District (right) will relieve some of the traffic along Parramatta Rd.

Plate 13.11 Parramatta Rd Urban Renewal Project – Granville Precinct (*Source* NSW 2014a)

Plate 13.12 Proposed Transport Upgrades Details (*Source* NSW Dept. of Planning Web 2016)

Having reviewed the Parramatta Rd urban transformation proposals, there are a number of significant effects on council areas within the Greater Parramatta district that these proposals are likely to have. These can be summarized under a need for: (a). more details on the urban design aspects of projects (i.e. height, bulk, overshadowing) planned within the redevelopment areas; impact of the additional population on services and Parramatta Rd transport; integration of open spaces lands and pedestrian linkages; greater housing affordability provisions; and, more environmental and heritage protection measures.

13.3.3.1 Conclusions on Parramatta Rd Corridor

Having reviewed the Parramatta Rd urban transformation proposals, there are a number of significant effects within the Sydney Inner West District that these proposals are likely to have. These can be summarized under a need for: (a). more details on the urban design aspects of projects (i.e. the height, bulk, overshadowing guidelines for new residential buildings within the redevelopment areas); impact of the additional population on services and current Parramatta Rd transport; integration of open spaces lands and pedestrian linkages; greater housing affordability provisions; and, details on environmental and heritage protection measures.

13.3.4 WestConnex

Finally, there has been considerable debate about the State's WestConnex project, with large sections of the community objecting to many aspects of the proposal. In rebuttal, the Government states the Westconnex will save 40 min on a typical journey between Parramatta and Sydney Airport and helping the city cope with an expected population increase of 1.5 million people over 20 years (*Source* WestConnex web 2016). There are a number of significant effects on council areas within the Greater Parramatta district that these proposals are likely to have. These can be summarized under a need for: (a). more details on the cost-benefits of these proposals (especially given the size of project expenditures anticipated); more details on road vs. public transport alternatives; more information on and timing aspects of proposals; extent that the proposals will be a catalyst for further redevelopment projects; long term effect of potentially inducing additional car orientated demands; impact of the additional population on services and housing that the road proposals will have; greater integration of open spaces lands and pedestrian linkages; and, more environmental (i.e. not alienating public open spaces such as in the Sydney Park impact) and heritage protection measures.

13.3.4.1 Conclusions on WestConnex

Having reviewed the WestConnex proposals, there are a number of significant effects within the Sydney Inner West district that these proposals are likely to have.

13.3.4.2 Conclusions on Developing Sustainable Transport

Having reviewed a number of Greater Parramatta District transport plans and projects and author (Rauscher) qualitative research field work (2014–2015), conclusions can be drawn. The transport projects as reviewed above (Long Term Transport Plans; Light Rail Development; Parramatta Rd Corridor; and, WestConnex) have seen major advances made in recent years, however many questions on transport planning at this level remain. In terms of meeting the planning principle *Developing Sustainable Transport* within the Greater Parramatta District, it is concluded that new planning arrangements with the State and councils need consideration. Options for these new planning arrangements are canvassed in Chap. 18 (Directions for Planning Sustainable Cities). Having examined transport projects, the planning principle Protecting the Natural Environment is examined next.

13.4 Protecting the Natural Environment

Having examined the first two urban planning principals (Developing Sustainable Urban Strategic Centers and Developing Sustainable Transport) within Greater Parramatta District, the third planning principal to be examined is Protecting the Natural Environment. Conservation protections within Greater Parramatta District (municipalities of Parramatta, Auburn and Holroyd) has to date been a cooperative effort among the three councils and the State. At times, on a common planning basis, councils outside this District have become involved (i.e. where waterways or green spaces affect these other councils). Two natural environment protection features of the District are selected to be examined, being: (1). Parramatta River; and, (2). Parramatta Greenway.

13.4.1 Parramatta River

The Parramatta River has a major catchment across 16 LGAs as noted earlier. The River straddles the Parramatta CBD, running north to south-east (Plate 13.13).

The River restoration is illustrated in Plate 13.14. A portion of Parramatta River foreshores is in its natural state, mangroves here (top-left). Pathways (top-center) have been developed along sections of the foreshores, especially close to Parramatta CBD. Picnic areas provided within parklands take advantage of the River's banks (top-right). The River Cat ferry (bottom-left) between Parramatta and Circular Quay at the Sydney CBD operate 7 days per week. Maritime works (bottom-center) continue along stretches of the River. Finally, there is one section of the River that houses refinery works (bottom-right) near the river's foreshores. The environment protection of Parramatta River overall, however, will need to be examined by the State for additional environmental protection measures. Examples of River restoration, from author field work (2014/2015), are examined next.

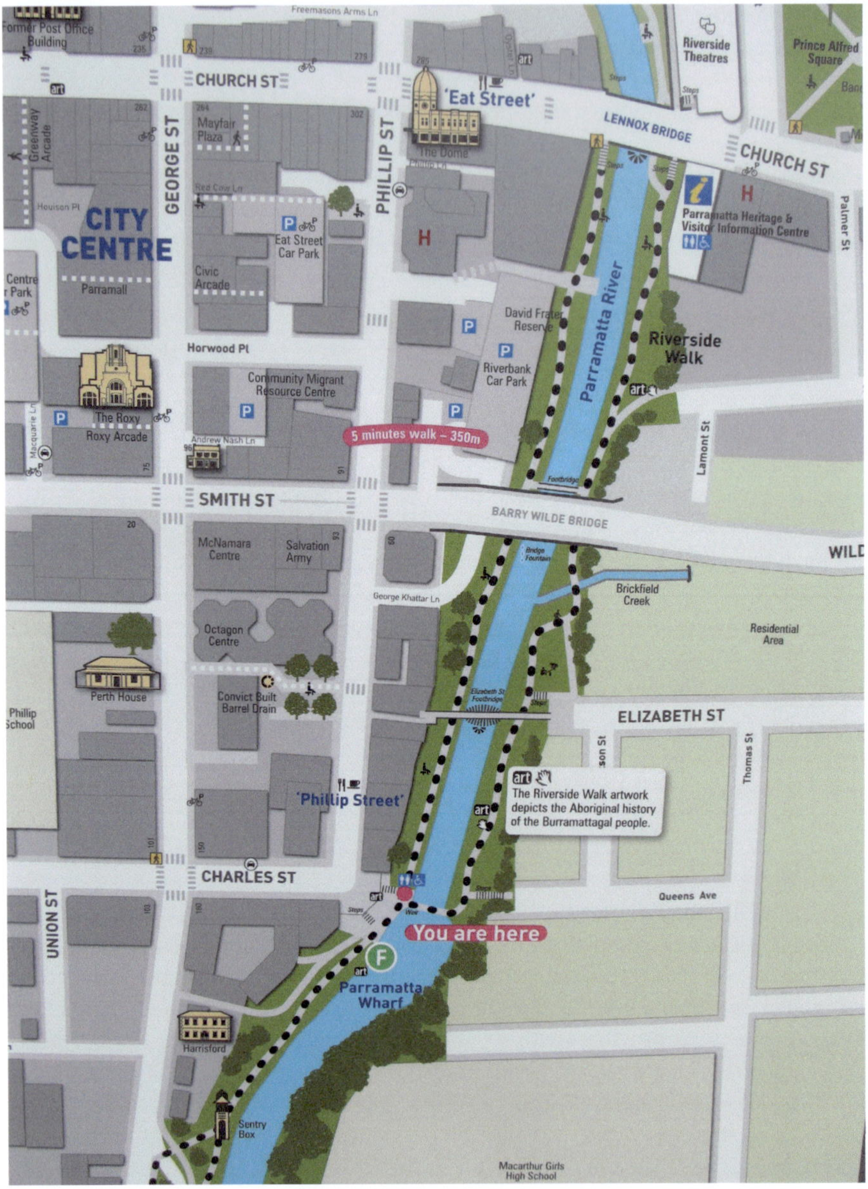

Plate 13.13 Schematic of Parramatta River Restoration (*Source* Parramatta Council 2016)

13.4.2 Parramatta Greenway

The Parramatta River 'GreenWay' is an urban green corridor in the Greater Parramatta District. The Greenway generally follows the route of the Parramatta River and its tributaries. Under *A Plan for Growing Sydney* (NSW 2014a, b) the State is proposing a Green Grid for Greater Parramatta. The State comments on the

Plate 13.14 Parramatta River Restoration. *Left to right* Mangroves under renewal; pathways along foreshore; Picnic areas in parklands; River Cat Ferry to Sydney CBD; Maritime Works; and, Refinery works downstream (*Source* Raymond Rauscher 2015)

broader Sydney Green Grid in that the State will: (a). include open spaces, parks, bushland, natural areas, waterway corridors and tree-lined streetscapes in a network that connects homes to centres, public transport, jobs and recreation; (b). build up the open space network already established the Metropolitan Greenspace Program and Regional Tracks and Tails Framework. The Parramatta Grid will provide environmental benefits such as better water and air quality, social benefits such as improved physical and mental health, and economic benefits such as higher property values and business retention (Source: *A Plan for Growing Sydney* 2014b). The environment protection of the Greenway overall, however, will need to be examined by the State for additional environmental protection measures. Examples of Greenway connections, from author field work (2014/2015), are examined next.

The Greenway connections (Plate 13.15) (Parramatta, Auburn and Holroyd) include a variety of natural and manmade spaces accessible to residents of the district. Greenway pathways here stretch along the Parramatta River (top-left). Here a formal memorial area has been created (top-center). Infrastructure has been provided through green spaces to provide public access (top-right). Provisions have been made for active recreation in this section of Parramatta River foreshore (bottom-left). There is increasing attention by the LGA councils to provide open space art fixtures as shown here (bottom-center). Finally, foreshore recreation areas connect through the district (bottom-right).

Plate 13.15 Greater Parramatta District Greenway. *Left to right* Greenway pathways and solar lighting; Formal memorial areas; Pedestrian access to open spaces; active play areas; open space art fixtures; and, Foreshore recreation areas (*Source* Raymond Rauscher 2015)

13.4.2.1 Conclusions on Protecting the Natural Environment

Having reviewed a number of a number of Parramatta River and Parramatta Greenway planning issues as above and author (Rauscher) qualitative research field work (2014–2015), conclusions can be drawn. The environment protection projects have seen major advances made in recent years, however many questions on environmental protection at this district level remain (aa noted in the review above). It can be concluded the State, in partnership with councils, needs to further address the planning principal Protecting the Natural Environment. Options for these new planning arrangements are canvassed in Chap. 18 (Directions for Planning Sustainable Cities). The subject of amalgamation of councils in the Greater Parramatta District is examined next.

13.5 Amalgamations of Councils and Greater Parramatta District

The subject of amalgamations of Greater Sydney councils (given implications for planning at municipal and Greater Sydney levels) was reviewed in Chap. 2. NSW Premier Mike Baird announced in Dec 2015, in examining the Greater Parramatta District (Plate 13.16), the State's preference: (a). for boundaries of Parramatta City to be adjusted to take in parts of Auburn (Olympic Park and Wentworth Point for example), Hornsby, and Baulkham Hills: and, (b). Auburn to merge with Holroyd and for the merged councils to absorb the Granville area of Parramatta.

The position on the councils' merger questions were reviewed in the respective chapter on each council. Every council adopted its preference on the merger question, with Holroyd wanting to stand-alone and Auburn wanting to merge with Burwood and Canada Bay. All councils remained opposed to forced amalgamations. The Mayors of all councils urged residents to have their say on the issue of amalgamations of councils. Councils noted that submissions to the new Council Boundary Review were being received (www.councilboundaryreview.nsw.gov.au). In the interim, the Councils noted, the Minister for Local Government had commenced the process of forcibly amalgamating the two groups of councils in accordance to State preference as noted earlier. Councils noted the Minister said he expected the new merged councils to be 'proclaimed' on 1 July 2016 (note the 12th May 2016 *Local Government (Council Amalgamations) Proclamation 2016* noted in Chap. 2). Results of the review of merger proposals would be referred to the

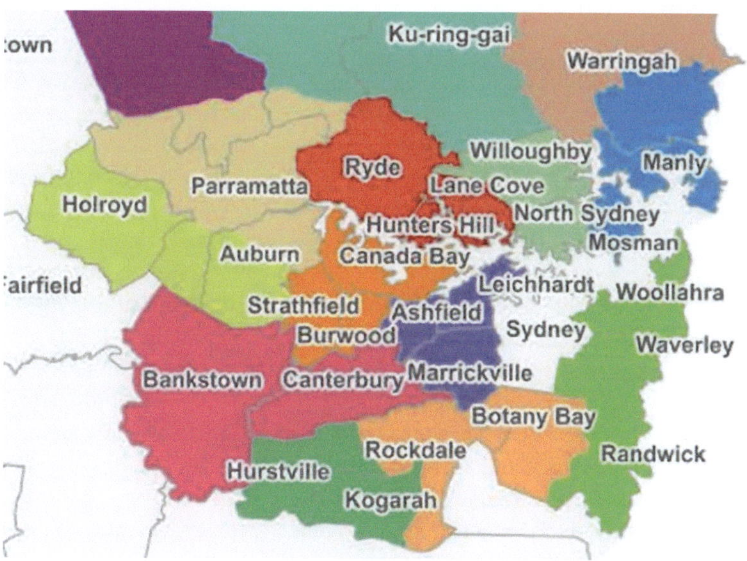

Plate 13.16 Amalgamations of Greater Sydney Councils – Greater Parramatta. *Left to right*: Holroyd, Parramatta, and Auburn (*Source* NSW Dept. of Local Government Web 2016)

Chief Executive of the Office of Local Government for examination and report under the Local Government Act (the Act). In turn, the Chief Executive had appointed a Delegate for the proposed merged Auburn and Holroyd councils and boundary adjustments of Parramatta Council and affected adjacent councils. The Delegates (2) would examine the proposals, conduct public inquiry, call for written submissions, and then report back to the Minister on the merger and boundary adjustment proposals.

13.5.1 Conclusion on Amalgamations

The State, in examining the Greater Parramatta District, expressed a preference: (a). boundary changes for Parramatta and adjacent councils; and, for Auburn and Holroyd councils to merge. Holroyd indicated to the State that it preferred to stand alone following Council's research and public engagement. Auburn preferred merging with Burwood and Canada Bay. At the end of the State review process (Council Boundary Review) the State would determine what amalgamations or boundary adjustments as proposed would proceed or not proceed. The 12th May 2016 *Local Government (Council Amalgamations) Proclamation 2016* proposed mergers and boundary changes to nominated Sydney Inner West councils was noted. This question will be further addressed in Chap. 18.

13.6 Overall Conclusions

Having completed an assessment of the development and planning of Greater Parramatta District (critiquing plans and developments against the adopted three planning principles) there are a number of overall conclusions reached. Firstly, examining conclusions reached in assessing development and planning in each of the three cities (Parramatta, Auburn, and Holroyd), it was concluded that each of these areas had been experiencing accelerated development and changes to the composition of their populations. Given these changes, and examining these council's responses, it was also concluded these councils were all moving in the right direction in increasing their sustainable urban planning (SUP) practices. It was also concluded, however, that there would be significant future impacts on all these cities from Greater Sydney growth and development. In reviewing potential councils and State responses to this future scenario, it was concluded that new partnerships between these councils and the State would be needed. A major initiative from the State (2009 onwards) was to investigate local government reform. This included a potential amalgamation of councils (e.g. Parramatta boundary changes; and, Auburn and Holroyd amalgamating). Holroyd council had during this State amalgamation process, however, expressed an intent to stand alone and not amalgamate. The State started a forced amalgamation process in early 2016, results not to date determined. From the city level to the district, the conclusions reached from the assessment of the development and planning of Greater Parramatta District as a whole will be examined next.

In examining the first planning principle, Developing Sustainable Urban Centers, two centers were examined: Parramatta District Centre (Parramatta) and the Sydney Olympic Park Growth Center (Auburn). Having reviewed the Parramatta District Center proposal, there are a number of significant effects on council areas within Greater Parramatta that this project is likely to have. These can be summarized under a need for: (a). more details on the urban design aspects of projects (i.e. height, bulk, overshadowing); impact of the additional population on services and transport; greater allocations of open space lands and pedestrian linkages; greater housing affordability provisions; and, more environmental and heritage protection measures. Secondly, having reviewed aspects of the Olympic Park Growth Centre, there are a number of significant issues (as with Parramatta District Center) for the councils within the Greater Parramatta. These can be summarized under needs to consider: (a). impacts of the urban design aspects of projects (i.e. height, bulk, and overshadowing); impact of the additional population on services and transport; greater allocations of open space lands and pedestrian linkages; greater housing affordability provisions; and, more environmental and heritage protection measures. Having reviewed these urban centers and examined author's (Rauscher) qualitative research field work (2014–2015) (Chaps. 10, 11 and 12), it can be concluded the State, in partnership with councils, needs to further address the impact of development under the State planning principal Developing Sustainable Urban Centers (Chap. 2). The urban centres (as reviewed above) have seen major advances made in recent years. Many questions, however, of the State and councils conferring on planning at this level remain. The planning principle Developing Sustainable Transport in Greater Parramatta District was assessed next.

Having reviewed a number of Greater Parramatta transport plans and projects and author (Rauscher) qualitative research field work (2014–2015), a number of conclusions can be drawn. The transport projects (light rail; bus corridors; and, WestConnex) have seen major advances made in recent years, however many questions on transport planning at this level remain. In terms of meeting the planning principle Developing Sustainable Transport within the Greater Parramatta District, it is concluded that new planning arrangements between the State and councils need consideration. Options for these new planning arrangements are canvassed in Chap. 18 (Directions for Planning Sustainable Cities). Having examined district transport projects, results of applying the planning principle Protecting the Natural Environment was examined next.

Having reviewed two Greater Parramatta environment protection projects (i.e. Parramatta River and Parramatta Greenway) and author (Rauscher) qualitative research field work (2014–2015), conclusions can be drawn. The environment protection projects (as reviewed above) have seen major advances made in recent years; however, many questions on environmental protection at this district level remain (as noted in the review). It is concluded the State, in partnership with councils, needs to further address new planning arrangements in addressing protection of the natural environment. Options for these new planning arrangements are canvassed in Chap. 18.

13.6.1 Overall Conclusion

Having addressed the question of planning in Greater Parramatta District (critiquing plans and developments against the adopted three planning principles) an overall conclusion is drawn that new planning arrangements among councils and the State (responsible for planning the District) are needed to address development and planning of the District. This conclusion will be further addressed within Chap. 18 (Directions for Planning Sustainable Cities). Having examined the Greater Parramatta District, the St George District is examined next (Chaps. 14, 15, 16 and 17).

13.7 Summary

Planning of Greater Parramatta District, the second of three case study districts of Greater Sydney, was examined in this chapter. Having examined the question of global transition and planning practices of the three cities within the District (Chaps.10, 11 and 12), the chapter addressed: (1). planning the Greater Parramatta District as a whole; and, (2). the extent the State (responsible for district planning) was meeting the planning issues. Key planning principles (Chap. 2) to apply to Greater Sydney (and this district) were reviewed. These key principles include: (1). Developing Sustainable Urban Centres; (2). Developing Sustainable Transport; and, (3). Protecting the Natural Environment. Sample plans and developments within Greater Parramatta were critiqued under each of these principles. After reaching a conclusion under each principle, an overall conclusion was drawn that new planning arrangements among Councils and the State (planning the District) were needed in planning the District. Finally, it was noted that these Greater Parramatta District conclusions would be assessed alongside conclusions reached in examining Sydney Inner West District (Chaps. 3, 4, 5, 6, 7, 8 and 9) and St George District (Chaps. 14, 15, 16 and 17). All these conclusions are examined (Chap. 18) in the context of SUP practices applied to Greater Sydney as a global city (affecting municipal areas within Greater Parramatta and the district as a whole). Having examined the Sydney Inner West District and the Greater Parramatta District, the St George District is examined next (Chaps. 14, 15, 16 and 17).

References

New South Wales (2014a) Draft Parramatta road urban renewal project. NSW, Sydney
New South Wales (2014b) A plan to grow Sydney. NSW, Sydney

Part IV
Planning in St George District

Part I introduced the planning of global cities and sustainable urban planning (SUP) (Chap. 1). A methodology for measuring the extent SUP practices are being applied at municipal and district levels was then presented in Chap. 2. The basis of this methodology is the selection and application of three key SUP principles (herein called planning principles). These planning principles are: Developing Sustainable Urban Centers; Developing Sustainable Transport; and, Protecting the Natural Environment. With these planning principles in mind, three questions are posed for each municipality (in separate chapters): (1) what is the pattern of urban transition the municipality is experiencing; (2) what is the extent SUP practices are being applied by that municipal council; and, (3) what is the impact of Greater Sydney planning on the development of the municipality, including proposed council amalgamations. Part II involved the application of this methodology to the Sydney Inner West District (Chaps. 3, 4, 5, 6, 7, 8 and 9) and Part III to the Greater Parramatta District (Chaps. 10, 11, 12 and 13). This Part IV now examines St George District cities (Hurstville, Kogarah and Rockdale) (Chaps. 14, 15 and 16), followed by an examination of the whole of the district (Chap. 17). Results from this district assessment (and results of the Sydney Inner West and Greater Parramatta district assessments) will contribute to addressing directions for planning a sustainable Greater Sydney and sustainable cities anywhere (Chap. 18).

Chapter 14
Planning Hurstville, Greater Sydney

Dedication *This chapter is dedicated to two past identities in the St George District, Hugh Patrick; and Jack Lindwall (depicted on heritage plaque in Hurstville) (Plate 14.1).*

Plate 14.1 Hugh Patrick; and Jack Lindwall

A snippet on each of the identities follows:

Hugh Patrick (1844–1922) served Hurstville as Mayor in 1895, 1902 and 1913

Jack Lindwall (1918–2000) was an honoured cricketer along with his brother Ray (Source: Wikipedia 2016)

R.C. Rauscher, S. Momtaz, *Cities in Global Transition*,
DOI 10.1007/978-3-319-39865-5_14

Abstract The chapter addresses the extent Hurstville Council (herein referred to as Hurstville) is applying sustainable urban planning (SUP) practices (see Chap. 1) in planning the municipal area. The chapter acknowledges the *Local Government (Council Amalgamations) Proclamation 2016* merging Hurstville and Kogarah councils noted in Chap. 2. Hurstville (in St George District) is one of twelve municipalities within three Greater Sydney Districts (Sydney Inner West, Greater Parramatta and St George), all considered to be in global transition (see Chap. 2). Examining SUP practices is based on the application of three key SUP principles (as adopted in Chap. 2), being: Developing Sustainable Urban Centers; Developing Sustainable Transport; and, Protecting the Natural Environment.

Firstly, a background to Hurstville development and population changes (e.g. global movement of population) is outlined. To examine the pattern of urban development, four phases of Hurstville development are examined: Early Hurstville; Traditional Hurstville; New Hurstville; and, Future Hurstville. Given conclusions reached here, key Hurstville Council planning documents are then critiqued to determine the extent Council applies SUP practices. In addition, given State planning of Greater Sydney affects Hurstville, a second question is posed: what is the impact of Greater Sydney planning and development on Hurstville? In examining this question, examples of State plans (including council amalgamations under the Proclamation noted above) likely to affect Hurstville are addressed. By chapter's end, conclusions are drawn on Council planning and Greater Sydney planning. The chapter notes that these conclusions will be addressed further in: Chap. 17 (examining the extent of SUP practices being applied over the whole of St George District); and, Chap. 18 (examining directions for planning a sustainable Greater Sydney or a sustainable city anywhere).

14.1 Background and Population Change

Hurstville City (herein called Hurstville) in the context of St George (as introduced in Chap. 2) is adjacent to the Georges River and south of Sydney Inner West to the south of the City of Sydney. Under the *Local Government (Council Amalgamations) Proclamation 2016* Hurstville was merged with Kogarah council as noted in Chap. 2. Suburbs of Hurstville include: Beverly Hills, Hurstville, Kingsgrove (partly within City of Canterbury & City of Rockdale), Lugarno, Mortdale, Narwee, Oatley, Peakhurst, Peakhurst Heights, Penshurst, Riverwood, Roselands. The area now known as Hurstville, home to the Wangal people who inhabited a major portion of the St George district. Prior to European settlement, the Wangal Aboriginal people lived around and beyond the St George District (cities of Hurstville, Kogarah, and Rockdale) (Chap. 2). European settlement began in the 1790s. Hurstville became a Municipality in 1887, eventually becoming a City in 1988. Hurstville prides itself as one of the most multicultural communities in Australia.

City of Hurstville, the second largest populated city (80,823 residents) of the three cities of St George District (Kogarah the smallest with 55,806 residents and Rockdale the largest with 97,340 residents). In respect to linguistically diverse communities Hurstville City has a significantly higher than average proportion (52.5 %) of family members speaking two or more languages at home (national average is 20.4 %). Finally, the proportion of Hurstville City residents who stated a religious affiliation with Islam was in excess of four times the national average respectively, and Hinduism was in excess of six times.

The Australian Bureau of Statistics comments on Hurstville City ancestry and languages spoken at home:

> According to the 2011 Australian Bureau of Statistics Census of Population, there were 26,040 residents in Hurstville. 68.4 % stated they were born overseas, with the top countries of birth being China (34.2 %), Hong Kong (4.6 %), and Nepal (2.7 %). English was stated as the only language spoken at home by 21.8 % of residents and the most common other languages spoken were Mandarin (26.7 %), Cantonese (20.4 %), Greek (3.7 %), and Arabic (3.6 %). The most common responses for religious affiliation were no religion (35.6 %), Catholic (15.0 %), Buddhism (10.0 %) and Orthodox (7.5 %). The largest ancestry group was Chinese, reported by 47.5 % of respondents, followed by English (7.6 %), Australian (6.8 %), and other or undeclared ancestry (38.1 %). (*Source* ABS 2011)

Finally, in author's (Rauscher) qualitative research (2014/2015), there is a poem written to reflect the cultural diversity of Hurstville (end of chapter under Notation 14.1) *Hurstville – Global Central* With this background introduction to Hurstville, the extent of Hurstville's development phases are examined next.

14.2 Phases of Development

An examination of the development phases of any locality provides a basis of understanding current trends and likely future directions of growth of that locality. Following a look at the background of Hurstville four phases of Hurstville development will be examined. These phases include: Early Hurstville; Traditional Hurstville; New Hurstville; and, Future Hurstville Directions. To assist in addressing this question, author (Rauscher) qualitative research field work was completed in Hurstville Municipality (2014–2015). A brief look at early Hurstville is shown Plate 14.2. The Hurstville Hotel (top-left) is shown here in Forest Rd. The long running Packham's bread delivery is shown here (top-right). Most shopping was done locally (Ashley's here in 1952) (middle-left). King Georges River Rd was a critical transport link (middle-right). The St George area was a popular district to build large estates as illustrated here (bottom-left). Finally, recreation boating (bottom-right) was a popular past time alongside Tom Ugly Bridge.

The transformation of Hurstville City downtown has been occurring over several decades (1980s onwards). These changes as illustrated in 'Traditional Hurstville' (Plate 14.3) include a number of commercial, service and cultural developments.

Plate 14.2 Early Hurstville. *Left to right* Early Hurstville Hotel; Packham's bread delivery; Ashley's shopping; King Georges River Rd; Early estate house; and Recreation boating at Tom Ugly Bridge

Footpaths in the main commercial area have been upgraded ((top-left). There is a popular civic memorial area near the train station (top-right). The commercial area downtown has expanded (middle-left). The Fire Station, still downtown, has a high rise residential block to the rear (middle-right). A mural at the rail station (bottom-left) depicts the Illawarra train coming into Hurstville Stn. Finally, the Council has been instrumental in the creation of the downtown Hurstville Gallery and Museum (bottom-right).

Plate 14.3 Traditional Hurstville Today. *Left to right*: Main shopping street; Statue of Miles Franklin, writer; Railway station entry; Fire station with high rise residential to the rear; Street wall art depicting the Illawarra train at Hurstville; and, the Hurstville Gallery and Museum

Plate 14.4 New Hurstville. *Left to right* Multistorey Waratah Pvt Hospital; Club Central; High rise residential; Six storey residential; High rise residential; and, Eight storey residential

The 'New Hurstville' (Plate 14.4) reflects a continuing change of the urban landscape. A section of the inner Hurstville area is zoned for multi- story residential complexes, some examples follow. The multistorey Waratah Pvt Hospital has been built in the CBD (top- left). The licensed Club Central (top-middle) occupies a key commercial downtown corner. There are numerous residential high rise buildings close to the CBD (top-right). Six storey residential buildings are integrated into the profile of the CBD (bottom-left). In contrast, high rise residential (bottom-centre) buildings present a higher profile and are built either in the CBD or in zones adjacent to the CBD. Finally, there is a scattering of mixed level developments of 3–6 stories that were built in an early phase of Hurstville CBD expansion (bottom-right).

Finally, Future Hurstville (Plate 14.5) reflects the continuing building boom in residential development at increasing densities throughout the municipality. The trend to build high rise residential (left), given the district centre that Hurstville represents, is likely to continue. The concentration of that high rise development (centre) is likely to find ever decreasing large redevelopment sites such as South Hurstville here. Finally, medium density demand is likely to continue (right) given transport advantages of Hurstville and ability to build on amalgamated residential older home sites. Factors of building design (such as height, density, housing afford-

Plate 14.5 Future Hurstville

ability, overshadowing and open space provisions) are likely to remain issues of development importance and contention within communities at the local government level into the future.

14.3 Conclusion

Given the review of development phases of Hurstville (above) and changes in population demographics, it can be concluded that Hurstville has been experiencing accelerated global transition (2000s) (Question 1). The planning of Hurstville (merged with Kogarah council as noted above under the Proclamation) in meeting development demands and applying sustainable urban planning (SUP) practices is examined next.

14.4 Hurstville and Sustainable Urban Planning

Planning of Hurstville City has been taking place since it became a local government body in 1861. A review of planning in Hurstville will pay particular attention to applications of sustainable urban planning (SUP) (as outlined in Chap. 1) practices (including use of indicators of sustainability) (Chap. 2). Hurstville Council key planning documents (a–z with date and web reference) are listed in Information Box 14.1. A summary of these documents and a brief critique assist with this review of Hurstville planning (documents can be sourced under the web link as noted in the right column).

The strategic planning focus of the Hurstville Council is contained within *Hurstville City Master Plan* (Hurstville 2004). Components of that Plan include

Information Box 14.1 Hurstville Municipality Planning

Hurstville City Development Control Plan (2012b)	www.hurstville.nsw.gov.au
Hurstville City Local Environment Plan (2012a)	www.hurstville.nsw.gov.au
Hurstville City Master Plan (Hurstville 2004)	www.hurstville.nsw.gov.au

Source Webs as noted February 2016

environment, transport, and urban design components. The Plan also links to the long-term future of the Sydney Metropolitan Region, noted above. The *Hurstville Local Environment Plan* (Hurstville City 2012a) *was adopted as the overriding plan for the City. Complimenting this plan is the Hurstville Development Control Plan* (Hurstville City 2012b). Hurstville City transport objectives affecting the implementation of these above plans are based on the continued expansion of the bus routes and greater access to the rail station heads. Removing traffic where feasible from local streets, in favour of through roads, is another prime target under Council's traffic management planning. The function and impact of the WestConnex motorway section through Hurstville City will be addressed in Chap. 17 (Planning St George District).

Having examined environmental and transport objectives of the Hurstville City Council the third planning component is urban design. There are a number of Hurstville City design projects under the *Hurstville City Master Plan* (Plate 14.6). One of the overriding goals of the Master Plan is to improve pedestrian access through the Super Centre and unite both sides of the town. This would be done through a number of strategically placed pedestrian north-south connections. Plans include the Woodville St bus interchange (top left) to upgrade transport connections. Council notes that a study will be undertaken on; (a). the capacity of the proposed bus interchange to carry both local and the new regional bus services; (b). the bus movements around the new interchange; and, (c). a concept design for the interchange. A concourse above rail tracks (top right) is planned to take advantage of air space. A new civic precinct (bottom left) will open up a central meeting place. Finally, more shade will be created with the introduction of pergolas (bottom right).

Hurstville Council will undertake a number of additional plans and studies in implementing the Master Plan. The Council will undertake (for example) Precinct Plans, with Council commenting:

> Council will undertake a detailed analysis of the City Centre to integrate urban design principles, community & cultural programs, retail and commercial trends. The precinct plan draws together the various elements of the Hurstville City Centre such as community, residential, commercial, public and private domain to identify a unique character, by area/precinct and improve their interrelationship within the whole city centre. The Precinct plans aim to identify a series of practical measures to enhance city center attraction, legibility and sustainability. (*Source* Hurstville Council Web 2015)

Plate 14.6 Hurstville Designs. *Left to right*: Woodville St bus interchange; Concourse above rail tracks; New Civic precinct; and introduction of pergolas (*Source* Hurstville Council Web 2015)

Under infrastructure planning the Council notes:

An infrastructure plan will examine the physical requirements and quality standards expected in a regional centre such as roads, drainage, traffic and parking, pedestrian access, recreation, public art, festivals, niche markets, noise minimization, lighting, landscaping, signage, cultural diversity, community safety, street furniture, community facilities, outdoor dining, access policy, shelter and wind tunnel reduction and whole of life maintenance of assets. A detailed study needs to be undertaken to establish the development potential for new commercial development in the area identified as the CBD core. This would feed into the financial analysis on the options for funding new public infrastructure. This study will also address improved pedestrian links between key destination points. A steering committee has been formed with membership of Hurstville City Council and Department of Planning & Environment for improved liaison with transport agencies and private transport operators. (*Source* Hurstville Council Web 2015)

14.4.1 Conclusion on Council Planning and SUP Practices

It can be concluded, having reviewed Hurstville City Council planning documents and their applications, it can be concluded that Council is moving in the direction of applying sustainable urban planning (SUP) practices (Question 2 above). The State's Greater Sydney planning (including infrastructure projects) and State council amalgamation proposals likely to affect Hurstville (Question 3) are reviewed next.

14.5 Greater Sydney Planning and Hurstville

There are planning and developments of Greater Sydney likely to impact on Hurstville City. In many instances the Hurstville City Council has had the task of providing: planning input on State proposals affecting Hurstville. These include making and reviewing public submissions; implementing State planning policies at the local level; joint partnership on service or infrastructure projects; and, monitoring State or private development projects affecting Hurstville City. Other Council planning responsibilities under State legislation that also faces the Council include: engaging the municipality community (e.g. residents, business, services, development interests, and institutions) on State planning policies and development proposals; statutory plan making (e.g. adopting local environment plans (LEPs); and, reporting back to the State on a range of planning matters (i.e. State proposed roads, state of environment reporting and infrastructure projects).

Greater Sydney planning requirements were set by the NSW State in 2006, with Hurstville Council adopting in 2011 xxx. Council's plans had to be consistent with the *Sydney Metropolitan Strategy* (NSW 2011). Council's complimentary town planning documents were the *Hurstville Local Environment Plan* (Hurstville 2012b) and the *Hurstville Development Control Plan* (2012b).

Moving from Hurstville City planning to State planning in Hurstville City, the NSW *A Plan for Growing Sydney* (NSW 2014), as noted earlier (Chap. 2), is the strategic document that outlines a vision for Sydney to 2031. The Plan forecasts a population growth of 1.6m (to be added to the Greater Sydney population of 4.84m at June 2014). The document sets out objectives and actions under seven themes or strategies, including: economy and employment; centres and corridors; housing; transport; environment and resources; parks and public places; and, implementation and governance. St George is within the South Subregion as designated in the above Plan and the *South Subregion Draft Subregional Strategy* (SSDSS) (NSW 2014).

The State designates this district within the South Subregion (Plate 14.7) under *A Plan for Growing Sydney* (NSW 2014). Note within this subregion: (a). circles representing 'strategic centers' of Kogarah and Hurstville; (b). WestConnex road project at top right to left; (c). the three blue bands right to left representing 'urban renewal corridors'; (d). the broken black dotted line right-top to bottom-left repre-

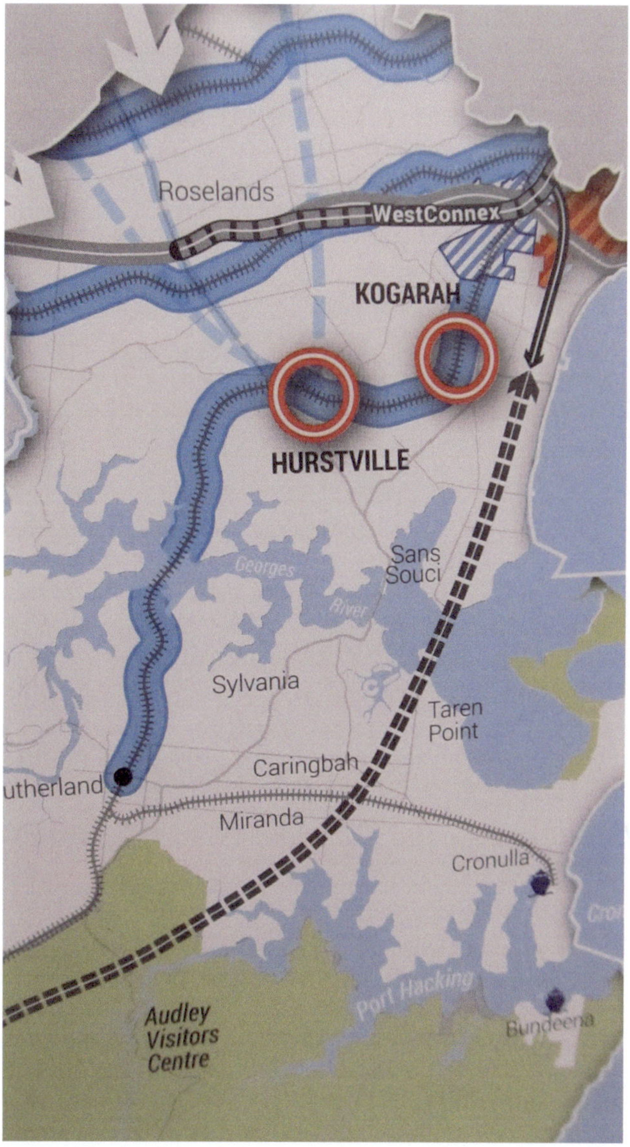

Plate 14.7 Sydney South Subregion (*Source*: *A Plan for Growing Sydney*, NSW 2014)

senting 'road/motorway investigation; and, blue crossed lines top-right (Rockdale) representing a 'priority precinct'; and, finally, red crossed top-right representing 'Sydney airport precinct'.

In examining this Plan and its impact on Marrickville, three key planning principles as adopted in Chap. 2 are addressed: (1). Developing Sustainable Urban Centres; (2). Developing Sustainable Transport; and, (3). Protecting the Natural

Environment. The planning principle Developing Sustainable Urban Centres is examined first. The Plan notes the State will work with Hurstville City Council to retain a commercial core in Hurstville, as required, for long-term employment growth. The State also intends to provide capacity for additional mixed-use development in Hurstville including offices, retail, services and new homes. The planning principle Developing Sustainable Transport is examined next. Hurstville Council, along with other St George councils, were asked in late 2014 to comment on the proposed State's WestConnex (NSW 2014) project. The Plan also notes other key transport projects (to be addressed further in reviewing the whole of St George District in Chap. 17) likely to affect Hurstville, including:

1. Building the M5 East Airport Link motorway as part of WestConnex.
2. Investigating the F6 motorway corridor.
3. Preserving potential future transport corridors for Sydney Rapid Transit with a possible extension of Sydney Rapid Transit to Hurstville.

Finally, the planning principle Protecting the Natural Environment is examined next. The Plan commits the State to 'protect the health of the subregions waterways and aquatic habitats'. This goal, for example, would apply to the prime conservation challenges in Hurstville. This would include the protection of bushlands, waterways and foreshores within Hurstville. The efforts of Hurstville Council in protecting, for example, waterways and foreshores of the Georges River and Botany Bay in cooperation with other councils in St George District will be outlined in more detail in Chap. 17.

14.5.1 Conclusion on Hurstville and Greater Sydney Planning

Having reviewed above the State's Greater Sydney planning (including infrastructure projects) it can be concluded the implementation of these plans and developments is likely to have a significant impact on Hurstville (Question 3). The extent of this impact will be better understood when these State plans are further examined in the context of the whole of the Sydney Inner West District (including Hurstville) in Chap. 9. Finally, in examining State proposed council amalgamations, options for councils (including Hurstville) working together and in partnership with the State to meet local and district needs will be addressed in Chap. 18 (Planning Reforms and Directions for Planning Sustainable Cities). The subject to amalgamations of councils and how this affects Hurstville examined next.

14.6 Amalgamation of Councils and Hurstville

The subject of amalgamations of Greater Sydney councils (given implications for planning at municipal and Greater Sydney levels) was reviewed in Chap. 2. NSW Premier Mike Baird announced in Dec 2015 the State's preference for: (a). Hurstville

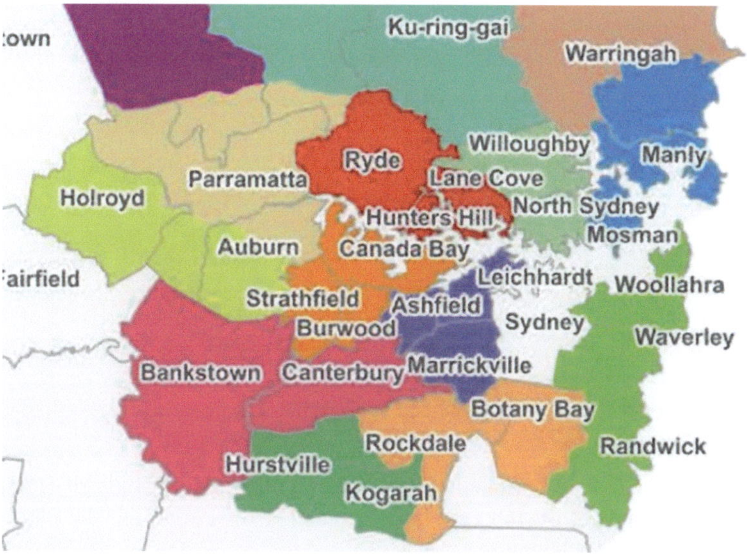

Plate 14.8 Amalgamation of Councils. *Bottom to Top*: Hurstville and Kogarah; and, Rockdale and Botany Bay (*Source*: NSW Dept. of Local Government Web 2016)

Council to merge with Kogarah; and (b). Rockdale Council to merge with Botany Bay Council (outside the St George District (Plate 14.8)).

Hurstville Council as well as Kogarah Council remained opposed to forced amalgamations. Kogarah City Council preferred to stand-alone as a council and not to merge. The Mayor of Hurstville City urged all residents to have their say on the issue of amalgamations of councils. Council noted that submissions to the new Council Boundary Review were open till the end of February (www.councilboundaryreview.nsw.gov.au). In the interim, the Council noted, the Minister for Local Government had commenced the process of forcibly amalgamating Hurstville and Kogarah councils. Council also noted the Minister said he expected the new council (with adjusted boundaries) to be 'proclaimed' on 1 July 2016 (note the 12th May 2016 *Local Government (Council Amalgamations) Proclamation 2016* noted in Chap. 2). The results of the review of proposed boundary adjustments and merger (Hurstville and Kogarah councils) would be referred to the Chief Executive of the Office of Local Government for examination and report under the Local Government Act. In turn, the Chief Executive had appointed a Delegate who would examine the proposals, conduct a public inquiry, call for written submissions, and then report back to the Minister. In turn, the Chief Executive had appointed a Delegate who would examine the proposals, conduct a public inquiry review, call for written submissions, and then report back to the Minister.

14.6.1 Conclusion on Amalgamations and Hurstville City Council

Hurstville City Council stated its case to the State, accepting that the State preferred Kogarah Council to merge with Hurstville. At the end of the State review process (Council Boundary Review) the State would determine what amalgamations as proposed would proceed or not proceed. The 12th May 2016 *Local Government (Council Amalgamations) Proclamation 2016* merging Hurstville and Kogarah councils was noted. This question will be further addressed in Chap. 18.

14.7 Overall Conclusions

Having completed an assessment of the development and planning of Hurstville, there are a number of overall conclusions reached. In examining the patterns of urban development of Hurstville it was concluded that Hurstville has been experiencing accelerated development and changes to the composition of its population. Given these changes, and examining Hurstville Council's response, it was also concluded Council is moving in the right direction in increasing its sustainable urban planning (SUP) practices. It was also concluded, however, that there would be significant future impacts on Hurstville from Greater Sydney growth and development. In reviewing potential Council and State responses to this future scenario, it was concluded that new partnerships between the Council and the State would be needed. A major initiative from the State (2009 onwards) was to investigate local government reform, including a potential amalgamation of councils (e.g. Hurstville merging with Kogarah; and Rockdale merging with Botany City). The 12th May 2016 *Local Government (Council Amalgamations) Proclamation 2016* that merged Hurstville and Kogarah councils was acknowledged. These overall chapter conclusions will be examined against results of assessing planning and development of the remainder of the three cities of St George (Chaps. 14, 15 and 16) and the district as a whole (Chap. 17). Results from the district assessment (and assessment of Sydney Inner West and Greater Parramatta districts in later chapters) will contribute to assessing how to plan for a sustainable Greater Sydney and a sustainable city anywhere (Chap. 18).

14.8 Summary

The chapter addressed the extent Holroyd Council was applying sustainable urban planning (SUP) practices (see Chap. 1) in planning the municipal area. It was noted that Hurstville, in the St George District, was merged with Kogarah council under the 12th May 2016 *Local Government (Council Amalgamations) Proclamation*

2016 (Chap. 2). It was concluded that Hurstville, one of twelve municipalities within three Greater Sydney Districts (Sydney Inner West, Greater Parramatta and St George) is considered to be in global transition. SUP practices (based on the application of three key SUP principles) were introduced, being: Developing Sustainable Urban Centers; Developing Sustainable Transport; and, Protecting the Natural Environment. A review of Hurstville background and population changes (i.e. global movement of population) was outlined. In examining the pattern of urban development, four phases of Hurstville development are examined: Early Hurstville; Traditional Hurstville; New Hurstville; and, Future Hurstville. With conclusions drawn, key Hurstville Council planning documents were then critiqued to determine the extent Council applies SUP practices. In addition, given State planning of Greater Sydney affects Hurstville, a second question was posed: what is the impact of Greater Sydney planning and development on Hurstville?. In examining this question, examples of State plans (including State proposed council amalgamations) likely to impact on Hurstville were addressed. By chapter's end, conclusions were drawn on Council planning and impact of Greater Sydney planning on Hurstville. The chapter noted that these conclusions would be addressed further in: Chap. 9 (examining the extent of SUP practices being applied over the whole of St George District); and, Chap. 18 (examining directions for planning a sustainable Greater Sydney and a sustainable city anywhere).

Notation 14.1 Cultural Diversity Poem – Hurstville
Poem (Rauscher 2015 unpublished) reflecting the diversity of Auburn:

Hurstville – Global Central (# 17)

Hurstville was the furthest a family would contemplate moving to in Sydney's early years
Today that center is a buzz of Chinese, Korean, Arabic new settlers
The history of Hurstville accords great sportspeople, politics and artists
Miles Franklin, Eddie Charlton, Fleur Mellor Dr Evatt and Christina Stead

The pristine environment of the Georges River and green spaces attracted escapees from city frenzy
Environment advocates Myles Dunphy and son Milo aligned with the willing to preserve nature
Today real estate push sees highest of towers and plazas of multicultural shops
The ground is sacred for the trails of history laid in Hurstville Global Central

References

Australian Bureau of Statistics (ABS) (2011) Australian census review. ABS, Canberra
Hurstville City Council (2004) Hurstville city master plan. NSW, Hurstville
Hurstville City Council (2012a) Hurstville city development control plan 1. NSW, Hurstville
Hurstville City Council (2012b) Hurstville city local environment plan. NSW, Hurstville
New South Wales Government (2014) A plan for growing Sydney. NSW, Sydney

Chapter 15
Planning Kogarah, Greater Sydney

Dedication *This chapter is dedicated to two past and current identities (Plate 15.1) in the Kogarah City and George District, (depicted on heritage plaque in Hurstville CBD).*

Plate 15.1 Myles Dunphy; and Eddie Charlton (*Source* Hurstville Council Archives 2015)

A snippet on each of the identities follows:

Myles Dunphy (1891–1985) was an Australian conservationist best known for the protection of parts of the Blue Mountains National Park.

Edward 'Eddie' Charlton (1929–2004) was an Australian professional snooker and English billiards player. (*Source* Wikipedia 2015)

R.C. Rauscher, S. Momtaz, *Cities in Global Transition*,
DOI 10.1007/978-3-319-39865-5_15

Abstract The chapter addresses the extent Kogarah Council (herein referred to as Kogarah) is applying sustainable urban planning (SUP) practices (see Chap. 1) in planning the municipal area. The chapter acknowledges the *Local Government (Council Amalgamations) Proclamation 2016* merging Kogarah and Hurstville councils noted in Chap. 2). Kogarah (in St George District) is one of twelve municipalities within three Greater Sydney Districts (Sydney Inner West, Greater Parramatta and St George), all considered to be in global transition (see Chap. 2). Examining SUP practices is based on the application of three key SUP principles (as adopted in Chap. 2), being: Developing Sustainable Urban Centers; Developing Sustainable Transport; and, Protecting the Natural Environment.

Firstly, a background to Kogarah development and population changes (e.g. global movement of population) is outlined. To examine the pattern of urban development, four phases of Kogarah development are examined: Early Kogarah; Traditional Kogarah; New Kogarah; and, Future Kogarah. Given conclusions reached here, key Kogarah Council planning documents are then critiqued to determine the extent Council applies SUP practices. In addition, given State planning of Greater Sydney affects Kogarah, a second question is posed: what is the impact of Greater Sydney planning and development on Kogarah? In examining this question, examples of State plans (including council amalgamations under the Proclamation noted above) likely to affect Kogarah are addressed. By chapter's end, conclusions are drawn on Council planning and impact of Greater Sydney planning on Kogarah. The chapter notes that these conclusions will be addressed further in: Chap. 17 (examining the extent of SUP practices being applied over the whole of St George District); and, Chap. 18 (examining directions for planning a sustainable Greater Sydney or a sustainable city anywhere).

15.1 Background and Population Change

Kogarah City (herein called Kogarah) in the context of St George (as introduced in Chap. 2) is east of Hurstville (and adjacent to Botany Bay) and south of Sydney Inner West and the City of Sydney. Under the *Local Government (Council Amalgamations) Proclamation 2016* Kogarah was merged with Hurstville council as noted in Chap. 2.

Suburbs of Kogarah include: Allawah (partly in City of Rockdale), Beverley Park, Blakehurst, Carlton (partly in City of Rockdale), Carss Park, Connells Point, Hurstville, Hurstville Grove, Kogarah (partly in City of Rockdale), Kogarah Bay, Kyle Bay, Mortdale, Oatley, Penshurst (partly in City of Hurstville), Sans Souci, and South Hurstville.

The area now known as Kogarah was home to the Wangal people, inhabiting a large portion of the St George district (Chap. 2). European settlement began in the 1790s in the Kogarah area. The municipality was formed in 1885, and eventually Kogarah became a City in 2008. The population of Kogarah is 55,806 (Census 2011) and the area contains a multicultural communityy. In respect to linguistically diverse communities (Census 2011) Kogarah City has a significantly higher than average proportion (xxx) of family members speaking two or more languages at home (national average is 20.4 %). Finally, a significant proportion of Kogarah City residents stated a religious affiliation with Islam. With this background introduction to Hurstville Municipality (herein referred to as Hurstville), the extent of Hurstville's global transition can be addressed. The Australian Bureau of Statistics comments on the suburb of Kogarah within Kogarah City ancestry and languages spoken at home:

> According to the 2011 Australian Bureau of Statistics Census of Population, there were 12,764 people usually resident in Kogarah. 63.2 % stated they were born overseas with the top countries of birth being China 12.0 %, Nepal 6.5 % and Bangladesh 4.4 %. English was stated as the only language spoken at home by 30.0 % of residents and the most common other languages spoken were Mandarin 10.2 %, Nepali 6.8 % and Cantonese 6.7 %. (*Source* ABS 2011)

Finally, in author's (Rauscher) qualitative research (2014/2015), there is a poem written to reflect the cultural diversity of Kogarah (end of chapter under Notation 15.1). Kogarah's development phases and population change are thus examined next.

15.2 Phases of Development

An examination of the development phases of any locality provides a basis of understanding current trends and likely future directions of growth of that locality. Following a look at the background of Kogarah four phases of Kogarah development will be examined. These phases include: Early Kogarah; Current Kogarah; and, New Kogarah Directions. To assist in addressing this question, author (Rauscher) qualitative research field work was completed in Kogarah City (2014–2015). Early Kogarah (Plate 15.2) was serviced by a tram, pulling into a station here (top-left). The main shopping centre (top-right) is laid out with the Post Office anchoring the main corner. The popularity of early shopping around the rail station is shown here (middle-left). The roads were basic in early days, as seen here at Rocky Pt. Road (middle-right). Another main road was Georges River Rd (bottom-left). Finally, the subdivision map of Kogarah (bottom-right) shows the early township layout.

The transformation of Kogarah City downtown has been occurring over several decades (1980s onwards). These changes include (Plate 15.3) a number of commercial developments. Note the commercial below and residential above (top-left). Kogarah Town Centre (top-right) is an indoor shopping complex with direct access from the Kogarah rail station. An outdoor sculpture (middle-left) is part of a small plaza at the street entry to the station. Kogarah Court House (middle-right) is one of only a few local court houses operating in suburbs of Sydney. Kogarah has retained a number of historic buildings (Leah Building here) (bottom-left), followed by Leah House alongside (bottom-right).

Plate 15.2 Early Kogarah City. *Left to right* Tram at Kogarah Railway; Main shopping street with Post Office in the distance; Kogarah early shopping area; Rocky Pt. Road, Sans Souci; Georges River Rd; and, early Kogarah township layout (Source City of Kogarah Archives)

Plate 15.3 Traditional Kogarah. *Left to right* Commercial and residential; Kogarah Town Centre; Outdoor sculpture; Kogarah Court House; Historic Leah Buildings; and, Leah House (*Source* Raymond Rauscher 2015)

The new Kogarah City (Plate 15.4) reflects a continuing change of the urban landscape. The Kogarah CBD has a modern library (top-left). A section of the inner Kogarah City area is zoned for multi-story residential complexes such as this one (top-middle). There has been considerable high rise residential. Note the residential balconies (top-right) afford a cosmopolitan view of the shopping area. A group of residential units with shopping below (bottom-middle) anchor a main street corner of the CBD. A new residential block (bottom-middle) has shops and businesses below. Finally, the Kogarah Railway square (bottom-right) provides a resting place.

Finally, Future Kogarah (Plate 15.5) reflects the continuing building boom in residential development at increasing densities throughout the municipality. The trend to build high rise residential is likely to continue along the main road arteries (left). A simpler construction method (centre) may find an increasing group of buyers looking for more affordable housing. Finally, as available sites for high rise is likely to reduce, older commercial and light industry sites (right) could become more attractive for redevelopment. Factors of building design (such as height, density, housing affordability, overshadowing and open space provisions) are likely to

Plate 15.4 New Kogarah *Left to right* Kogarah Library; High rise residential cluster; Residential balconies detail; Main Rd residential group; New residential block; and, Kogarah Railway square (*Source* Raymond Rauscher 2015)

Plate 15.5 Future Kogarah. *Left to right*: Developed and proposed developments (*Source*: Raymond Rauscher 2015 and Kogarah Council Web 2016)

remain issues of development importance and contention within communities at the local government level into the future.

15.3 Conclusion

Given the review of development phases of Kogarah (above) and changes in population demographics, it can be concluded that Kogarah has been experiencing accelerated global transition (2000s) (Question 1). The planning of Kogarah (merged with Hurstville council as noted above under the Proclamation) in meeting development demands and applying sustainable urban planning (SUP) practices is examined next.

15.4 Kogarah and Sustainable Urban Planning

Planning of Kogarah City has been taking place since it became a local government body in 1861. A review of planning in Kogarah will pay particular attention to applications of sustainable urban planning (SUP) (as outlined in Chap. 1) practices (including use of indicators of sustainability) (Chap. 2). Kogarah Council key planning documents (a–z with date and web reference) are listed in Information Box 15.1. A summary of these documents and a brief critique assist with this review of Kogarah planning (documents can be sourced under the web link as noted in the right column).

Information Box 15.1 Kogarah Municipality Planning

Kogarah City Local Environment Plan (2012)	www.kogarah.nsw.gov.au
Kogarah City Development Control Plan (2013)	www.kogarah.nsw.gov.au
draft Kogarah New City Plan (2015)	www.kogarah.nsw.gov.au

Source Webs as noted February 2016

The physical strategic planning of the Kogarah City Council is contained within the *draft Kogarah New City Plan* (Kogarah 2015) (www.kogarah.nsw.gov.au/new-cityplan). The Council states:

> Council is exhibiting changes to the Kogarah Local Environmental Plan 2012 (LEP2012). The Kogarah LEP 2012 is the primary planning document for development in the City of Kogarah. Those changes can only be made through a process that includes developing a Planning Proposal which is publicly exhibited for comment. The Planning Proposal that is now on public exhibition is called the New City Plan. The changes propose rezoning some land and putting in place rules for floor space ratio and height for development across the City of Kogarah. The aim of these changes is to help deliver housing and employment opportunities to meet the current and future needs of Kogarah's population. The New City Plan has been endorsed by the NSW Department of Planning & Environment allowing Council to place it on exhibition. The New City Plan is a draft document pending community input and a final Council decision on the proposal. (Source Kogarah Council 2015)

Council notes under environment a full range of environmental programs from bush care to water protection (George River and Botany Bay). Moving to transport, Council notes key strategies under transport planning. Both environment and transport programs can be viewed at www.kogarah.nsw.gov.au. Having examined environmental guidelines and transport of the City of Kogarah City the third planning factor of urban design is reviewed. The Council outlines its approach to designing the City via a project team the Council has established. There are a number of City design projects reflected earlier in viewing Kogarah New and Kogarah Future.

15.4.1 Conclusion on Council Planning and SUP Practices

It can be concluded, having reviewed Kogarah Council planning documents and their applications, it can be concluded that Council is moving in the direction of applying sustainable urban planning (SUP) practices (Question 2 above). The State's Greater Sydney planning (including infrastructure projects) and State council amalgamation proposals likely to affect Kogarah (Question 3) are reviewed next.

15.5 Greater Sydney Planning and Kogarah

There are planning and developments of Greater Sydney likely to impact on Kogarah City. In many instances Kogarah City Council has had the task of providing: planning input on State proposals affecting Kogarah. These include making and reviewing public submissions; implementing State planning policies at the local level;

joint partnership on service or infrastructure projects; and, monitoring State or private development projects affecting Kogarah City. Other Council planning responsibilities under State legislation that also faces the Council include: engaging the municipality community (e.g. residents, business, services, development interests, and institutions) on State planning policies and development proposals; statutory plan making (e.g. adopting local environment plans (LEPs); and, reporting back to the State on a range of planning matters (i.e. State proposed roads, state of environment reporting and infrastructure projects).

Greater Sydney planning requirements were set by the NSW State in 2006, with Kogarah City Council adopting in 2012 plans had to be consistent with the *Sydney Metropolitan Strategy* (NSW 2011). The Strategy), as noted in the previous chapter, is the strategic document that outlines a vision for Sydney to 2031. Council's complimentary town planning documents were the *Kogarah Local Environment Plan* (Kogarah Council 2012) and the *Kogarah Development Control Plan* (Kogarah Council 2013). The State in 2014 adopted its strategy for planning Greater Sydney to 2031 (*A Plan for Growing Sydney* (NSW 2014). The Plan forecasts a population growth of 1.6m (to be added to the Greater Sydney population of 4.84m at June 2014). In examining this Plan and its impact on Marrickville, three key planning principles as adopted in Chap. 2 are addressed: (1). Developing Sustainable Urban Centres; (2). Developing Sustainable Transport; and, (3). Protecting the Natural Environment. The planning principle Developing Sustainable Urban Centres is examined first. The Plan notes the State will work with Kogarah City Council to provide capacity for additional mixed-use development in Kogarah including offices, health, education, retail, services and new homes. The State also notes its support for health-related land uses and infrastructure around St George Hospital. The planning principle Developing Sustainable Transport is examined next. Kogarah Council, along with other St George councils, were asked in late 2014 to comment on the proposed State's WestConnex project. The Plan also notes other key transport projects likely to affect Kogarah (to be addressed further in reviewing the whole of St George District in Chap. 17), including:

1. Building the M5 East Airport Link motorway as part of WestConnex.
2. Investigating the F6 motorway corridor.
3. Preserving potential future transport corridors for Sydney Rapid Transit with a possible extension of Sydney Rapid Transit to Hurstville.

Finally, the planning principle Protecting the Natural Environment is examined next. The Plan commits the State to 'protect the health of the subregions waterways and aquatic habitats'. This goal, for example, would apply to the prime conservation challenges in Kogarah. This would include the protection of bushlands, waterways and foreshores within Kogarah. The efforts of Kogarah Council in protecting, for example, waterways and foreshores of the Georges River and Botany Bay in cooperation with other councils in St George District will be outlined in more detail in Chap. 17.

Having reviewed above the State's Greater Sydney planning (including infrastructure projects) it can be concluded the implementation of these plans and developments is likely to have a significant effect on Kogarah (Question 3). The extent of this effect will be better understood when these State plans are further examined in the context of the whole of the St George District (including Kogarah) in Chap. 17. Finally, options for councils (including Kogarah) working together and in partnership with the State to meet local and district needs will be addressed in Chap. 18 (Structures for Councils and State Planning Partnerships).

15.5.1 Conclusion on Greater Sydney Planning and Kogarah

Having reviewed above the State's Greater Sydney planning (including infrastructure projects) it can be concluded the implementation of these plans and developments is likely to have a significant impact on Kogarah (Question 3). The extent of this impact will be better understood when these State plans are further examined in the context of the whole of the St George District (including Kogarah) in Chap. 9. Finally, in examining State proposed council amalgamations, options for councils (including Kogarah) working together and in partnership with the State to meet local and district needs will be addressed in Chap. 18 (Planning Reforms and Directions for Planning Sustainable Cities). The subject to amalgamations of councils and how this affects Kogarah examined next.

15.6 Amalgamations of Councils and Kogarah

The subject of amalgamations of Greater Sydney councils (given implications for planning at municipal and Greater Sydney levels) was reviewed in Chap. 2. NSW Premier Mike Baird announced in Dec 2015 the State's preference for: (a). Kogarah and Hurstville Councils to merge; and (b). Rockdale Council to merge with Botany Bay Council (this council is outside the St George District). Kogarah City Council preferred to stand-alone as a council and not to merge. Hurstville Council as well as Kogarah Council remained opposed to forced amalgamations. The Mayor of Kogarah City urged all residents to have their say on the issue of amalgamations of councils. Council noted that submissions to the new Council Boundary Review were open till the end of February (www.councilboundaryreview.nsw.gov.au). In the interim, the Council noted, the Minister for Local Government had commenced the process of forcibly amalgamating Kogarah and Hurstville councils. Council also noted the Minister said he expected the new council (with adjusted boundaries) to be 'proclaimed' on 1 July 2016 (note the 12th May 2016 *Local Government (Council Amalgamations) Proclamation 2016* noted in Chap. 2). The results of the review of proposed boundary adjustments and merger (Kogarah and Hurstville councils) would be referred to the Chief Executive of the Office of Local Government for examination and report under the Local Government Act (the Act). In turn, the

Chief Executive had appointed a Delegate who would examine the proposals, conduct a public inquiry, call for written submissions, and then report back to the Minister. In turn, the Chief Executive had appointed a Delegate who would examine the proposals, conduct a public inquiry review, call for written submissions, and then report back to the Minister.

15.6.1 Conclusion on Amalgamations and Kogarah

Kogarah City Council stated its case to thc State that it preferred to stand-alone and not be merged with any other council. At the end of the State review process (Council Boundary Review) the State would determine what amalgamations as proposed would proceed or not proceed. The 12th May 2016 *Local Government (Council Amalgamations) Proclamation 2016* merging Kogarah and Hurstville councils was noted. This question will be further addressed in Chap. 18.

15.7 Overall Conclusions

Having completed an assessment of the development and planning of Kogarah, there are a number of overall conclusions reached. In examining the patterns of urban development of Kogarah it was concluded that Kogarah has been experiencing accelerated development and changes to the composition of its population. Given these changes, and examining Kogarah Council's response, it was also concluded Council is moving in the right direction in increasing its sustainable urban planning (SUP) practices. It was also concluded, however, that there would be significant future impacts on Kogarah from Greater Sydney growth and development. In reviewing potential Council and State responses to this future scenario, it was concluded that new partnerships between the Council and the State would be needed. A major initiative from the State (2009 onwards) was to investigate local government reform, including a potential amalgamation of councils (e.g. Kogarah merging with Hurstville; and Rockdale merging with Botany City). The 12th May 2016 *Local Government (Council Amalgamations) Proclamation 2016* that merged Kogarah and Hurstville councils was acknowledged. These overall chapter conclusions will be examined against results of assessing planning and development of the remainder of the three cities of St George (Chaps. 14, 15 and 16) and the district as a whole (Chapter 17). Results from the district assessment (and assessment of Sydney Inner West and Greater Parramatta districts in later chapters) will contribute to assessing how to plan for a sustainable Greater Sydney and a sustainable city anywhere (Chap. 18).

15.8 Summary

The chapter addressed the extent Kogarah Council was applying sustainable urban planning (SUP) practices (see Chap. 1) in planning the municipal area. It was noted that Kogarah, in the St George District, was merged with Hurstville council under the 12th May 2016 *Local Government (Council Amalgamations) Proclamation 2016* (Chap. 2). Firstly, a background to Kogarah development and population changes (i.e. global movement of population) was outlined. To examine the pattern of urban development, four phases of Kogarah development were examined: Early Kogarah; Traditional Kogarah; New Kogarah; and, Future Kogarah. Given conclusions reached here, key Kogarah Council planning documents were then critiqued to determine the extent Council applies SUP practices. In addition, given State planning of Greater Sydney affects Kogarah, a second question was posed: what is the impact of Greater Sydney planning and development on Kogarah?. In examining this question, examples of State plans (including State proposed council amalgamations) likely to impact on Kogarah are addressed. By chapter's end, conclusions are drawn on Council planning and impact of Greater Sydney planning on Kogarah. The chapter notes that these conclusions will be addressed further in: Chap. 17 (examining the extent of SUP practices being applied over the whole of St George District); and, Chap. 18 (examining directions for planning a sustainable Greater Sydney and a sustainable city anywhere).

Notation 15.1 Cultural Diversity Poem – Kogarah
Poem (Rauscher 2015 unpublished) reflecting the diversity of Kogarah:

Kogarah – A Fine Touch (# 17)

Kogarah was an escape to Botany Bay, Carr's Park and shorelines of Sans Souci
The Greek, Chinese, Latin American and Korean communities mix among long settled families
An urban story is penned and blue prints spread to create an enticing cityscape
The town layout is conceived to encourage walking, lanes are angular to amuse

The striving to create a city blending the past, embracing the new, and spawning new architecture
A stability of community hangs onto the familiar Post Office and Greek Orthodox Church downtown
The new high rise spaced to preserve the earlier terraces and rows of bungalows
Kogarah is still the escape, now to a city of refined touch and still the water calls

References

Australian Bureau of Statistics (ABS) (2011) Australian census review. ABS, Canberra
Kogarah City Council (2012) Kogarah city local environment plan. NSW, Kogarah
Kogarah City Council (2013) Kogarah development control plan. NSW, Kogarah
Kogarah City Council (2015) Draft Kogarah new city plan. NSW, Kogarah
New South Wales (2011) NSW government's metropolitan strategy. NSW, Sydney
New South Wales (2014) A plan for growing sydney. NSW, Sydney

Chapter 16
Planning Rockdale, Greater Sydney

Dedication *This chapter is dedicated to two past and current identities in the Rockdale City and George District, (depicted on heritage plaque in Hurstville CBD) (Plate 16.1).*

Plate 16.1 Norm Provan; and, Fleur Mellor (*Source* Hurstville Council Archives 2015)

A snippet on each of the identities follows:

Norm Provan (1932-) is an Australian former rugby league footballer and coach.

Fleur Mellor (1935-) is an Australian athlete and Olympic champion in relay running. (*Source* Wikipedia 2015)

R.C. Rauscher, S. Momtaz, *Cities in Global Transition*,
DOI 10.1007/978-3-319-39865-5_16

Abstract The chapter addresses the extent Rockdale Council (herein referred to as Rockdale) is applying sustainable urban planning (SUP) practices (see Chap. 1) in planning the municipal area. The chapter acknowledges the *Local Government (Council Amalgamations) Proclamation 2016* proposed merging Rockdale with Botany council noted in Chap. 2. Rockdale (in St George District) is one of twelve municipalities within three Greater Sydney Districts (Sydney Inner West, Greater Parramatta and St George), all considered to be in global transition (see Chap. 2). Examining SUP practices is based on the application of three key SUP principles (as adopted in Chap. 2), being: Developing Sustainable Urban Centers; Developing Sustainable Transport; and, Protecting the Natural Environment.

Firstly, a background to Rockdale development and population changes (e.g. global movement of population) is outlined. To examine the pattern of urban development, four phases of Rockdale development are examined: Early Rockdale; Traditional Rockdale; New Rockdale; and, Future Rockdale. Given conclusions reached here, key Rockdale Council planning documents are then critiqued to determine the extent Council applies SUP practices. In addition, given State planning of Greater Sydney affects Rockdale, a second question is posed: what is the impact of Greater Sydney planning and development on Rockdale? In examining this question, examples of State plans (including council amalgamations under the Proclamation noted above) likely to affect Rockdale are addressed. By chapter's end, conclusions are drawn on Council planning and the impact of Greater Sydney planning on Rockdale. The chapter notes that these conclusions will be addressed further in: Chap. 17 (examining the extent of SUP practices being applied over the whole of St George District); and, Chap. 18 (examining directions for planning a sustainable Greater Sydney or a sustainable city anywhere).

16.1 Background and Population Change

Rockdale City (herein called Rockdale) in the context of St George (as introduced in Chap. 2) is north of Hurstville and Kogarah (and adjacent to Botany Bay) and south of Sydney Inner West and the City of Sydney. Under the *Local Government (Council Amalgamations) Proclamation 2016* Rockdale was propocsed to merge with Botany council as noted in Chap. 2). Suburbs of Rockdale include: Arncliffe, Banksia, Bardwell Park, Bardwell Valley, Bexley, Bexley North, Brighton-Le-Sands, Carlton, Dolls Point, Kingsgrove, Kogarah, Kyeemagh, Monterey, Ramsgate, Ramsgate Beach, Rockdale, Sandringham, Sans Souci (partly in City of Kogarah), Sydney Airport. Turrella, and, Wolli Creek. The area now known as Rockdale was home to the Wangal people, inhabiting a major portion of St George district. European settlement began in Rockdale in the 1790s and the Municipality was formed in 1887, eventually becoming a City in 1995. The population of Rockdale is 97,340 (Census 2011). Rockdale prides itself as one of the most multicultural communities in Australia. In respect to linguistically diverse communities (Census 2011) Rockdale City has a significantly higher than average proportion (. %) of family members speaking two or more languages at home (national average is 20.4 %). Finally, the proportion of Rockdale City residents who stated a religious affiliation with Islam was significantly in excess of the national average. With this background introduction to Rockdale, the extent of Rockdale's development phases and changing demographics are examined next.

The Australian Bureau of Statistics (ABS 2011) comments on the Rockdale City ancestry and languages spoken at home:

> The proportion of residents who stated their ancestry was Greek was 6.5 times the New South Wales and national averages; the proportion of households where Greek is spoken at home is in excess of eight times the state and national averages; and the proportion of residents who stated an affiliation with Eastern Orthodox religion was in excess of six times the state and national averages. According to the 2006 Australian Bureau of Statistics Census of Population, there were 13,901 people usually resident in Rockdale. 53.2 % stated they were born overseas with the top countries of birth being China 10.6 %, Nepal 7.6 % Macedonia 4.1 %, and Bangladesh 4 %. English was stated as the only language spoken at home by 29.9 % of residents and the most common other languages spoken were Macedonian 15.7 %, Mandarin 7.2 % and Arabic 7.2 % (*Source* ABS 2011)

Finally, in author's (Rauscher) qualitative research (2014/2015), there is a poem written to reflect the cultural diversity of Rockdale (end of chapter under Notation 16.1): *Rockdale – Metropolis Unfolding.* The phases of Rockdale development are examined next.

16.2 Phases of Development

An examination of the development phases of any locality provides a basis of understanding current trends and likely future directions of growth of that locality. Following a look at the background of Rockdale four phases of Rockdale's development will be examined. These phases include: Early Rockdale; Traditional Rockdale; New Rockdale; and, Future Rockdale. To assist in addressing this question, author (Rauscher) qualitative research field work was completed in Rockdale City (2014–2015). Early Rockdale, first settlement, is shown in Plate 16.2. Lady Robinson's Beach at Brighton-le-Sands in 1875 is seen here (top-left). The tram serving this district is shown here (top-right). The popular Davis Butchery, at Bexley in the 1800s here (middle-left). There were many early estates developed at Rockdale, one shown here (middle-right). The Arncliffe railway station in 1873 is shown here (bottom-left). Finally, Tempe House (bottom-right) in its natural surrounds is shown here.

The restoration of early Rockdale buildings and features illustrates the emerging traditional Rockdale, as shown in Plate 16.3. The Castle Palace Hall (top-left) here has been refurbished. The Uniting Church (top-right) is located in the CBD, opposite the Council chambers. Tempe House (shown in early sketch) (middle-left) has been restored, with a sketch of the early house (middle-right). The St Magdalene Chapel (bottom-left) has also been restored. Finally, a tributary drain (bottom-right) running into the Cooks River at Tempe is shown here.

Plate 16.2 Early Rockdale. *Left to right* Lady Robinson's Beach; Brighton-le-Sands tram; Davis Butchery, Bexley; Early estate; Arncliffe rail station; and, Tempe House in natural surrounds (*Source* Rockdale Council Archives 2015)

Plate 16.3 Traditional Rockdale. *Left to right* St. George Girls High school; Catholic Church; Rockdale Mall; Traditional housing giving Way to high rise residential; Suburban Street; and, medium density complex

Plate 16.4 New Rockdale. *Left to right*: Six story residential; High rise residential; Proposed new library in CBD area; Civic Centre; Brighton Le Sands; and, Residential high rise, Rockdale CBD

The New Rockdale City (Plate 16.4) reflects a continuing change of the urban landscape. A large section of inner Rockdale City area is zoned for higher density residential complexes such as this one (top-left). There has been considerable high rise residential (top-center) around the periphery of the CBD. The commercial draw of Rockdale City centre has increased significantly over the last two decades as illustrated here in the artist's sketch of the proposed new library (top-right). The Rockdale City administration centre located within the business centre of Rockdale (bottom-left). The growing commercial positioning of the CBD is flowing onto the adjacent suburb of Brighton Le Sands illustrated here (bottom-centre) in this new residential and tourist complex overlooking Botany Bay. Finally, demand for residential premises is illustrated in this high rise complex adjacent to Rockdale CBD (bottom-right).

Finally, Future Rockdale (Plate 16.5) reflects the continue building boom in residential development at increasing densities throughout the municipality. There are also a number of design projects centered on Wolli Creek suburb. Here (left) is a residential high rise facing a planned civic space. A residential boulevard is shown here, running from the rail station (middle). Finally, a row of high rise stretches (right) towards the Princes Highway end of Wolli Creek.

Plate 16.5 Future Rockdale. *Left to right:* Current and proposed developments

16.3 Conclusion

Given the review of development phases of Rockdale (above) and changes in population demographics, it can be concluded that Rockdale has been experiencing accelerated global transition (2000s) (Question 1). The planning of Rockdale (proposed to be merged with Botany council as noted above under the Proclamation) in meeting development demands and applying sustainable urban planning (SUP) practices is examined next.

16.4 Rockdale and Sustainable Urban Planning

Planning of Rockdale City has been taking place since it became a local government body in 1887. A review of planning in Rockdale will pay particular attention to applications of sustainable urban planning (SUP) (as outlined in Chap. 1) practices (including use of indicators of sustainability) (Chap. 2). Rockdale Council key planning documents (a–z with date and web reference) are listed in Information Box 16.1. A summary of these documents and a brief critique assist with this review of Rockdale planning (documents can be sourced under the web link as noted in the right column).

Having examined environmental guidelines and transport of Rockdale City the third planning factor is urban design. There are a number of City design projects (Plate 16.6) being considered by Rockdale Council. Two are presented here. The new open spaces will be introduced within the *Rockdale Master Plan* (2014) (left). Secondly, more public areas would be integrated with new developments such as illustrated here at Bay St and the Princes Highway (right).

The planning of the suburb of Wolli Creek has been a challenge over the 1990-current 2015 period for Rockdale Council. Wolli Creek was named after the

Information Box 16.1 Rockdale Municipality Planning

Rockdale City Development Control Plan (2011a)	www.rockdale.gov.nsw.au
Rockdale City Local Environment Plan (2011b)	www.rockdale.gov.nsw.au
Rockdale Master Plan (2014)	www.rockdale.gov.nsw.au

Source Webs as noted February 2016

Plate 16.6 Details of Rockdale designs. *Left to right* Rockdale Master Plan; Bay St and Princes H'way; Wolli Creek high rise; Wolli Creek Railway Stn.; Boulevard from rail station; and row of high rise

waterway on its northern border and became predominantly industrial until redevelopment was planned in the late 1990s. Wolli Creek railway station opened in 2000 and the area of Wolli Creek was officially declared a suburb in 2002. The suburb has been developing since into a high density residential and commercial area (taking advantage of its proximity to the Sydney CBD and Sydney Airport.

Rockdale Council notes:

> The area around the Wolli Creek train station is also being transformed into a village square. The piazza-style Wolli Creek Village Square will be an outdoor shopping and dining precinct once fully completed. The village square area will contain an Asian Supermarket and a variety of restaurants, cafés and specialty stores. Part of the village square will include the transport interchange, for rail, bus and taxi. Wolli Creek railway station is a railway junction for the Illawarra line and Airport, Inner West & South Line on the Sydney Trains network. The train trip to Town Hall railway station takes approximately 16 minutes on the Illawarra line via Sydenham, while the trip to Museum railway station takes approximately 18 minutes on the Airport line via the Airport. There are a few different bus routes in Wolli Creek. The 348 bus route stops near Wolli Creek station at Magdalene Terrace where it continues to Bondi Junction, stopping at stops including: Tempe, St Peters, Kensington, UNSW and Randwick. (*Source* Rockdale Council web)

16.4.1 Conclusion on Council Planning and SUP Practices

It can be concluded, having reviewed Rockdale Council planning documents and their applications, that Council is moving in the direction of applying sustainable urban planning (SUP) practices (Question 2 above). The State's Greater Sydney planning (including infrastructure projects) and State council amalgamation proposals likely to affect Rockdale (Question 3) are reviewed next.

16.5 Greater Sydney Planning and Rockdale

There are planning and developments of Greater Sydney likely to impact on Rockdale City. In many instances Rockdale City Council has had the task of providing: planning input on State proposals affecting Burwood. These include making and reviewing public submissions; implementing State planning policies at the local level; joint partnership on service or infrastructure projects; and, monitoring State or private development projects affecting Rockdale City. Other Council planning responsibilities under State legislation that also faces the Council include: engaging the municipality community (e.g. residents, business, services, development interests, and institutions) on State planning policies and development proposals; statutory plan making (e.g. adopting local environment plans (LEPs); and, reporting back to the State on a range of planning matters (i.e. State proposed roads, state of environment reporting and infrastructure projects). Greater Sydney planning requirements were set by the NSW State in 2006, with Rockdale Council adopting in 2011 the *Rockdale City Local Environment Plan* (Rockdale Council 2011b). Council's plans had to be consistent with the *Sydney Metropolitan Strategy* (NSW 2011). Council's complimentary town planning document was the Rockdale *Development Control Plan* (Rockdale Council 2011a). The *NSW Government's Metropolitan Strategy* (NSW 2012), as noted in the previous chapter, is the strategic document that outlines a vision for Sydney to 2031. The South *Subregion Draft Subregional Strategy* (SSDSS) is the overarching strategic document that outlines a vision for the South Central Subregion. Sydney metropolitan planning requirements were set by the NSW State in 2006, with Rockdale City Council adopting in 2011 xxx. Council's plans had to be consistent with the *Sydney Metropolitan Strategy* (NSW 2011). Council's complimentary town planning documents were the *Rockdale Local Environment Plan* (xxx) and the *Rockdale Development Control Plan* (xxx).

Moving from Rockdale City planning to State planning in Rockdale, the NSW *A Plan for Growing Sydney* (NSW 2014), as noted earlier (Chap. 2 and previous two chapters), is the strategic document that outlines a vision for Sydney to 2031. St George (thus Rockdale) is within the South Subregion as designated in the above Plan and the *South Subregion Draft Subregional Strategy* (SSDSS). Several Greater Sydney State planning proposals currently affecting Strathfield are examined next.

As was noted in Chap. 14 (Hurstville) this subregion includes (a). the 'strategic centers' of Kogarah and Hurstville; (b). WestConnex road project; (c). 'urban renewal corridors'; (d). 'road/motorway investigation'; (e). a 'priority precinct' in Rockdale; and, finally, (f). the 'Sydney airport precinct'. The Plan forecasts a population growth of 1.6 m (to be added to the Greater Sydney population of 4.84 m at June 2014). In examining this Plan and its impact on Rockdale, three key planning principles as adopted in Chap. 2 are addressed: (1). Developing Sustainable Urban Centres; (2). Developing Sustainable Transport; and, (3). Protecting the Natural Environment. The planning principle Developing Sustainable Urban Centres is examined first. The Plan notes, given the capacity for additional mixed-use development in Hurstville and Kogarah as strategic centers (offices, health, education, retail, services and new homes), Rockdale is likely to also experience continued growth in residential and service areas. The planning principle Developing Sustainable Transport is examined next. Rockdale Council, along with other St George councils, were asked in late 2014 to comment on the proposed State's WestConnex (NSW 2014) project. The Plan also notes other key transport projects likely to affect Rockdale (to be addressed further in reviewing the whole of St George District in Chap. 17), including:

1. Building the M5 East Airport Link motorway as part of WestConnex.
2. Investigating the F6 motorway corridor.
3. Preserving potential future transport corridors for Sydney Rapid Transit with a possible extension of Sydney Rapid Transit to Hurstville. (*Source* WestConnex web 2015)

Finally, the planning principle Protecting the Natural Environment is examined next. The Plan commits the State to 'protect the health of the subregions waterways and aquatic habitats'. This goal, for example, would apply to the prime conservation challenges in Rockdale. This would include the protection of bushlands, waterways and foreshores within Rockdale. The efforts of Rockdale Council in protecting, for example, waterways and foreshores of the Georges River and Botany Bay in cooperation with other councils in St George District will be outlined in more detail in Chap. 17.

Having reviewed above the State's Greater Sydney planning (including infrastructure projects) it can be concluded the implementation of these plans and developments is likely to have a significant effect on Rockdale (Question 3). The extent of this effect will be better understood when these State plans are further examined in the context of the whole of the St George District (including Rockdale) in Chap. 17. Finally, options for councils (including Rockdale) working together and in partnership with the State to meet local and district needs will be addressed in Chap. 18 (Structures for Councils and State Planning Partnerships).

16.5.1 Conclusion on Rockdale and Greater Sydney Planning

Having reviewed above the State's Greater Sydney planning (including infrastructure projects) it can be concluded the implementation of these plans and developments is likely to have a significant impact on Rockdale (Question 3). The extent of this impact will be better understood when these State plans are further examined in the context of the whole of the Greater St George District (including Rockdale) in Chap. 9. Finally, in examining State proposed council amalgamations, options for councils (including Rockdale) working together and in partnership with the State to meet local and district needs will be addressed in Chap. 18 (Planning Reforms and Directions for Planning Sustainable Cities). The subject to amalgamations of councils and how this affects Rockdale examined next.

16.6 Amalgamations of Councils and Rockdale

The subject of amalgamations of Greater Sydney councils (given implications for planning at municipal and Greater Sydney levels) was reviewed in Chap. 2. NSW Premier Mike Baird announced in Dec 2015 the State's preference for: (a). Rockdale Council to merge with Botany Bay Council (outside the St George District); and, (b). Kogarah and Hurstville Councils to merge. Rockdale City preferred not to be separated from the two other St George councils, Kogarah and Hurstville. Kogarah City Council preferred to stand-alone as a council and not to merge. Rockdale, Hurstville Council and Kogarah councils remained opposed to forced amalgamations. The Mayor of Rockdale City urged all residents to have their say on the issue of amalgamations of councils. Council noted that submissions to the new Council Boundary Review were open till the end of February (www.councilboundaryreview.nsw.gov.au). In the interim, the Council noted, the Minister for Local Government had commenced the process of forcibly amalgamating Rockdale and Botany Bay councils. Council also noted the Minister said he expected the new council (with adjusted boundaries) to be 'proclaimed' on 1 July 2016 (note the 12th May 2016 *Local Government (Council Amalgamations) Proclamation 2016* noted in Chap. 2). The results of the review of proposed boundary adjustments and merger (Rockdale and Botany Bay councils) would be referred to the Chief Executive of the Office of Local Government for examination and report under the Local Government Act (the Act). In turn, the Chief Executive had appointed a Delegate who would examine the proposals, conduct a public inquiry, call for written submissions, and then report back to the Minister.

16.6.1 Conclusion on Amalgamations and Rockdale City Council

Rockdale City Council stated its case to the State that it preferred to remain within the St George district of councils and not be merged with Botany Bay council. At the end of the State review process (Council Boundary Review) the State would determine what amalgamations as proposed would proceed or not proceed. The 12th May 2016 *Local Government (Council Amalgamations) Proclamation 2016* proposed a merger of Rockdale and Botany councils was noted. This question will be further addressed in Chap. 18.

16.7 Overall Conclusions

Having completed an assessment of the development and planning of Rockdale, there are a number of overall conclusions reached. In examining the patterns of urban development of Rockdale it was concluded that Rockdale has been experiencing accelerated development and changes to the composition of its population. Given these changes, and examining Rockdale Council's response, it was also concluded Council is moving in the right direction in increasing its sustainable urban planning (SUP) practices. It was also concluded, however, that there would be significant future impacts on Rockdale from Greater Sydney growth and development. In reviewing potential Council and State responses to this future scenario, it was concluded that new partnerships between the Council and the State would be needed. A major initiative from the State (2009 onwards) was to investigate local government reform, including a potential amalgamation of councils (e.g. Hurstville merging with Kogarah; and Rockdale merging with Botany City). The 12th May 2016 *Local Government (Council Amalgamations) Proclamation 2016* proposed a merged Rockdale and Rockdale councils was acknowledged. These overall chapter conclusions will be examined against results of assessing planning and development of the remainder of the three cities of St George (Chaps. 14, 15 and 16) and the district as a whole (Chap. 17). Results from the district assessment (and assessment of Sydney Inner West and Greater Parramatta districts in later chapters) will contribute to assessing how to plan for a sustainable Greater Sydney and a sustainable city anywhere (Chap. 18).

16.8 Summary

The chapter addressed the extent Rockdale Council was applying sustainable urban planning (SUP) practices (see Chap. 1) in planning the municipal area. It was noted that Rockdale, in the St George District, was proposed to be merged with Botany council under the 12th May 2016 *Local Government (Council Amalgamations)*

Proclamation 2016 (Chap. 2). It was concluded that Rockdale, one of twelve municipalities within three Greater Sydney Districts (Sydney Inner West, Greater Parramatta and St George) is considered to be in global transition. SUP practices (based on the application of three key SUP principles) were introduced, being: Developing Sustainable Urban Centers; Developing Sustainable Transport; and, Protecting the Natural Environment. A review of Rockdale background and population changes (i.e. global movement of population) was outlined. In examining the pattern of urban development, four phases of Rockdale development are examined: Early Rockdale; Traditional Rockdale; New Rockdale; and, Future Rockdale. With conclusions drawn, key Rockdale Council planning documents were then critiqued to determine the extent Council applies SUP practices. In addition, given State planning of Greater Sydney affects Rockdale, a second question was posed: what is the impact of Greater Sydney planning and development on Rockdale?. In examining this question, examples of State plans (including State proposed council amalgamations) likely to impact on Rockdale were addressed. By chapter's end, conclusions were drawn on Council planning and Greater Sydney planning. The chapter noted that these conclusions would be addressed further in: Chap. 17 (examining the extent of SUP practices being applied over the whole of St George District); and, Chap. 18 (examining directions for planning a sustainable Greater Sydney and a sustainable city anywhere).

Notation 16.1 Cultural Diversity Poem – Rockdale
Poem reflecting the diversity of Rockdale:

Rockdale – Metropolis Unfolding (# 19)

They set up curb side shops of Macedonian, Turkish, Greek and Iraqi national backgrounds
The multi-coloured high rise march along straight lines to the shores of Botany Bay
Suburban homes near rail station corridor become victims of increasing developer purchases\
The short black coffees are popular in the Mall, a sea of cosmopolitan men swapping stories

Rockdale was the exit from Sydney to bigger estates and endless market gardens
The evolution cultural interest in Gilbert and Sullivan and Illawarra Theatre Guild widened horizons
Today an older home is under the gaze of change for a growing international population
The Princes Highway snakes through the commercial Rockdale, a hybrid metropolis unfolding

References

Australian Bureau of Statistics (ABS) (2011) Australian census review. ABS, Canberra
New South Wales (2011) Sydney metropolitan strategy. NSW, Sydney
New South Wales (2012) NSW government's metropolitan strategy. NSW, Sydney
New South Wales (2014) A plan for growing Sydney. NSW, Sydney
Rockdale City Council (2011a) Rockdale city development control plan. NSW, Rockdale
Rockdale City Council (2011b) Rockdale city local environment plan. NSW, Rockdale
Rockdale City Council (2014) Rockdale master plan. NSW, Rockdale

Chapter 17
Planning Greater Sydney Districts – St George Case Study

Dedication *This chapter is dedicated to four past identities in the St George District, Miles Franklyn, Sumner Elliott, Christina Stead, and H.V. Evatt (depicted on heritage plaque in Hurstville) (Plate 17.1).*

Plate 17.1 Past Citizens of St George District. *Left to right* Miles Franklyn; Sumner Locke; Christina Stead; and H.V. Evatt. (*Source* Hurstville Council Archives 2015)

R.C. Rauscher, S. Momtaz, *Cities in Global Transition*,
DOI 10.1007/978-3-319-39865-5_17

A snippet on each of the identities follows:

Miles Franklyn (1879–1954) was an Australian writer and feminist who is best known for her novel *My Brilliant Career*,

Sumner Locke Elliott (1917–1991) was an Australian (later American) novelist.

Christina Stead (1902–1983) was an Australian novelist and short-story writer acclaimed for her satirical wit and penetrating psychological characters.

Dr H.V. Evatt (1894–1965) was an Australian judge, lawyer, parliamentarian and writer. (*Source* Wikipedia 2015)

Abstract Planning of the St George District, the third of three case study districts of Greater Sydney, is examined in this chapter. The chapter acknowledges the *Local Government (Council Amalgamations) Proclamation 2016* proposed mergers and expanding the boundaries of nominated councils noted in Chap. 2. Having examined the question of global transition and planning practices of the three municipalities within the District (Chaps. 14, 15 and 16), the chapter addresses: (1). planning the St George District as a whole; and, (2). the extent the State (responsible for district planning) is meeting the planning issues of the District. Key planning principles as adopted in Chapter 2 to apply to Greater Sydney (and this district) are reviewed. These key principles include: (1). Developing Sustainable Urban Centres; (2). Developing Sustainable Transport; and, (3). Protecting the Natural Environment. Sample plans and developments within St George are critiqued under each of these principles. After reaching a conclusion under each principle, an overall conclusion will be drawn on extent the State (planning the District) is meeting the planning issues of the District. Finally, it's noted that these St George District conclusions will be assessed alongside conclusions reached in examining Sydney Inner West District (Chaps. 3, 4, 5, 6, 7, 8 and 9) and Greater Parramatta District (Chaps. 10, 11, 12 and 13). All these conclusions are examined in Chapter 18 (examining directions for planning a sustainable Greater Sydney or a sustainable city anywhere).

17.1 Background to St George District

The geography of St George district (Cities of Rockdale, Kogarah, and Hurstville) (as with Greater Parramatta) (Chap. 13) was favorable to early European settlers. St George, in the context of City of Sydney, is south east. The common geography, early history (Aboriginal and European) and common population ties reinforce St George as a metropolitan district. Looking early transport, the St George district serves has served from its first settlement as a connection between the City of Sydney and outlying areas further south. In time the district became a hub for train and bus services. The suburbs (from a real estate basis) contained in each of the

LGAs within St George are noted within the respective Chaps. 14, 15 and 16). A brief historical background of St George is gleaned from historical photos of the district (Plate 17.2). The coming of the English fleet entering Botany Bay in 1770 shown here (top-left). Botany Bay is a busy port by 1880, shown here (top-middle). Work on a sewer system is shown here (top-right) in 1915 at Wollongong Rd, Arncliffe. The Tom Ugly punt was an essential service to cross the George's River (bottom-left). The Hurstville Picture Palace (bottom-middle) served St George from the start of motion pictures. Finally, the commercial growth along Forest Rd in 1937 is illustrated here (bottom-right).

St George (Cities of Hurstville, Kogarah and Rockdale) is today the south-east geographical extension of the center of the Sydney Metropolitan Region. Population growth in St George has been significantly higher than the national average over the last two Census periods (2003–2011). Both K and R also had higher that the national average population growths over this period. St George prides itself as one of the most multicultural communities in Australia. St George, under the A *Plan for Growing Sydney* (NSW 2014) (Chap. 2), plays a strategic role in Sydney's growth to 2031 (Plate 17.3). St George is shown here (Hurstville and Kogarah) (bottom) with City of Sydney (top-right) and City of Parramatta (top-left).

Plate 17.2 Early St George district. *Left to right*: Botany Bay 1770; Botany Bay 1880; Wollongong Rd, Arncliffe; Tom Ugly Punt; Hurstville Picture Palace; and Forest Rd 1937 (*Source* Hurstville City Archives)

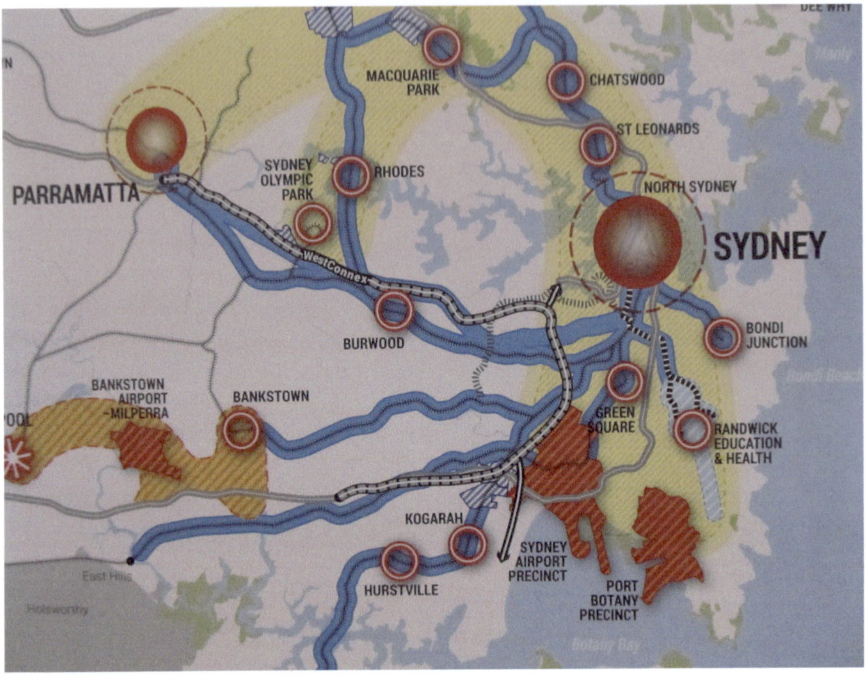

Plate 17.3 St George in context of Greater Sydney Centres (Source NSW 2014)

The Plan notes, in addressing the 'South Subregion':

> The South Subregion will be a significant focus for infrastructure investment and intensive growth over the next 20 years. Greater Parramatta will continue to be Sydney's second CBD and focus for jobs growth and services delivery Sydney's west. A growing and prosperous Grater Parramatta will be supported by a network of centres providing job and services closer to home for many of the subregions residents. This will improve liveability and contribute to strong, resilient communities throughout the subregion. (Source: NSW 2014)

The key urban planning principals as outlined in Chap. 2 are now addressed for St George district, including: (1). Developing Sustainable Urban Centres; (2). Developing Sustainable Transport; and, (3). Protecting the Natural Environment.

17.2 Developing Sustainable Urban Centers

The planning principal Developing Sustainable Urban Centers is the first principal to be examined. Case study examples include: (1). Kogarah Center; and (2). Hurstville Center; (3). Rockdale Center; and, (4). Wolli Creek Center.

17.2.1 Hurstville Center

The State has worked with Hurstville City Council to provide Hurstville with an updated district centre. This would provide Hurstville with the capacity for additional mixed commercial, service and residential uses to serve the St George District. Plans for the core of the Hurstville Center are contained in Plate 17.4. A central green square is planned within the middle of the Center (left). Details of the square are shown at right.

Examples of new development in Hurstville City reflecting centre renewal are illustrated in Plate 17.5. A high rise residential complex at South Hurstville, adjacent to the City Centre, under construction here (top-left). The allocation of open spaces within new developments is contrasted here with new complex in background (top-right). Hurstville is designated as an office centre as shown here with this new office building in South Hurstville (middle-left). Advantage is taken of the high elevation of Hurstville City Centre with generous balconies on these high-density 7-story apartments block here (middle-right). A portion of the new high rise developments have minimum setbacks on some frontages as illustrated here (bottom-left). Finally, the preservation of heritage buildings is a challenge to architects as shown here on this corner block (bottom-right).

Moving away from the immediacy of the District Center to areas adjacent to this Centre shows the proliferation of building styles and heights (Plate 17.6). A dramatic residential complex of two connected buildings here (top-left). Forest Rd is a main thorough fare on the outskirts of the District Center, shown here is a medium-rise commercial and residential building (top-middle). A series of 7–8 storey residential blocks (top-right) are being built on the fringe of the District Center, still within walking distance to the Hurstville rail station and bus routes. High rise residential buildings are prominent adjacent to the District Center (bottom-left).

Plate 17.4 Hurstville's planned district centre (*Source* Hurstville City Council Web 2016)

Plate 17.5 Hurstville's expanding centre (*Source* Ray Rauscher 2015)

Proposed residential complexes often are built on whole or part of residential street blocks as shown here (bottom-centre). Finally, back to South Hurstville, the assemblage of construction cranes here is an indication of the final residential density to be achieved in here (bottom-right).

Plate 17.6 Building styles and heights adjacent to Hurstville Center (*Sources* Ray Rauscher and Hurstville Council Web 2016)

17.2.1.1 Conclusions on Hurstville District Centre

Having reviewed the Hurstville District Center proposal, there are a number of significant effects on council areas within the St George district that this project is likely to have. These can be summarized under a need for: a. more details on the urban design aspects of projects (i.e. height, bulk, overshadowing); impact of the additional population on services and transport; greater allocations of open spaces lands and pedestrian linkages; greater housing affordability provisions; and, more environmental and heritage protection measures.

17.2.2 Kogarah Center

The State has worked with Kogarah City Council to provide Kogarah with an updated centre. This would provide Kogarah with the capacity for additional mixed commercial, service and residential uses to serve Kogarah and the wider St George district (built and proposed buildings in Plate 17.7).

The City center (top-left) is highly dense and generally surrounded by earlier built lower density residential developments. From an integrated transport perspective this higher central density is advantageous, with Kogarah rail station and bus interchange in the center. The Kogarah Square as built is shown in this artist's drawing (top-right). High-rise residential buildings have been and will continue to be

Plate 17.7 Kogarah's city centre development and proposals (*Source* Kogarah City Council Web 2016)

accommodated within the City Center (bottom-left). Finally, to cater for an increasing aged population new aged facilities are being proposed for the Center or in its vicinity where services can be provided (bottom-right).

Urban design in Kogarah is reaching higher investment levels and bigger projects (i.e. councils developing civic precincts). Kogarah's expanding center is illustrated in Plate 17.9 (left to right). The Kogarah City Council created a central square (top-left) in the commercial centre with public facilities such the City library at the rear of the square. Kogarah City has developed a significant cultural program with a statute of Bruce Lee (donated from China) in the town square (top-right). Small business offices have been allocated at ground level of residential complexes as shown here (middle-left). Shops below medium and high rise buildings is a popular trend in Kogarah as illustrated here (middle-right). The Council has developed a new administration centre within the town centre, maintaining easy public access (bottom-left). Finally, Kogarah has a large Greek population with their church located facing the town square (bottom-right).

Plate 17.8 Kogarah city built environment details (*Source*: Ray Rauscher and Kogarah City Council Web 2016)

Plate 17.9 Rockdale city centre development and proposals (*Source* Rockdale City Council Web 2016)

17.2.2.1 Conclusions on Kogarah Centre

Having reviewed the Kogarah Center commercial center development, there are a number of significant effects on council areas within the St George district that this project is likely to have. These can be summarized under a need for: a. more details on the urban design aspects of projects (i.e. height, bulk, overshadowing); impact of the additional population on services and transport; greater allocations of open spaces lands and pedestrian linkages; greater housing affordability provisions; and, more environmental and heritage protection measures.

17.2.3 Rockdale Center

The State has worked with Rockdale City Council to provide Rockdale with an updated centre. This would provide Rockdale with the capacity for additional mixed commercial, service and residential uses to serve Rockdale City and the wider St George district (Plate 17.9).

Urban design in Rockdale is reaching higher investment levels and bigger projects (i.e. councils developing civic precincts). Rockdale's expanding center is illustrated in Plate 17.10 (left to right). The City Council developed a popular outdoor mall with generous numbers of coffee houses (top-left). The mall is particularly well used 7 days a week by a cross sectional of members of ethnic communities. A new library is under construction (2016) adjacent to the town hall (top-right). A row of Rockdale heritage buildings is shown here at the Rockdale railway station (middle-left). Within walking distance to the rail station are high density developments are shown here (middle-right). The earlier Rockdale housing near the city centre is giving way to high rise residential buildings (middle-right). The assembling of large sites via demolition of earlier housing has seen significant high rise building erected (bottom-right). The fringes of the Rockdale city centre have seen major construction of high rise residential complexes. The buildings in general are being built along the Princes Highway and at Brighton Le Sands (tourist orientated).

17.2.3.1 Conclusions on Rockdale Town Centre

Having reviewed the Rockdale Center commercial development, there are a number of significant effects on council areas within the St George district that this project is likely to have. These can be summarized under a need for: a. more details on the urban design aspects of projects (i.e. height, bulk, overshadowing); impact of the additional population on services and transport; greater allocations of open spaces lands and pedestrian linkages; greater housing affordability provisions; and, more environmental and heritage protection measures.

17.2.4 Wolli Creek Center

The State has worked with Rockdale City Council (covering the Wolli Creek Suburb) to provide Rockdale City with an updated Wolli Creek Town Centre. This would provide Rockdale with the capacity for additional residential housing to serve the City and in general the St George district.

Wolli Creek's expanding center is illustrated in Plate 17.11 (left to right). The medium rise residential buildings here are located in a semi-circular pattern taking advantage of Discovery Point (top-left). Note the wide pedestrian way passing this complex). Tempe House, first house on the original estate, has been preserved and is open to the public on nominated days (top-right). Tempe House can be viewed here midst the high rise developments in the background (left-middle). The extent of the high rise developments is illustrated here (middle-right) in this 16 story complex of two matching buildings. The town square of Wolli Town Centre is illustrated here (bottom-left). The town square currently has a variety of shops (for example real estate, cafes, and super-market), with expansion plans under way

Plate 17.10 Rockdale's city centre details (*Source*: Ray Rauscher and Rockdale City Council Web 2016)

Plate 17.11 Wolli Creek's Town Centre (*Source*: Ray Rauscher 2015 and Rockdale Council Web 2016)

(2016). A walkway to the town square leads into a green space, shaded here by the surrounding buildings (that also act as a wind break for square users).

17.2.4.1 Conclusions on Wolli Creek Town Centre

Having reviewed the Wolli Creek Center development, there are a number of significant factors Rockdale Council and the State will need to monitor. These can be summarized under a need for: a. close coordination of the urban design aspects of projects (i.e. height, bulk, overshadowing); impact of the additional population on services and transport; greater allocations of open spaces lands and pedestrian linkages; greater housing affordability provisions; and, more environmental and built heritage protection measures.

17.2.4.2 Conclusions on Developing Sustainable Urban Centers

Having reviewed the main development centres within the three municipalities (Hurstville, Kogarah and Rockdale) and author (Rauscher) qualitative research field work (2014–2015), a number of conclusions can be drawn. The three cities' centers (as reviewed above) have seen major advances made in recent years, however many questions on planning at this level remain. It can be concluded the State, in partnership with councils, need to address the new planning arrangements needed to meet the planning principal *Developing Sustainable Urban* Centers. Options for these new planning arrangements are canvassed in Chap. 18.

17.3 Developing Sustainable Transport

Having examined the first urban planning principal (*Developing Sustainable Urban Centers*) within St George District, the second principal is D*eveloping Sustainable Transport*. With an extra 1.6 million people expected to be living in Sydney by 2036, there are many opportunities for transport to meet sustainability principles. All of the local government areas within St George District have been experiencing several decades of growth. As a result all the councils have been working to accommodate this growth and to apply the most effective transport provisions. At the district level the two State agencies responsible for developing plans for transport are the Dept. of Planning and Environment and the Dept. of Transport and Maritime Services. A number of transport projects either planned or underway are now reviewed under: light rail; bus corridors; and, WestConnex.

17.3.1 Light Rail

EcoTransit has researched a number of public transport alternatives to road transport proposals. One proposal aimed at assisting transport needs in the St George district is the Bay Light Express Line J (Plate 17.12). This proposal compliments the south east light rail currently under construction (with George St CBD start in 2015). Localities served would be: University Technology Sydney; Sydney University; Newtown; Central Industrial Area; Sydney Airport; Rockdale; Kogarah and Hurstville multiple stops; and, over the Georges River to Sutherland's key commercial and industrial areas.

Having reviewed the light rail planning proposals, there are a number of significant effects on council areas within the St George district that these proposals are likely to have. These can be summarized under a need for: a. increased opportunity for expanding the public transport system; stimulation of greater housing opportunities along the corridors of the light rail system; and, opportunities for expansion of connecting open space connectors.

17.3.2 Bus Corridors

Bus services within the three cities of St George are essential to the whole of transport planning of the district. There are bus corridors in St George linking major centres outside the district including: Bankstown; Burwood; Botany Bay bayside communities; Sutherland Shire and further south; Parramatta City; and the Sydney CBD. The buses operate locally and well as offering express services. The bus interchanges at all centres are integral for easy modal transport changes. Most of the interchanges have been upgraded (i.e. Hurstville, Kogarah, Rockdale and Wolli Creek) to provide easier access to transport services. These centres have also expanded commercial and service uses, thus increasing further the importance of these transport links. The State policy of integrated transport planning built around higher densities around town centres and rail or light rail hubs places an additional priority on the allocation of bus services. The buses provide an increasingly important service to workers who travel to other centres within the Sydney metropolitan region. The T-way rapid bus corridor, operating from the south west and west of the metropolitan region, is utilised by some of the St George express services to centres such as Parramatta City. A major road construction project, WestConnex, is examined next.

Plate 17.12 Proposed eco-transit bay light rail (*Source*: Eco-Transit Web 2016)

Plate 17.13 WestConnex proposal – St George district and connections (*Source*: Dept. Roads and Maritime Services Web 2016)

17.3.3 WestConnex

The State Government has committed to a major road proposal (Infrastructure NSW budget of about $15b over 10 years) called WestConnex (Plate 17.13) (left-right). This project involves: a. building 33 km of road; b. extending and widening the M5 (Sydney East and South) and the M4 (Sydney West); and, c. providing a link between them, via a long tunnel under the inner west of Sydney. The motorway proposal (left) provides a link to the Sydney Airport adjacent to the St George District. A consultant study examined the impacts of the motorway proposal from Wolli Creek to Rockdale Town Center. To get to Wolli Creek a controversial flyover interchange at St Peters (adjacent and affecting Sydney Park) is shown here (middle). The M5 section of the motorway affects St George, shown here at the interchange with King Georges Rd. (right). Advocates of more public transport argue for an alternative to the WestConnex, such as further extensions of light rail (noted above) from Dulwich Hill.

The Government created a web site for public input http://engage.haveyoursay.nsw.gov.au/westconnex and developed a video on the project video-summary-of-westconnex. As the WestConnex project is likely to many parts of the St George district the Wolli Creek Preservation Society, has assessed the project. The Society writes in summarizing two concerns about the project:

1. The proposals for the M5 East WestConnex section will have implications for Bexley North (more traffic) and Wolli Creek (its bushland).
2. There is a major investment imbalance between rail and road, thus the Society supports targeted investment in public transport alternatives that would have greater and more sustainable benefit to inner west, south-west and western Sydney residents at a fraction of the cost of WestConnex. *Source* Wolli Creek Preservation Society 2015).

Having reviewed the WestConnex proposals, there are a number of significant effects on council areas within the St George district that these proposals are likely to have. These can be summarized under a need for: a. more details on the cost-benefits of these proposals (especially given the size of project expenditures anticipated); more details on road vs. public transport alternatives; more information on and timing aspects of proposals; extent that the proposals will be a catalyst for further redevelopment projects; long term effect of potentially inducing additional car orientated demands; impact of the additional population on services and housing that the road proposals will have; greater integration of open spaces lands and pedestrian linkages; and, more environmental (i.e. not alienating public open spaces) and heritage protection measures.

17.3.3.1 Conclusions on Developing Sustainable Transport

Having reviewed a number of St George transport plans and projects and author (Rauscher) qualitative research field work (2014–2015), a number of conclusions can be reached. The transport projects (light rail; bus corridors; and, WestConnex) have seen major advances made in recent years, however many questions on transport planning at this level remain. In terms of meeting the planning principle *Developing Sustainable Transport* within St George District, it is concluded that new planning arrangements between the State and councils need consideration. Options for these new planning arrangements are canvassed in Chap. 18. The next planning principal to be examined is *Protecting the Natural Environment.*

17.4 Protecting the Natural Environment

Having examined the first two urban planning principals (developing strategic centers and developing networked transport) within St George District, the third planning principal to be examined is *Protecting the Natural Environment.* Conservation protections within the St George District has to date been a cooperative effort among the three councils and the State (refer to Chaps. 14, 15 and 16). At times, on a common planning basis, councils outside St George District have become involved (i.e. where waterways or green corridors affect these other councils). Three conservation protection features of the district are selected to be examined, being: (1). the Georges River; (2). Botany Bay; and, Wolli Creek National Park.

17.4.1 Georges River

The Georges River (Plate 17.14) is the main tributary of Botany Bay (Plate 1). A secondary tributary of the Georges River is the Cooks River as addressed under 'Sydney's Inner West' (Chap. 9). A background on the Georges River follows.

Plate 17.14 Georges River, St George District (*Source* Wikipedia 2015)

The Georges River, formerly known as Tucoerah River, is an intermediate tide dominated drowned valley estuary. The land adjacent to the Georges River was occupied for many thousands of years by the Tharawal and Eora Aboriginal peoples. They used the river as an important source of food and a place for trade. The total catchment area of the river is approximately 930.9 square km (359.4 sq. mi) and the area surrounding the river is managed by a large number of local government authorities and NSW Government agencies. The banks of the river along the lower reaches are marked by large inlets and indentations overlooked by steep sandstone ridges and scarps. (Source: Wikipedia 2015)

An examination of the River and its catchment up close (Plate 17.15) reflects the central role this water system is to a number of South West Sydney districts, including the St George district. The catchment of the river (noted above) covers a number of local government areas. The Lower Georges River Sustainability Initiatives is one of several programs to ensure the health of the river (addressed below). There are a number of recreation places established along the foreshores of the river. These include for example: Burrawang Reach picnic area (top-left); Cattle Duffers Flat (top-right); and, Yeramba Lagoon (bottom-left). Finally, some of the estuary areas of the river are accessible to the public (canoeing here) (bottom-right).

The Georges River and its catchment is the responsibility of the NSW Government's Coastal Management Program and local councils. This program provides support to local councils to manage the risks from coastal hazards, such as coastal erosion. A secondary objective of the program is to restore degraded coastal habitats. In addition, the NSW Government's Estuary Management Program provides support to councils to improve the health of NSW estuaries, such as those in the Georges River catchment. The support provided to councils under both these programs includes: technical and financial assistance to prepare coastal zone management plans; managing risks associated with coastal hazards; and, protecting the

Plate 17.15 Georges river and foreshores conservation. *Left to right* Burrawang Reach picnic area; Cattle Duffer's Flat boat launch; Yeramba Lagoon; and Canoe activities in wetlands (*Sources* Catchment Management Committee and Google Images 2016)

estuary health. In 2008, for example, a number of councils (including those under the Georges River Combined Councils Committee) completed the *Georges River Comprehensive Strategic Plan* (NSW 2008). This plan focuses on the rehabilitation of the catchment area of the river. Finally, the State has responsibility for floodplain management of the Georges River catchment.

Having reviewed the Georges River restoration and protection, there are a number of significant effects on council areas within the St George district that these measures are likely to have. These can be summarized under a need for: a. wider State agreement to expand funding of these measures; increasing restoration and protection programs; impact of the additional population on these projects as planned for the district under State plans; further integration of open spaces lands and pedestrian linkages; and, more environmental and heritage protection measures.

17.4.2 Botany Bay

The Georges River flows into Botany Bay, another large water body of significance to the St George district. Botany Bay is an open oceanic embayment (Plate 17.16), most of whose foreshores extend over many local government areas, including areas within St George (left side of map). The background of the Bay follows:

> Botany Bay has its source in the confluence of the Georges River and the Cooks River and flows 10 km (6 mi) to the east before meeting its mouth, the Tasman Sea. The total catchment area of the bay is approximately 55 km^2 (21 sq. mi). Despite its relative shallowness, the bay serves as greater metropolitan Sydney's main cargo seaport. Two runways of Sydney Airport extend into the bay. Botany Bay National Park is located on the northern and southern headlands of the bay. The land adjacent to Botany Bay was occupied for many thousands of years by the Tharawal and Eora Aboriginal peoples and their associated clans. On 29 April 1770, Botany Bay was the site of James Cook's first landing of HMS *Endeavour* on the continent of Australia, after his extensive navigation of New Zealand. Later the British planned Botany Bay as the site for a penal colony. (Source: Wikipedia 2015)

The geography and history of the Bay is depicted in Plate 17.17. A watercolour by Charles Gore shows the English fleet arriving in 1788 (top-left). The fleet (headed by Capt. Cook) encountered Aboriginal peoples (top-right) of the district (commented on earlier in Chapter). The occupation of the Bay area is depicted here (bottom-left). Finally, the map (bottom-right) shows the Bay and St George district on the left.

The environment Botany Bay, including its diverse marine population and sensitive foreshores, has been impacted upon over the history of the Bay. In recent times, it's reported: 'Weedy sea dragons are just one of hundreds of territorial marine crea-

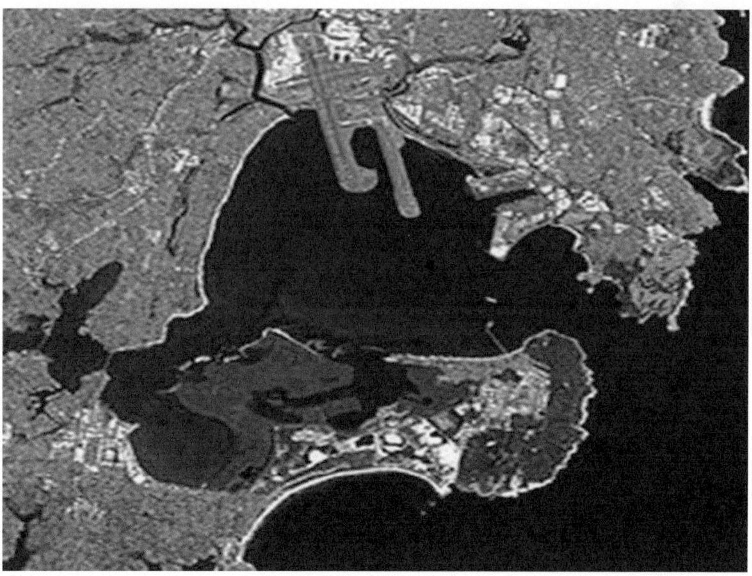

Plate 17.16 Botany bay aerial. *Note* St George on *left-middle* (Source: Wikipedia 2016)

Plate 17.17 Botany bay geography. *Left to right Watercolor* (Charles Gore) on English fleet arrival in 1788; Encounter with Aboriginal peoples of the district; Occupation of Bay by Aboriginal Clans; and map on left, the Bay and St George district on *left* (*Source* NSW Archives 2016)

tures found within Botany Bay. The eastern blue grouper is the state fish of New South Wales; and are commonly found following divers along the shore line of Botany Bay.' (Source: Wikipedia 2015). Recent reports on environment threats to Botany Bay indicate:

> Botany Bay and its catchment waterways are subject to ongoing threats due to nutrient and sediment-laden run-off from various non-agricultural land uses. A substantial part of the catchment is highly developed with almost 40 % of its area being used for urban, industrial or commercial purposes. Pollutants of concern are nitrogen, phosphorus and total suspended solids. Development pressure is contributing to degradation of bushland and recreational areas which is increasing sediment loads within the catchment. (Source: Wikipedia 2015)

Key water quality improvements projects have been targeted for Botany Bay. The key State agencies responsible for these projects include: Sydney Metropolitan Catchment Management Authority; NSW Office of Environment and Heritage and NSW Office of Water. These projects come under the Local Land Services (Greater Sydney) (http://greatersydney.lls.nsw.gov.au/home) include: Botany Bay and

Catchment Water Quality Improvement Plan; Water Sensitive Urban Design Adoption Strategy for the Botany Bay Catchment; Opportunities and barriers to WSUD adoption in the Botany Bay Catchment; Ecological Response Model; Modelling pollutants; and, Environmental Values and Water Quality Objectives

There are additional programs undertaken by Councils of St George (and surrounding councils) to improve the environment of Botany Bay. Rockdale, Kogarah, Canterbury and Marrickville Councils, for example, received a grant from the Cooks River Foreshore Improvement Program (2007/2008) to undertake an ecological study in relation to water quality of the Cooks River. The project's objective is to establish a cost-effective ecological monitoring program that will develop information and guidelines for the management of tidal vegetation and inform future capital works projects along the riverbanks. Specific projects have included installing gully pit traps, trash racks, and mesh screens to prevent gross pollutants from eventually washing up on the beaches at the northern end of Botany Bay. Also, outlet pipelines have been installed to improve water quality, ventilation and solar access to promote fish access between the ponds and Botany Bay. Finally, wetlands corridors have been constructed to treat water flowing from reserves into Botany Bay.

There are many linked natural areas and open spaces within the St George district. Two examples of Councils, State and community cooperation in planning and protecting these open spaces are illustrated next. First is the cooperation in the planning and part construction of a 57.2 km (Plate 17.18) cycle route. The route is

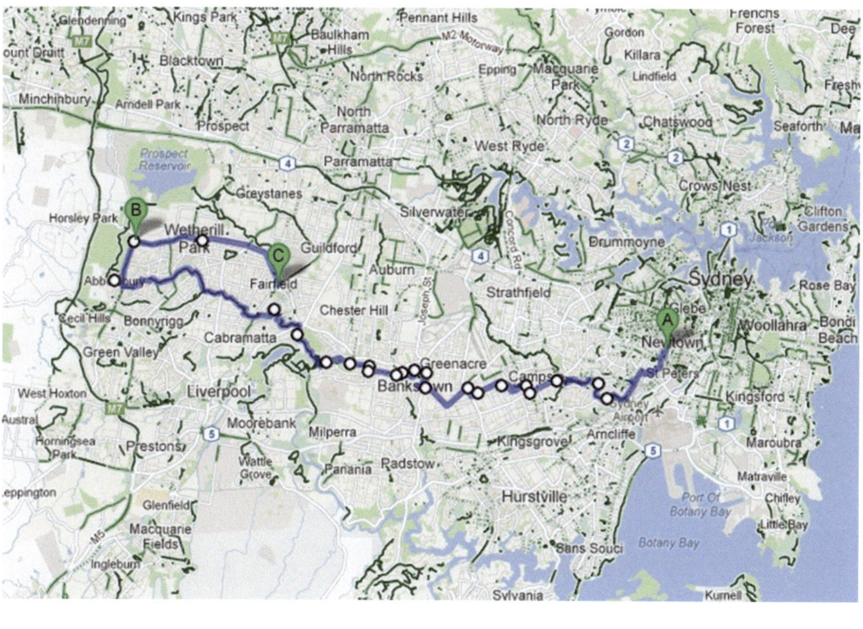

Plate 17.18 Georges River and Cooks River Cycle Route (*Source* Hurstville City Council Web 2016)

planned to run from Newtown (Marrickville Council) to Fairfield) and follows the Georges River (to left) and the Cooks River (to the right).

Having reviewed the Botany Bay restoration and protection, there are a number of significant effects on council areas within the St George district that these measures are likely to have. These can be summarized under a need for: a. wider State agreement to expand funding of these measures; increasing restoration and protection programs; impact of the additional population on these projects as planned for the district under State plans; further integration of open spaces lands and pedestrian linkages; and, more environmental and heritage protection measures.

17.4.3 Wolli Creek Regional Park

A second example of cooperation among councils, the State and the community is the open space (50 ha) planning of the Wolli Creek Regional Park (Plate 17.19). The park is a stretch of bushland in St George (Rockdale) with varied flora and fauna. The park contains heathland, woodland, rain forest trees, wetlands, salt marsh and parklands. The Wolli Creek Preservation Society (http://www.wollicreek.org.au/) reports:

> The Wolli Creek Valley contains the only bushland of substantial size left in inner south-west Sydney. It is also the only large, high-quality open space that remains in a heavily developed residential and industrial region. The Society was originally formed in 1983 to actively oppose the construction of the M5 East motorway through the Wolli Creek Valley. Over the years, the Valley has faced some dire planning proposals, particularly the proposal

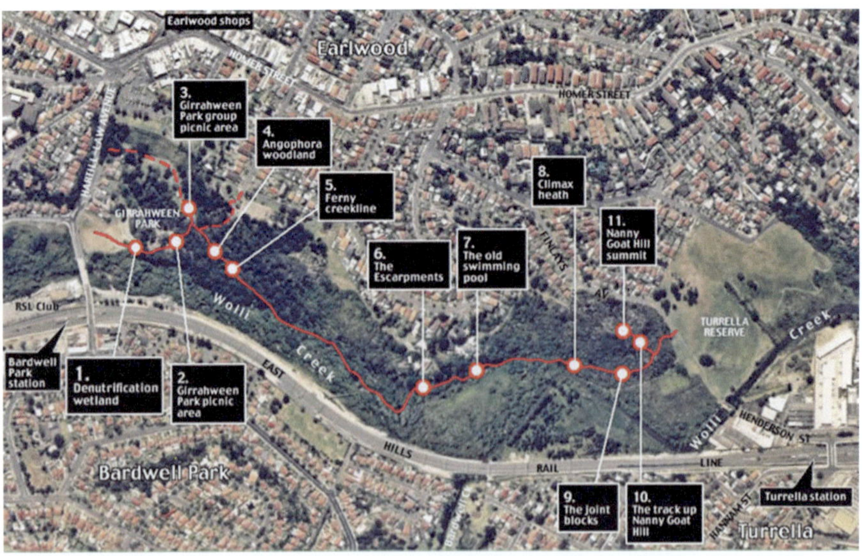

Plate 17.19 Wolli Creek National Park (*Source* Wolli Creek Preservation Society Web 2016)

for an eight-lane highway above ground through the whole Valley. We have fought hard to retain and enhance the Valley. There is still a long way to go in restoring this area, but bringing the Valley under the protection afforded by Regional Park status is a step in the right direction. The Wolli Creek Preservation Society aims to ensure the preservation of the natural and cultural heritage of the Wolli Creek Valley, its tributaries, associated corridors and their ecological communities, and the ecologically sensitive restoration, expansion and maintenance of these areas.Finally, the Wolli Creek Preservation Society acknowledges that Wolli Creek flows through the cultural landscape of the Gadigal people. (*Source* Wolli Creek Preservation Society 2015)

Having reviewed a number of St George District projects aimed at protecting the natural environment and author (Rauscher) qualitative research field work (2014–2015), it can be concluded the St George District Councils are addressing State planning principal *Protecting the Natural Environment* under *Planning for Sydney's Growth* (NSW 2014) (Chap. 2). These Councils, in addressing this principal, are applying the individual council plans (Chaps. 14, 15 and 16) and acknowledging the State 'subregional plan guidelines' for their district (Chap. 2). In many instances it has been the local area resident groups, in cooperation with Councils and the State, which has raised environment protection and enhancement proposals. Finally, in terms of advancing planning within the St George District, new arrangements among Councils and the State need consideration. The environment planning affecting the St George District to date have seen major advances made, though questions have arisen (as reviewed above). To advance further environment sustainability principles, the authorities (local, State and Federal) will need to develop the governance structures that meet the challenge. These options facing the municipalities within St George District (i.e. structural changes in partnership between councils and the State) are canvassed in Chap. 18.

Having reviewed the Wolli Creek restoration and protection, there are a number of significant effects on council areas within the St George district that these measures are likely to have. These can be summarized under a need for: a. wider State agreement to expand funding of these measures; increasing restoration and protection programs; impact of the additional population on these projects as planned for the district under State plans; further integration of open spaces lands and pedestrian linkages; and, more environmental and heritage protection measures.

17.4.3.1 Conclusions on Protecting the Natural Environment

Having reviewed a number of St George natural areas and author (Rauscher) qualitative research field work (2014–2015), it can be concluded the State needs to further address natural environment issues under Protecting the Natural Environment. The St George Councils, it was concluded earlier in Chaps. 14, 15 and 16, were addressing: (1). environmental protection within each council's planning practices; and, (2). the planning principle Protecting the Natural Environment (Chap. 2). The environment protection projects (as reviewed above) have seen major advances made in recent years, however many questions on environmental protection at this level remain (as noted in the review). In terms of meeting the State principle Protecting the Natural Environment within the St George District, it is concluded that new planning arrangements among Councils and the State need consideration. Options for these new planning arrangements are canvassed in Chap. 18. The subject to amalgamations of councils and how this affects the St George District examined next.

17.5 Amalgamations of Councils and St George District

The subject of amalgamations of Greater Sydney councils (given implications for planning at municipal and Greater Sydney levels) was reviewed in Chap. 2. NSW Premier Mike Baird announced in Dec 2015, in examining the St George District (Plate 17.20) (and Botany Bay Council), the State's preference: a. for Hurstville and Kogarah councils to merge; and b. for Rockdale and Botany Bay councils.

The position on the councils' merger questions were reviewed in the respective chapter on each council. Each council adopted its preference on the merger question, with: a. Hurstville agreeing to a merger with other St George councils; b. Kogarah wanting to stand-alone; and c. Rockdale to merge with Botany Council (outside the St George District) but with a preference (under any forced amalgamations) for Rockdale to merge with Hurstville and Kogarah to keep St George as a

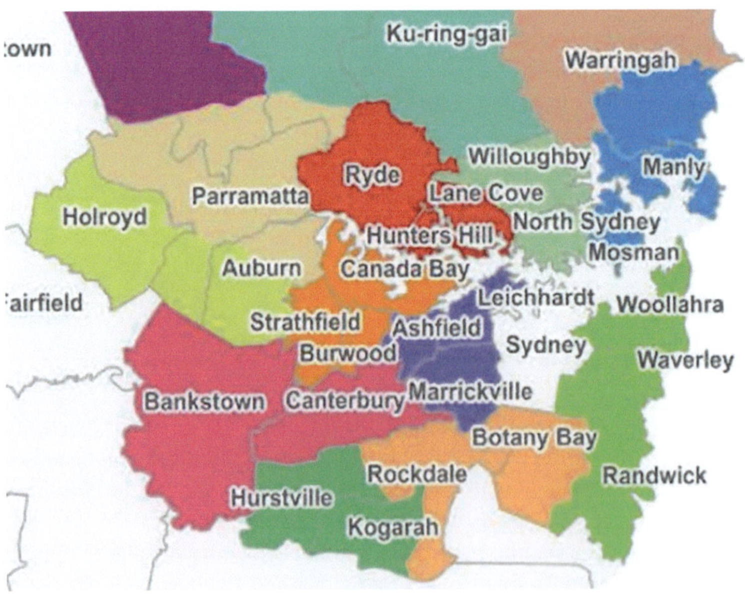

Plate 17.20 St George District Councils' Amalgamations. *Left to right at bottom*: Hurstville and Kogarah; and Rockdale and Botany Bay to east (*left*) (*Source*: NSW Dept. of Local Government Web 2016)

district entity. The Mayors of all councils urged residents to have their say on the issue of amalgamations of councils. Councils noted that submissions to the new Council Boundary Review were open till the end of February 2016 (www.council-boundaryreview.nsw.gov.au). In the interim, the Councils noted, the Minister for Local Government had commenced the process of forcibly amalgamating the two groups of councils (Hurstville and Kogarah; and Rockdale and Botany Bay) in accordance to State preference as noted earlier. Councils noted the Minister said he expected the new merged councils to be 'proclaimed' on 1 July 2016 (note the 12th May 2016 *Local Government (Council Amalgamations) Proclamation 2016* noted in Chap. 2). Results of the review of merger proposals would be referred to the Chief Executive of the Office of Local Government for examination and report under the Local Government Act (the Act). In turn, the Chief Executive had appointed a Delegate for each of the proposed merges. The Delegates (2) would examine the proposals, conduct public inquiry, call for written submissions, and then report back to the Minister on the mergers.

17.5.1 Conclusion on Amalgamations

The State, in examining the St George District, expressed a preference for Hurstville and Kogarah councils to merge and for Rockdale and Botany Bay councils to merge. At the end of the State review process (Council Boundary Review) the State would determine what amalgamations as proposed would proceed or not proceed. The 12th May 2016 *Local Government (Council Amalgamations) Proclamation 2016* proposed mergers and boundary changes to nominated Sydney Inner West councils was noted. This question will be further addressed in Chap. 18.

17.6 Overall Conclusions

Having completed an assessment of the development and planning of St George District (critiquing plans and developments against the adopted three planning principles) there are a number of overall conclusions reached. Firstly, examining conclusions reached in assessing planning and development in each of the three cities (Hurstville, Kogarah and Rockdale), it was concluded that each of these cities had been experiencing accelerated development and changes to the composition of their populations. Given these changes, and examining these councils' responses, it was also concluded these councils were all moving in the right direction in increasing their sustainable urban planning (SUP) practices. It was also concluded, however, that there would be significant future impacts on all these cities from Greater Sydney growth and development. In reviewing potential Councils and State responses to this future scenario, it was concluded that new partnerships between these councils and the State would be needed. A major initiative from the State (2009 onwards) was to investigate local government reform. This included a potential amalgamation of councils (e.g. Hurstville and Kogarah; and, Rockdale and Botany City). Kogarah and Rockdale councils had during this State amalgamation process, however, expressed an intent to stand alone and not amalgamate. The State started a forced amalgamation process in early 2016, results not to date determined. The overall conclusion in applying the three planning principles is that new planning arrangements between the State and councils are needed to meet these principles. Options for these new planning arrangements are canvassed in Chap. 18.

17.7 Summary

Planning of St George District, the third of three case study districts of Greater Sydney, was examined in this chapter. Having examined the question of global transition and planning practices of the six municipalities within the District (Chaps. 14, 15 and 16), the chapter addressed: (1). planning the St George District as a whole; and, (2). the extent the State (responsible for district planning) was meeting the planning issues. Key planning principles as adopted (Chap. 2) to apply to Greater Sydney (and this district) were reviewed. These key principles include: (1). Developing Strategic Centres; (2). Developing Networked Transport; and, (3). Protecting the Natural Environment. Sample plans and developments within St George were critiqued under each of these principles. After reaching a conclusion under each principle, an overall conclusion was drawn that new planning arrangements among Councils and the State (planning the District) were needed in planning the District. Finally, it's was noted that these St George District conclusions would be assessed alongside conclusions reached in examining Sydney Inner West District (Chaps. 3, 4, 5, 6, 7, 8 and 9) and Greater Parramatta (Chaps. 10, 11, 12 and 13). All these conclusions are examined Chap. 18 (examining directions for planning a sustainable Greater Sydney and a sustainable city anywhere). Having examined the three districts of Sydney Inner West District, Greater Parramatta and St George, Chap. 18 is addressed next.

References

New South Wales (2008) Georges river comprehensive strategic plan. NSW, Sydney
New South Wales (2014) A plan for growing Sydney. NSW, Sydney

Part V
Conclusions on Planning Sustainable Cities

Parts II to IV assessed the extent of sustainable urban planning (SUP) practices being applied at the municipal and districts levels over three Greater Sydney Districts (Sydney Inner West, Greater Parramatta and St George) (Chaps. 3, 4, 5, 6, 7, 8, 9, 10, 11, 12, 13, 14, 15, 16 and 17). Part V now examines the conclusions reached from those assessments. From these conclusions, optional types of planning arrangements for planning partnerships between municipalities and the State are offered. The aim here is to ensure the districts of Greater Sydney (and Greater Sydney overall) continue to move in the direction of becoming sustainable. These options are outlined in Chap. 18. In addition the chapter addresses how governments anywhere in the world can adopt SUP practices, in particular to meet global transition factors impacting on cities.

Chapter 18
Directions for Planning Sustainable Cities

Abstract This chapter examines the conclusions from the analyses on planning within the three case study districts of Greater Sydney (Sydney Inner West, Greater Parramatta and St George) (Chaps. 3, 4, 5, 6, 7, 8, 9, 10, 11, 12, 13, 14, 15, 16, and 17). The chapter acknowledges the *Local Government (Council Amalgamations) Proclamation 2016* proposed mergers and expanding the boundaries of nominated councils noted in Chap. 2. The first conclusion reached was that the State needs to further address the impact of development under the three planning principles (as assessed in earlier chapters) with specific reference to *A Plan for Sydney's Growth* (NSW 2014). The second conclusion reached, that new planning arrangements among Councils and the State are needed in planning the Greater Sydney districts, is also addressed. Suggested new planning arrangements focus on six options for planning at municipal and district levels. Comments are then offered on State planning directions for planning a sustainable Greater Sydney. The chapter, finally, addresses how sustainable urban planning (SUP) practices can be used to achieve a sustainable city anywhere. A sample range of planning guidelines are offered for any city in the world to consider to move in the direction of achieving sustainability.

18.1 Planning Greater Sydney to Date

The analyses of planning at municipal and district levels within the three case study districts of Greater Sydney (Sydney Inner West, Greater Parramatta and St George) (Chaps. 3, 4, 5, 6, 7, 8, 9, 10, 11, 12, 13, 14, 15, 16, and 17) provided many conclusions. In examining the municipal councils (12 in all over the 3 districts), three questions were addressed for each municipality (in separate chapters): (1). is that municipal area experiencing accelerated global transition in its development; (2). what is the extent Sustainable Urban Planning (SUP) practices are being applied by that municipal council; and, (3). what is the extent that council area is affected by Greater Sydney Region planning. To addressing #1 it was concluded that all of the councils had been experiencing accelerated global transition (especially in the 2000s). The #2 question, the extent of SUP being practiced, was then examined. Having reviewed key planning documents within each of the councils (Chaps. 3, 4, 5, 6, 7, 8, 9, 10, 11, 12, 13, 14, 15, 16, and 17), it was concluded the councils were

R.C. Rauscher, S. Momtaz, *Cities in Global Transition*,
DOI 10.1007/978-3-319-39865-5_18

overall moving in the direction of SUP practices being applied. The State's Greater Sydney planning (including infrastructure projects) likely to affect those councils was then reviewed (#3). This review concluded that the implementation of these State plans and State approved developments were likely to have significant effects on councils (12) within the three districts. Having drawn these conclusions to the three questions at the local government level, it was noted that the analysis would then move to assessing planning within the three districts as whole units (Chaps. 9, 13, and 17). The planning principles were (as adopted in Chap. 2): (1). Developing Sustainable Urban Centres; (2). Developing Sustainable Transport; and, (3). Protecting the Natural Environment. Conclusions drawn from applying these applications is addressed next.

The analysis of the planning principle Developing Sustainable Urban Centres was applied to each of the three case study districts (Sydney Inner West, Greater Parramatta and St George). Both selected planning and development projects and author (Rauscher) qualitative research field work (2014–2015) were examined. It was concluded, firstly, that the State needed to further address the impact of development under the planning principle Developing Sustainable Urban Centres. Attention was drawn here (for comparison bases to State plans and on-the-ground development), to the main State planning reference *A Plan for Sydney's Growth* (NSW 2014) (Chap. 2). In terms of meeting this principle within the three districts, it was concluded that new planning arrangements between the State and councils would need consideration. These new planning arrangements are addressed later in this chapter.

The next planning principal that was examined was Developing Sustainable Transport. A number of the districts' transport projects and author (Rauscher) qualitative research field work (2014–2015) were examined. It was concluded, firstly, that the State would need to further address transport issues under the planning principle Developing Sustainable Transport. It was noted that the transport projects (as reviewed for the three districts) have seen major advances made in recent years, however many questions on transport planning within the three districts remained. It was also concluded that new planning arrangements between the State and councils would need consideration (again, addressed later in this chapter).

Having reviewed the first two principles as applied to the three districts, the third and final principle Protecting the Natural Environment was assessed. It was concluded, firstly, that the State needs to further address the protection of the natural environment issues under this principle. It was noted that the environment protection projects (as reviewed within the three districts) had seen major advances made in recent years, however many questions on environmental protection at the district level remain. It was also concluded, that In terms of meeting this principle within the three districts, new planning arrangements between the State and councils would need consideration (canvassed later in chapter).

Having addressed planning within the case study districts (Sydney Inner West, Parramatta, and St George) (i.e. critiquing plans and developments against the three planning principles) two overriding conclusions (both addressed next) were **(and** addressed in the next Sect. 18.2 Directions for Planning a Sustainable Greater Sydney):

1. *The State needed to further address the three planning principles: Developing Sustainable Urban Centers; Development Sustainable Transport; and, Protecting the Natural Environment.*
2. The State and councils needed to consider new planning arrangements to create a sustainable Greater Sydney

18.2 Directions for Planning a Sustainable Greater Sydney

There are many options for the State and councils (or governments in any country) to address in adopting sustainable planning principles and planning decision making to create a sustainable city (in this case Greater Sydney). A crucial aspect of these planning principles and planning decision making is their application to all levels of governance (e.g. municipal, district and Greater Sydney wide). A successful application at the municipal and district levels will normally have a flow on effect to the Greater Sydney level. How the three planning principles referred to throughout the chapters can be realized and new planning decision making adopted (i.e. State in partnership with councils) is addressed next. Offering planning option models to achieve both these goals reflects the complexity of planning sustainably for a city and achieving an inclusive decision making. These planning option models will need also to acknowledge the current State's local government reform agenda. This reform ranges from: (a). creating (at State level) the Greater Sydney Commission (Chap. 2 and addressed later); and, (b). amalgamations of councils (Chaps. 2 3, 4, 5, 6, 7, 8, 9, 10, 11, 12, 13, 14, 15, 16, and 17). The overall voice of councils (above and from Chaps. 2 3, 4, 5, 6, 7, 8, 9, 10, 11, 12, 13, 14, 15, 16, and 17) is that they welcome helpful reforms and improved partnerships with the State. Councils have stated, however, that they oppose State planning that will impact on their areas or force amalgamations. These councils, to date, have stated many reasons that forced amalgamations don't work (i.e. removes independence of elected local government councils and defeats democratic principles). There are six options (including a no change option) offered here for State, in partnership with councils, planning decision making at a district level. These options are:

18.2.1 District Planning Groups

State bodies and councils often form 'planning groups' to cooperate on specific planning or public works projects within one or more local government areas. A number of these types of planning groups were outlined in Chaps. 3, 4, 5, 6, 7, 8, 9, 10, 11, 12, 13, 14, 15, 16, and 17 (e.g. planning urban centers or light rail projects).

There is an opportunity here for the State and councils to expand their partnerships in district level projects (i.e. transport or protecting natural environments as outlined in earlier chapters). Where a District Planning Group was established it could participate in State decisions, such as nomination of 'public infrastructure benefit levies' (PIBL). A PIBL is a proposal to levy developer owned land that is set to benefit from a State public infrastructure project (i.e. new light rail station, transport interchange, or train station). Many district level projects outlined in Chaps. 9, 13 and 17 would be the type of projects the State and local councils could work in partnership within a District Planning Group. Such a group, in being created, would be responsible for engaging the participating councils and the communities likely to be affected by planning decisions at district level. As noted above, the State's adoption of *A Plan for Growing Sydney* (NSW 2014) designates Greater Sydney sub-regions (in 2016 the State was calling these 'districts'). Related to this, the newly created (late 2015) *Greater Sydney Commission* announced it will develop a district plan for each of the Greater Sydney six districts. A second option of *District Planning Boards* is examined next.

18.2.2 District Planning Boards

District Planning Boards is another option for the State and councils to work in an ongoing partnership on specific planning and infrastructure works projects. These planning and infrastructure projects could include, for example: (a). environmental projects; (b). climate change planning and related programs; (c). alternative energy projects; (d). waterways protection; and, (e). sustainable transport initiatives. The planning of the Inner West Greenway and Inner West Light Rail Corridor are prime examples of council initiatives being endorsed and later partnered with the State. This option of a District Planning Board could be created (and the State is taking steps in that direction in early 2016) within the newly formed Greater Sydney Commission. The Board would be more formal that a District Planning Group (#1) and could also be incorporated within any legislation stemming from the State's local government reform proposals (Chap. 2). The main objective of a District Planning Board would be to guide district level planning and infrastructure provision (including economic advice). The Board would engage the participating councils and be responsible for public consultation processes on district level planning and developments. A Board could operate under a sun-set timed agreement (or renewable every 4 years) as written within any legislation. A municipality participating within a District Planning Board would still maintain its integrity at municipal level planning and infrastructure provisions under the State acts. Such councils would also continue their local planning and service responsibilities, for example: traffic management; environmental works; planning and delivery of local public works; and, parks and open spaces.

Moving further, the functions of a District Planning Board could be defined in the current (2016) State local government reform program in cooperation with the

NSW Local Government Association (representing councils). The Board could have final accountability to the State Minister for Planning and Environment. The composition of the Board could consist of: general manager and mayor (representatives of the municipal citizens) of each council; an appointee from each of Dept. of Planning and Environment and NSW Public Works and Infrastructure; and, District Commissioner under Greater Sydney Commission. The meeting place of the District Planning Board would be within the district (likely a council administration building). The decision making meetings would be open to the public (unless declared commercial-in-confidence). The budget of the District Planning Board could be under the State via the Greater Sydney Commission. A rolling 4 year District Planning Board program (and budget) of planning and infrastructure works could be adopted by the Board (and approved by the State). This program could be updated annually to enable participating councils to note designated planning and infrastructure works (i.e. including in councils' annually updated Community Strategic Plans and reports to ratepayers). Of final note, the Premier's Office could arbitrate any disputes. Participating councils would have the right to ask that disputed items go to an Independent Disputes Arbitrator, assigned under the District Planning Boards legislation. Having examined the option of District Planning Boards, the third option for the State and councils to consider is District Planning Committees under partnerships with current Regional Organizations of Councils' (ROCs).

18.2.3 District Planning Committees (Under Regional Organizations of Councils)

Regional Organizations of Councils (ROCs) were established by the Whitlam Government in 1973, with ROCs currently operating in all States of Australia. The Southern Sydney Regional Organisation of Councils (SSROC), for example, represents more than 1.4 million residents in the districts of Sydney Inner West and Sydney South. This ROC represents sixteen councils. The role of SSROC is defined as:

> To provide a forum for the exchange of ideas between our member councils, and an interface between governments, other councils and key bodies on issues of common interest. The organization acts as a facilitator of joint activities between councils which provide benefits through economies of scale. Member councils can reduce the costs of procurement, share the costs of engaging external assistance if needed, and can expose their staff and councillors to a wide network of expertise and views. SSROC is also a means of resolving issues and challenges that cross our municipal boundaries, such as the environment and transport. SSROC's work is structured to respond to the priorities identified and agreed by the members in the annual work plan, and ultimately to contribute to the social, environmental and economic improvement of the southern Sydney region. (*Source*: SSROC **2015** Web)

The State could ask a Regional Organization of Council to assist in the creation of a District Planning Committee. Such a body would undertake planning in cooperation with the State. The councils of that district (within the larger ROC area) could participate in such a Committee. The budget of these committees could operate similar to the District Planning Boards, with a rolling 4 year planning program as agreed by the State (and endorsed by the Committee). The Committee would be responsible for engaging the public under any guidelines as negotiated at the initial setting up of a District Planning Committee. As with the Planning Boards, any dispute could be arbitrated by an Independent Disputes Arbitrator. Having examined three options, a fourth option, District Planning Advisory Bodies is considered next.

18.2.4 *District Planning Advisory Bodies*

The chapter acknowledges the *Local Government (Council Amalgamations) Proclamation 2016* proposed mergers and expanding the boundaries of nominated councils noted in Chap. 2 and subsequent chapters. The Proclamation was accompanied by: (a). an endorsement (and time schedule) for amalgamation of councils that nominated as such; and, (b). a State decision on amalgamation of other councils that chose not to amalgamate. The State proposed amalgamation of nominated councils is indicated below (Plate 18.1) as reviewed in Chaps. 3, 4, 5, 6, 7, 8, 9, 10, 11, 12, 13, 14, 15, 16, and 17.

Two or more amalgamated councils are likely to see all current planning and infrastructure provisions reviewed under the one new council (under an Administrator without mayor or councillors) until local government elections in September 2017.

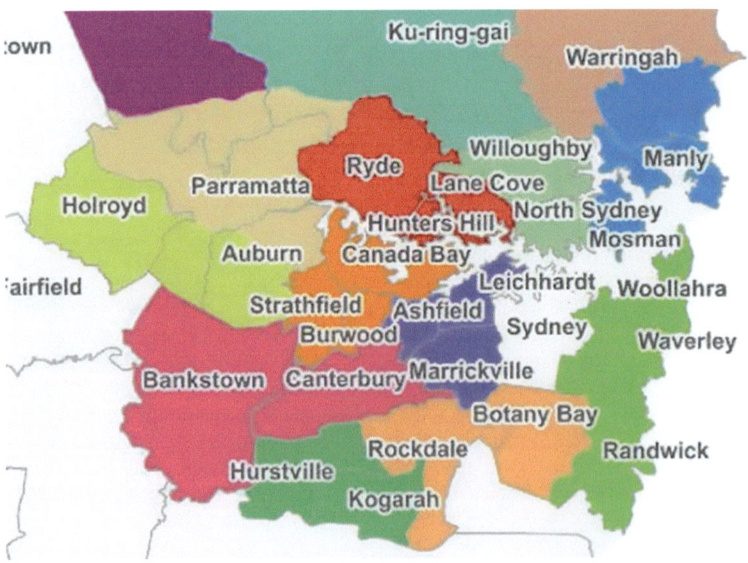

Plate 18.1 Amalgamations of Councils. Clockwise from *left*: Parramatta District (*left*); Inner West (*center*); and St George (*bottom*) (*Source*: NSW Dept. of Local Government Web 2016)

The subject of planning and infrastructure provisions is also likely to be reviewed with the State (potentially under new State legislation and to reflect the local government reform process (Chap. 2). The State would likely work directly with the newly amalgamated council (and other councils) within a Greater Sydney district (discussed earlier). The issue here remains (as reviewed earlier in Chaps. 3, 4, 5, 6, 7, 8, 9, 10, 11, 12, 13, 14, 15, 16, and 17 and above) many councils have expressed their desire to stand alone as an ongoing municipal unit. At the time of this book in 2016 there were a number of court cases involving councils disputing mergers of boundary changes as related to the Proclamation above. The fifth option of the State expanding its district planning role is examined next.

18.2.5 District Planning Units (Under State)

Having outlined four options for State and councils partnerships in planning and infrastructure provisions, the option of the State expanding its role in this area is addressed next. Under this option District Planning Units (possibly under the NSW Department of Planning and Environment) (herein called the Department) could be created for district planning. This would be the instance, for example, in implementing *A Plan for Growing Sydney* (NSW 2014). Currently, under the Department, major urban center strategic plans have been completed within districts of Greater Sydney (Chaps. 9, 13 and 17). This urban centre planning task has also been provided to regional councils outside Greater Sydney, such as Gosford City (on the Central Coast), Wollongong City (on the South Coast) and Newcastle City (in the Hunter Valley). The district (or regional) planning work of the Department (in conjunction with other State departments) could be expanded within designated districts of Greater Sydney (and regions). An expanded role in district plan making and infrastructure provisions could be initiated by the State within Department designated District Planning Units. The proposed changes to the EPA Act (1979) (Chap. 2) and the final recommendations of the State local government reform process (Chap. 2) could include a program for District Planning Units**.** Within a District Planning Unit, councils could be given a more defined (and agreed to) partnership planning role than currently exists. Also, under this option (as in other options above) there could greater State commitment (including budgeting) to planning and infrastructure commitments within districts. Finally, these District Planning Units could also be a model to be taken up in regions outside Greater Sydney (as noted above). All these above points need examination and debate at council, community and State levels. The no change option to current council and State working arrangements is examined next.

18.2.6 No Change to Councils Planning

A range of options (1–5) have been addressed above for councils working in partnership with other councils and the State. Examples of how these options could be taken up, as case studies, by councils in the metropolitan districts of Sydney Inner West, Greater Parramatta and St George were reviewed. Finally, there is also the option of 'no change' to current local council and State working arrangements. Given the extent of issues canvassed in Chaps. 1, 2, 3, 4, 5, 6, 7, 8, 9, 10, 11, 12, 13, 14, 15, 16, and 17 and the options outlined above, the 'no change' option is unlikely to be attractive to most parties. These parties want expanded opportunities, for example, in better 'urban planning and infrastructure provision' at all levels of government. The aspects of the State and councils working in partnerships is further explored in reviewing other State planning and infrastructure bodies, including the Greater Sydney Commission next.

18.2.6.1 State Planning Bodies and Greater Sydney Commission

Given the options of State and councils partnerships in planning and infrastructure provisions (as reviewed above), it is useful to review related State governance in planning. There are a range of other State bodies with responsibilities in planning and infrastructure provisions (in addition to the Department of Planning and Environment as discussed under District Planning Units above). The current work of State bodies engaged in planning and infrastructure provisions include for example Department of Roads and Maritime Services; NSW Public Works; UrbanGrowth; and, State Development Corporations. These bodies are often criticized (and often justified) by councils and the communities within municipal areas for not adequately engaging councils and the community in State planning and infrastructure provisions. In the exercise of any of the options above (or other State district planning for Greater Sydney) the question of engaging councils and the community is a valid issue (Chaps. 3, 4, 5, 6, 7, 8, 9, 10, 11, 12, 13, 14, 15, 16, and 17). In addition, the current work of NSW bodies such as State appointed 'planning panels' (created to make recommendations on developments beyond the decision making of a council) could be reviewed. The State's *A Plan for Growing Sydney* (NSW 2014) contains within it many opportunities for a review of the way councils and the community can be more effectively engaged. The intent to complete district plans (noted above) within the work of the Greater Sydney Commission (GSC) (herein called the Commission) presents a new opportunity for the State to recognise the roles of citizens and local councils within the districts of Greater Sydney. The State (NSW 2014) notes that the role of the Commission (Plate 18.2) also includes overseeing: action delivery; subregional planning (now called districts); infrastructure investment; and, local planning. Thus the Commission will need the full cooperation of local councils (representing their communities).

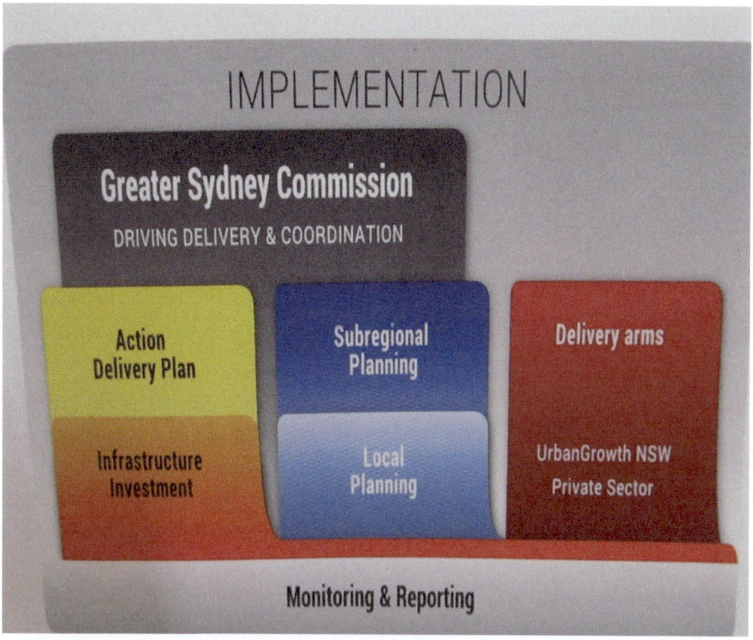

Plate 18.2 Greater Sydney Commission (*Source* NSW 2014)

The question thus arises how councils will be engaged in the operations of the Commission, especially in planning and infrastructure provisions within a district. What mechanisms, for instance, will be created by the State to enable councils to scrutinize proposals affecting a council area? Will the State look at a range of State-councils partnership options such as noted above? One scenario from the NSW Property Council (representing the property industry) on the implementation of *A Plan for Sydney's Growth* (NSW 2014) was expressed in 2015. The Council (Chris Johnson, CEO) writing in the Sydney Morning Herald (SMH 19 May 2015) takes up implementation the Commission and district planning (summarizing) as follows: the Commission could run metropolitan planning and governance; district offices would contain local government planning staff and provide other planning services such as road planning; and, the above could take place with or without merged councils.

In examining the Commission, it is also noted that the Commission: (1). is based on the Greater London Authority; (2). local councils within a district will select one representative; (3). The Commission will have a Chair and CEO and commissioners representing economic, environment and social areas (these were appointed in early 2016); (4). the Chair and CEO would sit on NSW Cabinet Infrastructure Committee; will create plans for each district, including housing targets; and review LEPs of councils and be the ultimate decision makers in rezoning proposals (currently under the Minister for Planning). In summary, there is every possibility the NSW councils and the State will be operating under different planning and infrastructure provisions by the end of 2016 and early 2017. The options of the State and councils working in partnership presents an array of options (as above). The ideal would be to formulate the State and councils partnerships that would satisfy community engagement and apply sustainable urban planning (SUP) practices in municipal and district level planning (Chaps. 3, 4, 5, 6, 7, 8, 9, 10, 11, 12, 13, 14, 15, 16, and 17). With the analysis of planning Greater Sydney now complete, the subject of planning sustainable cities anywhere is addressed next.

18.3 Directions for Creating Sustainable Cities Anywhere

As illustrated in examining planning in Greater Sydney (above) each city needs to find the best practice path to incorporating SUP practices (thus ensuring a sustainable city into the future). Key planning guidelines (and reflecting overall conclusions reached in earlier chapters) can be drawn up for any city in the world to consider to move in the direction of sustainability. Here are some examples of guidelines (drawn from Chaps. 1, 2, 3, 4, 5, 6, 7, 8, 9, 10, 11, 12, 13, 14, 15, 16, and 17) that a city government could consider adopting to move towards sustainability of that city:

1. Responding to environmental issues, including renewing older parts of a city. Current priority environment issues for many cities over the world that could be examined include: impact of climate change; carbon emissions; pressures to accommodate new immigrant residents; and, depleting resources (e.g. peak oil).
2. Ensuring current planning practices are moving towards a more coordinated and holistic framework in incorporating sustainability principles in planning of the city.
3. Working closely with that portion of the development industry accepting development proposals coming under SUP practices, including adopting ecologically sustainable development (ESD) principles.
4. Developing education, bonuses and tax breaks for industry within city urban development policies.
5. Acknowledging the fact that 50 % of world's oil resources ('peak oil') has been expended. Thus, car dependent cities will need to examine alternatives to motor transport, including more cycling and walking.
6. Investigating and acting on the causal factors within urban growth creating environmental problems.
7. Making advances in sustainability applications and translating these into ESD related documents (in the form of protocols, acts and reports). A city can tap into these documents as they exist today at all levels of government (municipal, district, regional, state, national, and international).
8. Tapping into documents drawn up by non-government environment and planning organizations. Generally, the NGO sector has been increasingly pro-active

in alerting governments to act on sustainability planning. Key NGO documents, for example, have contributed to the debate around national and city actions on climate change

9. Interpreting international ESD related documents (commenced in the early 1970s). Of particular importance for an area's sustainability are those documents applying to the natural and built environments.
10. Adopting ESD principles and goals to ensure renewal of the city and preservation of resources for future generations. Equipped with ESD goals, the city is then in a position to adopt the key ESD components for measurement (e.g. indicators of sustainability). These indicators will enable the city to measure ESD goals and monitor outcomes.
11. Working with other cities (often around the world) on urban issues that have international implications. Chapter 1 noted that by the early 2010s the mayors of a number of cities, agreed to cooperate on CO2 emissions. The mayors adopted a maximum future 2° world temperature change to avert the predicted greater disasters if temperature changes are greater than 2°. This action was then taken up in the Paris Agreement (UN 2015) on climate change as adopted in late 2015.
12. Examining key urban planning schools of thought that have assisted cities to adopt ESD criteria into city planning. Some of these schools of thought (as reviewed in Chap. 1 and noted next) lead to the adoption of city planning goals.
13. Planning compact cities, with the city becoming more compact using higher density development around transport nodes.
14. Planning cities around the new urbanism school, such as building neighborhoods with a diversity of residential, commercial and light industrial land uses in close proximity
15. Using the smart growth, such as aiming to limit urban sprawl through improved land use and transport policies. As noted in Chap. 1, older inner city areas often reflect the aims of smart growth.
16. Using eco-city planning such as adopting a range of environmental design strategies (e.g. biodiversity protection).
17. Adopting mechanisms for more cooperative planning between other municipalities and districts and the State government.
18. Challenging the State to develop new frameworks for planning cities and city districts, going beyond the main city CBD.
19. Using more master planning of the city and its districts, as this approach could ensure sustainability principles are taken up. These master plans need to be functional and still result in urban designs that create the sustainable city and sustainable districts.
20. Ensuring metropolitan communities have adequate open spaces and natural lands. This can be achieved by developing city policies that emphasize ecological assets and green urbanism.
21. Considering policies that enable engineering and architecture disciplines to carry through the designs under SUP principles.

22. Responding to the factors of cultural complexities in building the city. The use of adaptive ecologies could be considered as an approach (Chap. 1). City planners need to appreciate that urban places should reflect these cultural aspirations, which is turn lead to specific architectural designs
23. Planning for urban areas to be distinctly different from each other and still incorporate sustainability principles.
24. Considering the classical idea that a city expresses its own distinctive ethos or values. Gleeson (2014) (Chap. 1) in his book *The Public City* argues the importance of civic and institutional spaces in a city in creating the 'public city'.
25. Instigating wide community input in planning, for example incorporating community ownership of urban spaces. The city planners need to design urban places to first meet the social needs of the community.
26. Ensuring the development market is engaged in creating urban public spaces for the wider community benefit, not solely for developer interests. The city needs thus to ensure the public sector has not been relegated to the edges of the space-making process. Often developer created spaces fail to meet the needs of a wider public and of the diversity of users that make up that community.
27. Ensuring that urban design reflects learning from nature's patterns and to compliment these patterns (not opposing them) (Chap. 1).
28. Planning the city using a tool of eco-design as contained, for example, in the reference (referred to in an earlier chapter) *Dictionary of Ecodesign: An Illustrated Reference* (Ken Yeang 2010).
29. Considering low-energy architectural solutions in housing and commercial design.
30. Giving priority to urban areas where pedestrians congregate, as this is a key factor to bring about sustainable cities. The administrators in cities have learned that it is the pedestrians who need to have right of way over vehicles. Many cities around the world have turned over sections of downtowns to open plazas.
31. Considering, when planning for bigger populations, that the issue of ethnic diversity be given special importance. This factor is apparent when planning for multi-cultural populations in older metropolitan districts (Chaps. 3, 4, 5, 6, 7, 8, 9, 10, 11, 12, 13, 14, 15, 16, and 17).

There were a number of questions, raised in Chap. 1 on the subject of cities in global transition. It is anticipated that the Chaps. (1, 2, 3, 4, 5, 6, 7, 8, 9, 10, 11, 12, 13, 14, 15, 16, and 17 answered these questions, which were as follows: (1). What is a global city? (2). What is the meaning of cities in global transition? (3). What is the extent of impacts of global cities (social, environmental and economic)? (4). Can urban development in global cities be assessed against sustainability principles (e.g. UN environmental protocols)? (5). Is there some universal practices of sustainable urban planning (SUP) that can be applied to cities and city districts? (6). Can specific plans and developments likely to impact on a city district be measured and appropriate government policies be adopted to prevent (or minimize) any impact? (7). Can other cities learn lessons from the experience of one city and translate these lessons into improved governance planning policies? These questions were all addressed throughout the book, with options offered above on how best to achieve the governance structure to address the above questions.

In closing, the potential for any city or district to adopt SUP practices is within the grasp of elected offices, administrators and citizens of that city or district. It is hoped the conclusions reached on the applications of SUP principles to this study of Greater Sydney districts (and municipalities within) (Chaps. 3, 4, 5, 6, 7, 8, 9, 10, 11, 12, 13, 14, 15, 16, and 17) has provided useful models and lessons learned. A future scenario would see all cities worldwide advancing toward becoming sustainable. This will be paramount given world population growth (Chap. 1) and its ever increasing social, environmental and economic impacts (and challenges they present) both locally and globally.

18.4 Summary

This chapter examined the conclusions from the analyses on planning within the three case study districts Greater Sydney (Sydney Inner West, Greater Parramatta and St George) (Chaps. 3, 4, 5, 6, 7, 8, 9, 10, 11, 12, 13, 14, 15, 16, and 17). The first conclusion commented on was that the State needed to further address the impact of development under the three planning principles (as assessed in earlier chapters). Specific reference was drawn to *A Plan for Sydney's Growth* (NSW 2014). The second conclusion was then reviewed, being that new planning arrangements among Councils and the State were needed in planning the Greater Sydney districts. Suggested new planning arrangements under six options for planning at municipal and district levels were offered. Comments were then made on State directions for planning a sustainable Greater Sydney. The chapter then addressed how sustainable urban planning (SUP) practices can be used to achieve a sustainable city anywhere. Finally, a sample range of planning guidelines were offered for any city in the world to consider to move in the direction of achieving sustainability.

References

Gleeson B (2014) The public city. Melbourne University Press, Melbourne
NSW (2014) A plan for Sydney's growth. NSW Government, Sydney
United Nations (2015) Paris agreement (United Nations framework convention on climate change). United Nations, New York
Yeang K (2010) Dictionary of ecodesign: an illustrated reference. Routledge, New York

Appendix 1: Cultural Diversity Poems Titles – Districts of Sydney Inner West, Greater Parramatta, and St George

These poems (Raymond Charles Rauscher, author) were completed as a result of the qualitative research (2014/15) on population changes within the case study districts (Sydney Inner West, Greater Parramatta and St George). The poems are by order of appearance under a heading *Notation* at end of the respective chapter. The general theme heading is *Cultural Diversity* as that is the outstanding factor all the districts had in common. The poems cover a range of topics, including: municipality history; locality features; and revival of inner city living. The poems also include an insight into the trades and skills of the early settlers who contributed to the development of each district. At the same time, other poems acknowledge early residents (including Aboriginal clans) and new residents moving into the districts. A common element here in all districts is the impact of global population movements and expanding multi-cultural communities. The renewal of the districts by residents, governments and developers is acknowledged in a number of poems.

Chapter (that poem is in, with number of poem in brackets)

3. *Grains of Memory in Marrickville* (#1)
   White Cockatoos and Marrickville (#2)
   Newtown's Global Realm (#3)
4. *Ashfield's National Blends* (#4)
   Return to Your Summer Hill (#5)
5. *Annandale Crimson* (#6)
   Balmain Remnants (# 7)
   Growing City Cultures in Norton St (#8)
   Heartland Leichhardt Revives (# 9)
   Taverners Hill Lights Up (#10)
6. *Burwood Finds Metropolis' Heart* (#11)
7. *Who Took His Strathfield Away* (#12)
8. *Rhodes' New Order Modernism* (#13)
10. *Parramatta's Second City Reach* (#14)

R.C. Rauscher, S. Momtaz, *Cities in Global Transition*,
DOI 10.1007/978-3-319-39865-5

11. *Tripoli Rises in Auburn* ($15)
12. *Multicultural Meets Suburban Edge in Holroyd* (#16)
14. *Hurstville as Global Central* (#17)
15. *Kogarah as City Fine* (#18)
16. *Rockdale as Metropolis Unfolding* (#19)

Appendix 2: Sustainable Urban Planning Web Sites (Updated to Book Printing Date, Please Inform Authors of Any Change Via Email)

Agenda 21	www.unchs.org
Agenda 21 Europe	www.iclei.org/la21/eurola21.htm
American Planning Association	www.planning.org/
American Society of Civil Engineers	www.asce.org
Architects Australia	www.raia.com.au/
Austin, Texas	www.ci.austin.tx.us
Australian Conservation Foundation	www.acfonline.org.au
Australian Housing and Urban Research	www.ahuri.edu.au
British Columbia (Climate Change)	www.sdri.ubc.ca
Brookings Institute (Urban Policy), US	www.brook.edu.es/urban
Canada Planning	www.cip-icu.ca
Car free	www.carfree.com/
Centre for Neighborhood Technology (Chicago)	www.cnt.org/
City-states Group, US	www.citistates.com
Citizens for Better Environments, US	www.cbezambia.org
Citizens Network for Sustainable Development, US	http://orgs.tigweb.org
Community Environment Council, Santa Barbara	www.cecsb.org/
Community Environment Network, NSW Australia	www.cccen.org.au
Community Initiatives, US	www.communityinitiatives.com
Community Sustainability Resource Institute, US	www.sustainable.org
Community Viz (Vision), US	www.communityviz.com
Congress for New Urbanism, US	www.cnu.org
Cyburbia, US	www.cyburbia.org
Cyberhood, US	www.thecyberhood.net
Dept. of Planning and Infrastructure, NSW, Australia	www.dopi.nsw.gov.au
Earthsharing, Australia	www.earthsharing.org.au
Earth Council Summit	www.ecouncil.ac.cr
EcoAction 2000, US	www.ec.gc.ca/ecoaction
Ecodesign, US	www.ecodesign.com

(continued)

R.C. Rauscher, S. Momtaz, *Cities in Global Transition*,
DOI 10.1007/978-3-319-39865-5

Ecodesign Foundation/Australia	www.edf.edu.au
Ecopolitics, Australia	www.ecopolitics.org.au
Ecosystem Anthologies, US	www.ecoiq.com
Ecotransit, NSW Australia	www.ecotransit.org.au
Electronic Cultural Atlas Initiated, US	http://ecai.org
Environment Australia (Commonwealth)	www.environment.gov.au
Environment Defenders Office (EDO), Australia	www.edo.org.au
Environment Protection NSW (see Envalue)	www.epa.nsw.gov.au
Environment Protections Authority, US	www.epa.org
Europe Expert Group	www.iclei.org/europe/expert.htm
Futures Preferred, Australia	www.preferredfutures.org
Geocities, US	www.geocities.com/athens/2962/bos
Geotools, US	www.geotools.org
Global Urban Observatory (intranet), UN	www.urbanobservatory.org/indicators
Greenbelt, UK	www.greenbelt.org
Greendesign, US	www.greendesign.net
Greenmap, NYC, US	www.greenmap.org
Green Mountain Institute for Environ. Democracy, US	www.gmied.org
Green Network, Australia.	www.greennet.org.au
Ground Swell	www.progress.org.cg
Hamilton-Wentworth Regional Council, Canada	www.hamilton-went.on.ca/vis2020
Hawaii Ho'okipa Network	www.hawaiian.net/cbokauai
Healthy City Office (Toronto) (Healthy City Index)	www.city.toronto.on.ca/
Healthy Cities, UN	www.who.org
Innovative Urban Planning, NYC US	www.columbia.edu/~jws150/urban
Institute for Local Self-Reliance, US	www.ilsr.org
Institute for Sustainable Futures, NSW Australia	www.isf.edu.net.au
International Institute of Sustainable Development, US	www.iisd.org
International Soc. For Ecological Economics	www.csf.colorado.edu/ISEE
Jacksonville Community Council, US	www.sustainablemeasures.com
Lansing Sustainable, US	www.urbanoptions.org/sustain
Lincoln Land Use Institute, US	www.lincolninst.edu/main.html
Local Environment Initiatives, US	www.iclei.org/iclei.htm
Local Government and Shires Assoc, Australia	www.lgsa.org.au
Maine Development Foundation, US	www.mdf.org/meip.htm
Mapquest, US	www.mapquest.com
Metro Virtual (books/webs), US	www.metrovirtual.org
Millennium Ecosystem Assessment	www.maweb.org
Millennium Institute	www.igc.org/millennium
Minnesota Planning, US	www.mnplan.state.mn.us/mm/index.html
Mountain Assoc. for Community Econ. Dev., US	www.maced.org
Natural Resources Atlas (NSW aspects of NR), NSW	www.nr.nsw.gov.au

(continued)

Neighborhood Preservation Centre, NYC US	www.neighborhoodpreservation.org
New Urbanism, US	www.newurbanism.org
New Urbanism News (Image Bank Mapping), US	www.newurbannews.com
New Urbanism Resource Site, US	www.netsense.net/terry/newurban.htm
Newcastle City Council, NSW Australia	www.newcastle.nsw.gov.au
Olympia WA, US	www.olywa.net/roundtable
Ontario Healthy Communities Coalition, Canada	www.opc.on.ca/ohcc
Ontario Roundtable on Environment and Economy	www.web.net.ortee/scrp
Oregon Progress Board, US	www.econ.state.or.us/opb
Penrith Council, NSW, Australia	www.penrith.nsw.gov.au
Planners Network, NYC US	www.plannersnetwork.org
Planners Network, UK	www.plannersnetwork.co.uk
Planners Web, US	www.plannersweb.com
Port Stephens Council, NSW, Australia	www.portstephens.nsw.gov.au
Preserve Net, US	www.preserve.org
Positive Futures Network, UK Transport	www.futurenet.org
Project for Public Spaces, US	www.pps.org
Office of Environment and Heritage, NSW, Australia	www.environment.nsw.gov.au/
Quality Planning, New Zealand	www.qualityplanning.org.nz
Redefining Progress (progress indicators), US	www.rprogress.org
Resilience Alliance	www.resalliance.org
Rocky Mountain Institute, Colorado US	www.rmi.org
Santa Monica Council, US	www.ci.santa-monica.ca.us/environment
Seattle, US	www.ci.seattle.wa.us
Smart Growth America	www.smartgrowthamerica.com
Smart Growth Network, US	www.smartgrowth.org
Smart Growth Rhode Island, US	www.growsmartri.com
Smart Land Development, US	www.landuse.org
Society for Responsible Design, Australia	www.green.net.au/srd
Sprawl City, US	www.sprawlcity.org
Sprawl Watch Clearinghouse, US	www.sprawlwatch.org
Sustain Western Maine, US	www.mainewest.com/swm
Sustainability Project, US	www.cyberus.ca/choose.sustain
Sustainable Boston, US	www.ci.boston.ma.us/environment/ sustain.asp
Sustainable City, San Francisco	www.sustainable-city.org
Sustainable Communities Network, US	www.sustainable.org
Sustainable Development Centre (US DO Energy)	www.sustainable.doe.gov
Sustainable Development Research Institute, Canada	www.sdri.ubc.ca
Sustainable Earth Electronic Library, US	www.envirolink.org/pubs/seel/about. html
Sustainable Industries, Australia	www.sustainindustries.org.au
Sustainable Measurements, US	www.sustainablemeasurements.com
Ted Trainer (Environmental Teacher/Writer)	www.arts.unsw.edu.au/socialwork/ trainer.h

(continued)

Thoreau Centre for Sustainability, US	www.naturalstep.org
Tom Farrell Institute	www.newcastle.edu.au/research-centre/tfi
Transportation for Livable Communities Net, US	www.tlcnetwork.org
UN Commission of Sustainable Development	www.un.org/esa/sustdev
UN Habitat	www.unchs.org
UN Sustainable Cities Programme	www.undp.org/un/habitat/scp/index.html
UNESCO Man and the Biosphere (MAB) Urban Group	www.unesco.orb/mab
Urban Advantage (digital), US	www.urban-advantage.ocm
Urban Ecology, US	www.urbanecology.org
Urban Futures, US	www.urbanfutures.org
Urban Regeneration Companies, UK	www.urcs-online.co.uk
Urban Photo, US	www.urbanphoto.org
Urbanization Science Project IHDP	www.ihdp.uni-bonn.de
US Dept. of Housing and Urban Development	www.huduser.org
Vermont, US	www.vitalcommunities.org
Virtual Library on Sustainable Development, US	www.ulb.ac.be/ceese/meta/sustvl.html
Walkable Communities, US	www.walkable.org
Waterfront Regeneration Trust, US	www.waterfronttrust.com
Wiley Books (Sustainable Design)	www.interscience.wiley.com
World Business Council on SD	www.wbcsd.ch
World Watch Institute, UN	www.worldwatch.org
World Wildlife Foundation (footprint calculator)	www.wwf.org.au

Glossary

Act An act is "an order, law or judgment as of Parliament" (Macquarie Dictionary 2015).

Agenda 21 Agenda 21 is a detailed plan of actions dealing with all aspects of ecologically sustainable development (ESD) and desirable national policies. The concept was agreed to by the national representatives at the United Nations Conference on Environment and Development (UNCED) at Rio de Jeneiro in June 1992.

Area A region (or other size) such as a *settled area* (Macquarie Dictionary 2015).

Agreement Coming to an arrangement; a condition of agreeing (Macquarie Dictionary 2015).

Biodiversity The variety of life forms, including the different plants, animals and microorganisms, the genes they contain and the ecosystems they form (Macquarie Dictionary 2015).

Catchment The area of land drained by a river and its tributaries (Macquarie Dictionary 2015).

Catchment Management Coordination of land use and resource development within natural catchment areas rather than artificially defined areas or districts.

Cities in Global Transition (CGT) Cities that undergo a transition responding to the global impact factors (GIF's), including: increased overseas migration into those cities; expanding global financial institutions (i.e. international banks); and, changing residential housing reflecting global trends such as apartment housing and higher densities.

Climate change Global changes in climate associated with the greenhouse effect, including the overall effects on climate of human made and natural changes.

Community housing Assisted housing provided by and managed by community based groups (often subsidized by governments).

Cultural land Relating to tradition of Aboriginal lands (Macquarie Dictionary 2015).

R.C. Rauscher, S. Momtaz, *Cities in Global Transition*,
DOI 10.1007/978-3-319-39865-5

Development Control Plan A local government plan prepared to provide detail on particular policies for certain types of development or geographical areas.

District An area of land delineated by geographical features, and can also be an administrative area such as a sub-part of a city, district or region (Macquarie Dictionary 2015).

Ecology Branch of biology dealing with organism and their environment.

Environment The aggregate of surrounding things, conditions or influences. Broad natural surrounding conditions, such as the bush, rivers, air, sea in which human and natural elements exist (Macquarie Dictionary 2015).

ESD Ecologically Sustainable Development – using, conserving and enhancing the community's natural resources so that ecological processes on which life depends are maintained and the total quality of life, now and into the future, can be increased.

Global Impact Factors (GIF) Global factors that impact on urban areas (e.g. cities, districts, regional areas), including population growth; new infrastructure needs; and expanded urban centers.

Indicator Recording variations, reactions or changes affecting a system.

Kyoto Agreement under the United Nations (as signed by nations) effective when Russia signed in September 2004; requires nations to reduce greenhouse gases.

Limits to growth Title of a book published in 1972 by the Club of Rome with predictions of severe consequences if the world's population and resource use continued to grow.

Local Characterized by place as in local situation (Macquarie Dictionary 2015).

Local environment Eenvironment limited to a particular place or small area. Also relates to parts of an area as of a system.

Locality A local urban, regional or rural place.

Metropolis The full extent of a city beyond the official city boundaries.

Neighborhood A local living area that is usually part of a suburb or city and defined by boundaries. An area can also be a 'locality', normally not as big as a neighbourhood.

Precinct A space of definite or understood limits and its environs (Macquarie Dictionary 2015).

Principle A principle is 'a rule or law exemplified in natural phenomena' (The Macquarie Dictionary 2015).

Protocol A protocol is 'rules of behavior to be agreed upon by heads of organizations' (Macquarie Dictionary 2015).

Public housing Housing provided by the local council or state for rental.

Redevelopment The removal of buildings (i.e. residential, commercial or industrial) to create a new development in an urban area.

Rehabilitation The upgrading of urban areas by improving housing, buildings and services.

Renewal The introduction of new development and renewal of existing buildings in urban areas.

Social housing Subsidized housing provided by government, city, company or community group that caters for special needs.

Standard A basis of comparing things of a similar nature. Serving as a basis or measure or value (Macquarie Dictionary 2015).

Suburb A part of a local government area.

Sustainability Managing our natural resources in a way that maintains their environment, economic and cultural values, so that they continue to be available in the long-term.

Urban Relation to a city or town (Macquarie Dictionary 2015).

Urban Conglomerate A series of cities and urban centers stretching as a contiuous band of settlement.

Index

R.C. Rauscher, S. Momtaz, *Cities in Global Transition*,
DOI 10.1007/978-3-319-39865-5

MIX
Papier aus verantwortungsvollen Quellen
Paper from responsible sources
FSC® C105338

If you have any concerns about our products,
you can contact us on
ProductSafety@springernature.com

In case Publisher is established outside the EU,
the EU authorized representative is:
Springer Nature Customer Service Center GmbH
Europaplatz 3, 69115 Heidelberg, Germany

Printed by Libri Plureos GmbH
in Hamburg, Germany